LRFD
STEEL DESIGN

THE PWS SERIES IN ENGINEERING

ANDERSON, *Thermodynamics*

ASKELAND, *The Science and Engineering of Materials*, Third Edition

BOLLUYT, STEWART, AND OLADIPUPO, *Modeling for Design Using Silverscreen*

BORSE, *FORTRAN 77 and Numerical Methods for Engineers*, Second Edition

CLEMENTS, *Microprocessor Systems Design*, Second Edition

CLEMENTS, *Principles of Computer Hardware*, Second Edition

CLEMENTS, *68000 Family Assembly Language*

DAS, *Principles of Foundation Engineering*, Second Edition

DAS, *Principles of Geotechnical Engineering*, Third Edition

DAS, *Principles of Soil Dynamics*

FLEISCHER, *Introduction to Engineering Economy*

GERE AND TIMOSHENKO, *Mechanics of Materials*, Third Edition

GLOVER AND SARMA, *Power Systems Analysis and Design*, Second Edition

JANNA, *Design of Fluid Thermal Systems*

JANNA, *Introduction to Fluid Mechanics*, Third Edition

KEEDY, *Introduction to CAD using CADKEY*, Second Edition

KNIGHT, *The Finite Element Method in Machine Design*

LOGAN, *A First Course in the Finite Element Method*, Second Edition

MCGILL AND KING, *Engineering Mechanics: Statics*, Second Edition

MCGILL AND KING, *Engineering Mechanics, An Introduction to Dynamics*, Second Edition

MCGILL AND KING, *Engineering Mechanics: Statics and An Introduction to Dynamics*, Second Edition

RAINES, *Software for Mechanics of Materials*

REED-HILL AND ABBASCHIAN, *Physical Metallurgy Principles*, Third Edition

REYNOLDS, *Unit Operations and Processes in Environmental Engineering*

SACK, *Matrix Structural Analysis*

SCHMIDT AND WONG, *Fundamentals of Surveying*, Third Edition

SEGUI, *Fundamentals of Structural Steel Design*

SEGUI, *LRFD Steel Design*

SHEN AND KONG, *Applied Electromagnetism*, Second Edition

STRUM AND KIRK, *BookWare Companion: Contemporary Linear Systems Using MATLAB*

SULE, *Manufacturing Facilities*, Second Edition

VARDEMAN, *Statistics for Engineering Problem Solving*

WEINMAN, *VAX FORTRAN*, Second Edition

WEINMAN, *FORTRAN for Scientists and Engineers*

LRFD
STEEL DESIGN

WILLIAM T. SEGUI

The University of Memphis

PWS PUBLISHING COMPANY

**Boston • Albany • Bonn • Cincinnati • Detroit • London
Madrid • Melbourne • Mexico City • New York • Paris
San Francisco • Singapore • Tokyo • Toronto • Washington**

I(T)P **An International Thomson Publishing Company**

PWS Publishing Company
20 Park Plaza
Boston, MA 02116-4324

Editor: Jonathan Plant
Assistant Editor: Mary Thomas
Production Editor: Kerry Avin
Manufacturing Coordinator: Andrew Christensen
Interior and Cover Design: Abigail M. Heim
Composition and Illustration: Santype International Limited
Cover Printer: John P. Pow Company, Inc.
Text Printer and Binder: Quebecor/Martinsburg

International Thomson Publishing
The trademark ITP is used under license

Printed in the United States of America
2 3 4 5 6 7 8 9 —98 97

This book is printed on
acid-free, recycled paper.

Library of Congress Cataloging-in-Publication Data

A catalog record for this book is available from the Library of Congress
ISBN 0-534-95478-2

CONTENTS

PREFACE

This book, which reflects the 1993 revisions to the AISC LRFD Specification, serves as a basic text in structural steel design for undergraduate civil engineering students at the junior or senior level. While the primary function of *LRFD Steel Design* is that of a textbook, practicing civil engineers who need a review of current practice and the current AISC Specification will find it useful as a reference. Students using this book should have a background in mechanics of materials and analysis of statically determinate structures.

Depending upon the level of competence of the students, *LRFD Steel Design* can be used for one or two courses of three semester hours each. A suggested two-course sequence is as follows: a first course covering Chapters 1 through 7 and a second course covering Chapters 8 through 10, supplemented by comprehensive design assignments. This division of topics has been used successfully for several years at The University of Memphis.

The emphasis of this book is on the design of building components in accordance with the provisions of the 1993 AISC LRFD Specification and the second edition of the *LRFD Manual of Steel Construction*. Although the text makes occasional reference to AASHTO and Area Specifications, no examples or assigned problems are based on these documents.

Prior to the introduction of the Load and Resistance Factor Design Specification by AISC in 1986, the dominant design approach for structural steel was allowable stress design. The trend today is toward load and resistance factor design, but because allowable stress design is still in use, students should have some familiarity with it. To that end, Appendix B provides a brief introduction to that topic.

It is absolutely essential that the student have a copy of the *Manual of Steel Construction*. In order to promote familiarity with it, material from the *Manual* is not reproduced in this book so that the reader will be required to refer to the *Manual*. All notation in this book is consistent with that in the *Manual*, and AISC equation numbers are used in tandem with sequential numbering of other equations according to the textbook chapter.

United States customary units are used throughout, with no introduction of S.I. units. Metric units can be incorporated easily into a basic mechanics text, but it is difficult to justify their use in a design text geared to industry standards that are based on customary units. Although metrication is inevitable and will be the basis of future AISC specifications and manuals, the change has not yet taken place in the steel construction industry.

As far as design procedures are concerned, the application of fundamental principles is encouraged. Although this book is oriented toward practical design, sufficient theory is included to avoid a "cookbook" approach. Direct design methods are used where feasible, but no complicated design formulas have been developed. Instead, trial and error with "educated guesses" is the rule. Tables, curves, and other design aids from the *Manual* are used, but they have a role that is subordinate to the use of basic equations. Assigned problems provide practice with both approaches, and where appropriate, the required approach is specified in the statement of the problem.

In keeping with the objective of providing a basic text, a large number of assigned problems are provided at the end of each chapter. Computer programming assignments are also included. These problems specify required input and output as well as the general algorithm. The requirements are completely machine-independent to give the instructor maximum flexibility. Most of these assignments can be done with spreadsheet software, and the instructor may wish to require this. Because of the wide range of hardware and programming languages available, program listings are not included in this book. This decision is based on the belief that students will benefit more from writing short programs to apply basic procedures than from running "canned" programs. Answers to selected problems are given at the back of the book, and a *Solutions Manual* is available.

A fairly comprehensive treatment of roof trusses is provided in Chapter 3, since components of trusses are dealt with in subsequent chapters. Column base plates are considered in the discussion of beams in Chapter 5 rather than in Chapter 4, which covers the topic of compression members. Because base plate design requires a consideration of bending strength, coverage of the topic is deferred until bending has been discussed.

I would like to express my appreciation to the following people who reviewed the manuscript for this book and provided helpful comments:

Stuart S. Chen
State University of New York at Buffalo

W. Samuel Easterling
Virginia Polytechnic Institute and State University

Duane S. Ellifritt
University of Florida

Michael D. Folse
University of New Orleans

Louis F. Geschwindner
The Pennsylvania State University

Frank J. Hatfield
Michigan State University

Warner Lansing
Florida Atlantic University

Eric M. Lui
Syracuse University

Donald E. Milks
Ohio Northern University

William H. Walker
University of Illinois at Urbana-Champaign

In addition, several people at the American Institute of Steel Construction were helpful in providing updates on recent AISC Specification revisions as well as other assistance: Nestor R. Iwankiw, Abraham J. Rokach, and Lewis B. Burgett.

Finally, I wish to thank my wife, Angela, for her support and for her valuable suggestions and help in proofreading the manuscript for this book

William T. Segui

1 INTRODUCTION

1.1 ☰ STRUCTURAL DESIGN

The structural design of buildings, whether of structural steel or reinforced concrete, requires the determination of the overall proportions and dimensions of the supporting framework and the selection of the cross sections of individual members. In most cases the functional design, including the establishment of the number of stories and the floor plan, will have been done by an architect, and the structural engineer must work within the constraints imposed by this design. Ideally, the engineer and architect will collaborate throughout the design process so that the project is completed in an efficient manner. In effect, however, the design can be summed up as follows: The architect decides how the building should look; the engineer must make sure that it doesn't fall down. Although this is an oversimplification, it affirms the first priority of the structural engineer: safety. Other important considerations include serviceability (how well the structure performs in terms of appearance and deflection) and economy. An economical structure requires an efficient use of materials and construction labor. Although this can usually be accomplished by a design that requires a minimum amount of material, savings can often be realized by using slightly more material if it results in a simpler, more easily constructed project.

A good design requires the evaluation of several framing plans — that is, different arrangements of members and their connections. In other words, several alternate designs should be prepared and their costs compared. For each framing plan investigated, the individual components must be designed. This requires the structural analysis of the frame(s) of the building and the computation of forces and bending moments in the individual members. Armed with this information, the structural designer can then select the appropriate cross section. Before any analysis, however, a decision must be made on the building material to be used; it will usually be reinforced concrete, structural steel, or both. Ideally, alternate designs should be prepared with each.

The emphasis in this book will be on the design of individual structural steel members and their connections. Although the structural engineer must select and evaluate the overall structural system in order to produce an efficient and economical design, this cannot be done without a thorough understanding of the design of the components (the "building blocks") of the structure. Component design is the focus of this book.

Before proceeding with a discussion of structural steel, we will examine the various types of structural members to be covered. Figure 1.1 shows a truss

FIGURE 1.1 ≡

with vertical concentrated forces applied at the joints along the top chord. In keeping with the usual assumption of pinned connections, each member of the truss will be a two-force member, subject to either axial compression or tension. For simply supported trusses loaded as shown — and this is a typical loading condition — each of the top chord members will be in compression, and the bottom chord members will be in tension. The web members will either be in tension or compression, depending on their location and orientation and on the location of the loads.

Other types of members can be illustrated with the rigid frame of Figure 1.2a. The members of this frame are rigidly connected by welding and can be assumed to form a continuous structure. At the supports, the members are welded to a rectangular plate that is bolted to a concrete footing. If several of these frames are placed in parallel and connected with additional members that are then covered with roofing material and walls, we have a typical building system. Many important details have not been mentioned, but this is essentially the manner in which many small commercial buildings are constructed. The design and analysis of each frame in the system begins with the idealization of the frame as a two-dimensional structure as shown in Figure 1.2b. Because the frame has a plane of symmetry parallel to the page, we are able to treat the frame as two-dimensional and represent the frame members by their centerlines. (Although it is not shown in Figure 1.1, this same idealization is made with trusses, and the members are usually represented by their centerlines.) Note that the supports are represented as hinges (pins) and not as fixed. If there is a possibility that the footing will be able to undergo a slight rotation, the support must be considered to be pinned. One assumption made in the usual methods of structural analysis is that deformations are very small, and this means that only a slight rotation of the support is needed to qualify it as a pinned connection.

Once the geometry and support conditions of the idealized frame have been established, the loading must be determined. This is usually done by apportioning a share of the total load to each frame. If the hypothetical structure under consideration is subjected to a uniformly distributed roof load, the portion carried by one frame will be a uniformly distributed line load measured in force per unit length as shown in the figure. Typical units would be kips per foot.

FIGURE 1.2 ≡

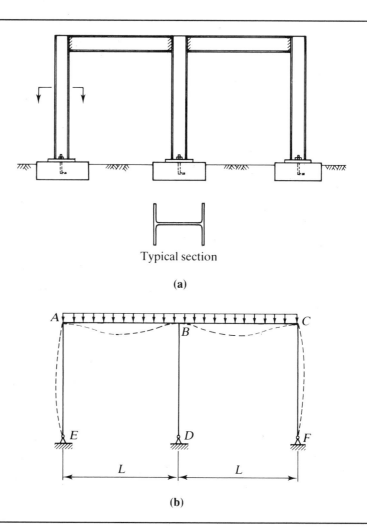

Typical section

(a)

(b)

For the loading shown in Figure 1.2, the frame will deform as indicated by the dashed line (drawn to a greatly exaggerated scale). The individual members of the frame can be classified according to the type of behavior represented by this deformed shape. The horizontal members *AB* and *BC* are subjected primarily to bending, or flexure, and are called *beams*. The vertical member *BD* is subjected to couples transferred from each beam, but for the symmetrical frame shown, they are equal and opposite, thereby canceling each other. Thus, member *BD* is subjected only to axial compression arising from the vertical loads. In buildings, vertical compression members such as these are referred to as *columns*. The other two vertical members, *AE* and *CF*, must resist not only axial compression from the vertical loads, but also a significant amount of bending. Such members are called *beam-columns*. In reality, all members, even those classified as beams or columns, will be subjected to both bending and axial load, but in many cases, one of the effects is minor and can be neglected.

In addition to the members described above, this book covers the design of connections and the following special members: composite columns, composite beams, and plate girders.

1.2 ≡ LOADS

The forces that act on a structure are called *loads*. They belong to one of two broad categories: *dead load* and *live load*. Dead loads are those that are permanent, including the weight of the structure itself, which is sometimes called the *self-weight*. Other dead loads in a building include the weight of nonstructural components such as floor coverings, suspended ceilings with light fixtures, and partitions. All of the loads mentioned thus far are forces due to gravity and are referred to as *gravity loads*. Live loads, which can also be gravity loads, are those that are not as permanent as dead loads. This type may or may not be acting on the structure at any given time, and the location may not be fixed. Examples of live load include furniture, equipment, and occupants of buildings. In general, the magnitude of a live load is not as well defined as that of a dead load, and it usually must be estimated. In many cases, a given structural member must be investigated for various positions of the live load so that a potential failure situation is not overlooked.

If the live load is applied slowly and is not removed and reapplied an excessive number of times, the structure can be analyzed as if the loads were static. If the load is applied suddenly, as would be the case when the structure supports a moving crane, we must account for the effects of impact. If the load is applied and removed many times over the life of the structure, fatigue stress becomes a problem, and we must account for its effects. Impact loading occurs in relatively few buildings, notably industrial buildings, and fatigue loading is rare, with thousands of load cycles over the life of the structure required before fatigue becomes a problem. For these reasons, all loading conditions in this book will be treated as static, and fatigue will not be considered.

Wind exerts a pressure or suction on the exterior surfaces of a building; because of its transient nature, it properly belongs in the category of live loads. Because of the relative complexity of determining wind loads, however, wind is usually considered a separate category of loading. Since lateral loads are most detrimental to tall structures, wind loads are usually not as important for low buildings, but uplift on light roof systems can be critical. Although wind is present most of the time, wind loads of the magnitude considered in design are infrequent and are not considered to be fatigue loads.

Earthquake loads are another special category and need to be considered only in those geographic locations where there is a reasonable probability of occurrence. A structural analysis of the effects of an earthquake requires an analysis of the structure's response to the ground motion produced by the earthquake. Simpler methods are sometimes used in which the effects of the earthquake are simulated by a system of horizontal loads, similar to those resulting from wind pressure, acting at each floor level of the building.

Snow is another live load that is treated as a separate category. Adding to the uncertainty of this load is the complication of drift, which can cause much of the load to accumulate over a relatively small area.

Other types of live load are often treated as separate categories, such as hydrostatic pressure and soil pressure, but the cases enumerated above are the ones ordinarily encountered in the design of structural steel building frames and their members.

1.3 ≡ BUILDING CODES

Buildings must be designed and constructed according to the provisions of a building code, which is a legal document containing requirements related to such things as structural safety, fire safety, plumbing, ventilation, and accessibility to the physically disabled. A building code has the force of law and is administered by a governmental entity such as a city, a county, or, for some large metropolitan areas, a consolidated government. Building codes do not give design procedures, but they do specify the design requirements and constraints that must be satisfied. Of particular importance to the structural engineer is the prescription of minimum live loads for buildings. Although the engineer is encouraged to investigate the actual loading conditions and attempt to determine realistic values, the structure must be able to support these specified minimum loads.

Although some large cities write their own building codes, many municipalities will adopt a "model" building code and modify it to suit their particular needs. Model codes are written by various nonprofit organizations in a form that is easily adopted by a governmental unit. Currently, there are three national model codes: the *BOCA National Building Code*,[1] the *Uniform Building Code*,[2] and the *Standard Building Code*.[3] A related document, similar in form to a building code, is ASCE 7-88, *Minimum Design Loads for Buildings and Other Structures*.[4] This standard is intended to provide load requirements in a format suitable for adoption by a building code.

1.4 ≡ DESIGN SPECIFICATIONS

In contrast to building codes, design specifications give more specific guidance for the design of structural members and their connections. They present the guidelines and criteria that enable a structural engineer to achieve the objectives mandated by a building code. Design specifications represent what is considered to be good engineering practice based on their latest research. They are periodically revised and updated by supplements or by completely new editions. As with model building codes, design specifications are written in a legal format by nonprofit organizations. They have no legal standing on their own, but by presenting design criteria and limits in the form of legal mandates and prohibitions, they can easily be adopted, by reference, as part of a building code.

The specifications of most interest to the structural steel designer are those published by the following organizations:

1. **American Institute of Steel Construction (AISC):**[5] This specification provides for the design of structural steel buildings and their connections. It is the one of primary concern in this book and will be discussed in detail.

2. **American Association of State Highway and Transportation Officials (AASHTO):**[6] This specification covers the design of highway bridges and related structures, and it provides for all structural materials normally used in bridges, including steel, reinforced concrete, and timber.

3. **American Railway Engineering Association (AREA):**[7] This document covers the design of railway bridges and related structures.

4. **American Iron and Steel Institute (AISI):**[8] This specification deals with cold-formed steel (see Section 1.6).

1.5 ≡ STRUCTURAL STEEL

The earliest use of iron, the chief component of steel, was for small tools in approximately 4000 B.C.[9] This material was in the form of wrought iron, produced by heating ore in a charcoal fire. In the latter part of the eighteenth century and in the early nineteenth century, cast iron and wrought iron were used in various types of bridges. Steel, an alloy of primarily iron and carbon, with fewer impurities and less carbon than cast iron, was first used in heavy construction in the nineteenth century. With the advent of the Bessemer converter in 1855, steel began to displace wrought iron and cast iron in construction. In the United States, the first structural steel railroad bridge was the Eads bridge, constructed in 1874 in St. Louis, Missouri.[10] In 1884 the first building with a steel frame was completed in Chicago.

The characteristics of steel that are of the most interest to structural engineers can be examined by plotting the results of a tensile test. If a test specimen is subjected to an axial load P as shown in Figure 1.3a, the stress and strain can be computed as follows:

$$f = \frac{P}{A} \quad \text{and} \quad \varepsilon = \frac{\Delta L}{L}$$

where

> f = axial tensile stress
>
> A = cross-sectional area
>
> ε = axial strain
>
> L = length of specimen
>
> ΔL = change in length

If the load is increased in increments from zero up to the point of fracture, and stress and strain are computed at each step, a stress–strain curve such as the one in Figure 1.3b can be plotted. This curve is typical of a class of steel known as *ductile*, or *mild*, *steel*. The relationship between stress and strain is linear up to the proportional limit; the material is said to follow *Hooke's law*. A peak value, the upper yield point, is quickly reached after that, followed by a leveling off at the lower yield point. The stress then remains constant, even though the

FIGURE 1.3 ≡

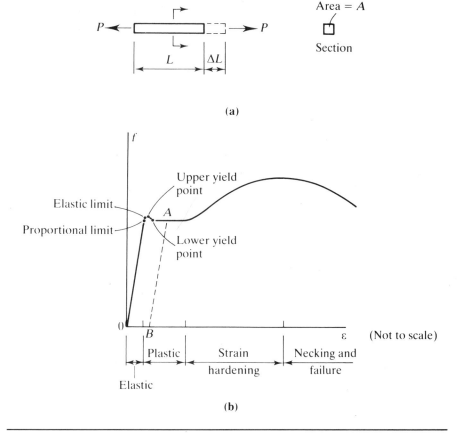

(a)

(b)

FIGURE 1.4 ☰

strain continues to increase. At this stage of loading, the test specimen continues to elongate as long as the load is not removed, even though the load cannot be increased. This constant stress region is called the *yield plateau,* or *plastic range.* At a strain of approximately 12 times the strain at yield, strain hardening begins, and additional load (and stress) is required to cause additional elongation (and strain). A maximum value of stress is reached, after which the specimen begins to "neck down" as the stress decreases with increasing strain, and fracture occurs. Although the cross section is reduced during loading (the Poisson effect), the original cross-sectional area is used to compute all stresses. Stress computed in this way is known as *engineering stress.* If the original length is used to compute the strain, it is called *engineering strain.*

Steel exhibiting the behavior shown in Figure 1.3b is called *ductile* because of its ability to undergo large deformations before fracturing. Ductility can be measured by the elongation, defined as

$$e = \frac{L_f - L_0}{L_0} \tag{1.1}$$

where

e = elongation (usually expressed as a percent)

L_f = length of the specimen at fracture

L_0 = original length

The elastic limit of the material is a stress that lies between the proportional limit and the upper yield point. Up to this stress, the specimen can be unloaded without permanent deformation; the unloading will be along the linear portion of the diagram, the same path as followed during loading. This part of the stress–strain diagram is called the *elastic range.* Beyond the elastic limit, unloading will be along a straight line parallel to the loading path, and there will be a permanent strain. For example, if the load is removed at point A in

Figure 1.3b, the unloading will be along line *AB*, resulting in the permanent strain *OB*.

Figure 1.4 shows an idealized version of this stress–strain curve. The proportional limit, elastic limit, and the upper and lower yield points are all very close to one another and are treated as a single point called the *yield point*, defined by the stress F_y. The other point of interest to the structural engineer is the maximum value of stress that can be attained, called the *ultimate tensile strength* F_u. The shape of this curve is typical of all mild structural steels, which are different from one another primarily in the values of F_y and F_u. The ratio of stress to strain within the elastic range, denoted by E and called *Young's modulus*, or *modulus of elasticity*, is the same for all structural steels and has a value of 29,000,000 psi (pounds per square inch) or 29,000 ksi (kips per square inch).

Figure 1.5 shows a typical stress–strain curve for high-strength steels, which are less ductile than the mild steels discussed thus far. Although there is a linear elastic portion and a distinct tensile strength, there is no well defined yield point or yield plateau. To use these higher strength steels in a manner consistent with the use of ductile steels, some value of stress must be chosen as a value for F_y so that the same procedures and formulas can be used with all structural steels. Although there is no yield point, we need to define one. As previously shown, when a steel is stressed beyond its elastic limit and then unloaded, the path followed to zero stress will not be the original path from zero stress; it will be along a line having the slope of the linear portion of the path followed during loading — that is, a slope equal to E, the modulus of elasticity. Thus, there will be a residual strain, or permanent set, after unloading. The yield stress for steel with a stress–strain curve of the type shown in Figure 1.5 is called the *yield strength* and is defined as the stress at the point of unloading that corresponds to a permanent strain of some arbitrarily defined amount. A strain of 0.002 is usually selected, and this method of determining the yield strength is called the

FIGURE 1.5 ≡

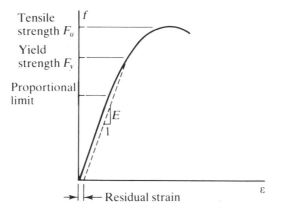

0.2% offset method. As mentioned above, the two properties usually needed in structural steel design are F_u and F_y, regardless of the shape of the stress–strain curve and regardless of how F_y was obtained. For this reason, the generic term *yield stress* is used, and it can mean either yield point or yield strength.

The various properties of structural steel, including strength and ductility, are determined by its chemical composition. Steel is an alloy, its principal component being iron. Another component of all structural steels, although in much smaller amounts, is carbon, which contributes to strength but reduces ductility. Other components of some grades of steel include copper, manganese, nickel, chromium, molybdenum, and silicon. Structural steels can be grouped according to their composition as follows:

1. **Plain carbon steels:** mostly iron and carbon, with less than 1% carbon

2. **Low-alloy steels:** iron and carbon plus other components (usually less than 5%). The additional components are primarily for increasing the strength, which is accomplished at the expense of a reduction in ductility.

3. **High-alloy or specialty steels:** similar in composition to the low-alloy steels, but with a higher percentage of the other components — that is, the components added to iron and carbon. These steels are also higher in strength than the plain carbon steels and also have some special quality such as resistance to corrosion.

Different grades of structural steel are identified by the designation assigned them by the American Society for Testing and Materials (ASTM). This organization develops standards for defining materials in terms of their composition, properties, and performance, and it prescribes specific tests for measuring these attributes.[11] The most commonly used structural steel today is a mild steel designated as ASTM A36, or A36 for short. It has a stress–strain curve of the type shown in Figures 1.3b and 1.4 and has the following tensile properties:

Yield stress: $F_y = 36,000$ psi (36 ksi)

Tensile strength: $F_u = 58,000$ psi to $80,000$ psi (58 ksi to 80 ksi)

A36 steel is classified as a plain carbon steel, and it has the following components (other than iron):

Carbon: 0.26% (maximum)

Phosphorous: 0.04% (maximum)

Sulfur: 0.05% (maximum)

These percentages are approximate, the exact values depending on the form of the finished steel product. A36 is a ductile steel, with an elongation as defined by Equation 1.1 of 20% based on an undeformed original length of 8 inches.

Steel producers who provide A36 steel must certify that it meets the ASTM standard. The values for yield stress and tensile strength given above are minimum requirements; they may be exceeded and usually are to a certain extent. The tensile strength is given as a range of values because this property cannot be achieved to the same degree of precision as the yield stress.

A steel with a yield stress of more than 36 ksi is usually considered to be a high-strength steel. The most frequently used high-strength steels are those with a yield stress of 50 ksi and a tensile strength of either 65 ksi or 70 ksi, although steel with a yield stress of 100 ksi is available. As an example, ASTM A242 is a low-alloy, corrosion-resistant steel available in yield stresses of 42, 46, and 50 ksi with corresponding tensile strengths of 63, 67, and 70 ksi. Its composition is as follows:

Carbon: 0.15% (maximum)

Manganese: 1.00% (maximum)

Phosphorus: 0.15% (maximum)

Sulfur: 0.05% (maximum)

Copper: 0.20% (minimum)

A242 steel is not as ductile as A36; the elongation, based on an original length of 8 inches, is 18% as compared with 20% for A36.

1.6 ≡ STANDARD CROSS-SECTIONAL SHAPES

In the design process outlined earlier, one of the objectives — and the primary emphasis of this book — is the selection of the appropriate cross sections for the individual members of the structure being designed. Most often, this will entail selecting a standard cross-sectional shape that is widely available rather than requiring the fabrication of a shape with unique dimensions and properties. The selection of an "off-the-shelf" item will almost always be the most economical choice, even if it means using slightly more material. The largest category of standard shapes includes those produced by *hot-rolling*. In this manufacturing process, which takes place in a mill, a heated block of steel is passed through a series of rollers that squeeze the material into the desired cross-sectional shape. The block of steel to be rolled is obtained by reheating (to more than 2000° F) the originally cast block, known as an *ingot*, and rolling it into smaller blocks. These smaller blocks are called either *billets* or *slabs*, depending on what is to be rolled from them. By heating the material before rolling, the material can be deformed with no resulting loss of ductility, as would be the case with cold-working. The rolling process greatly lengthens the original block of steel, and the member is cut to standard lengths, usually a maximum of 65 to 75 feet, which are subsequently cut (in a fabricating shop) to the lengths required for a particular structure.

Cross sections of some of the more commonly used hot-rolled shapes are shown in Figure 1.6. The dimensions and designations of the standard available shapes are defined in the ASTM standards.[12] The *W-shape*, also called a *wide-*

FIGURE 1.6 ≡

W-shape
(W18 × 50 shown)

American Standard, S
(S18 × 70 shown)

Equal-leg
angle, L
(L6 × 6 × ¾ shown)

Unequal-leg
angle
(L6 × 3 × ⅝ shown)

American Standard
Channel, C
(C9 × 20 shown)

Structural tee, ST or WT
(WT18 × 115 shown)

flange shape, consists of two parallel flanges separated by a single web. The orientation of these elements is such that the cross section has two axes of symmetry. A typical designation would be "W18 × 50," where W indicates the type of shape, 18 is the nominal depth parallel to the web, and 50 is the weight in pounds per foot of length. The nominal depth is the approximate depth expressed in whole inches. For some of the lighter shapes, it is equal to the depth to the nearest inch, but this is not a general rule for the W-shapes. All of the W-shapes of a given nominal size can be grouped into families that have the same depth from inside-of-flange to inside-of-flange but with different flange thicknesses.

The *American Standard*, or *S-shape*, is similar to the W-shape in having two parallel flanges, a single web, and two axes of symmetry. The difference is in the proportions: The flanges of the W are wider in relation to the web than are the flanges of the S. In addition, the outside and inside faces of the flanges of the W-shape are parallel, whereas the inside faces of the flanges of the S-shape are sloping with respect to the outside faces. An example of the designation of an

S-shape is "S18 × 70," with the S indicating the type of shape, and the two numbers giving the depth in inches and the weight in pounds per foot. This shape was formerly called an *I-beam*.

The angle shapes are available in either equal-leg or unequal-leg versions. A typical designation would be "L6 × 6 × 3/4" or "L6 × 3 × 5/8." The three numbers are the lengths of each of the two legs as measured from the corner, or heel, to the toe at the other end of the leg, and the thickness, which is the same for both legs. In the case of the unequal-leg angle, the longer leg dimension is always given first. While this designation provides all of the dimensions, it does not provide the weight per foot.

The *American Standard Channel*, or *C-shape*, has two flanges and a web, with only one axis of symmetry; it carries a designation such as "C9 × 20." This notation is similar to that for W- and S-shapes, with the first number giving the total depth in inches parallel to the web and the second number the weight in pounds per linear foot. For the channel, however, the depth is exact rather than nominal. The inside faces of the flanges are sloping, just as with the American Standard shape. Miscellaneous Channels — for example, the MC10 × 25 — are similar to American Standard Channels.

The *Structural Tee* is produced by splitting a W-, M-, or S-shape at mid-depth. This shape is sometimes referred to as a *split-tee*. The prefix of the designation is either WT, MT, or ST, depending on which shape is the "parent." For example, a WT18 × 115 has a nominal depth of 18 inches and a weight of 115 pounds per foot, and is cut from a W36 × 230. Similarly, an ST10 × 32.7 is cut from an S20 × 65.4, and an MT3 × 10 is cut from an M6 × 20.

Not shown in Figure 1.6 are two hot-rolled shapes similar to the W-shape: the HP- and M-shapes. The HP shape, used for piles, has parallel flange surfaces, approximately the same width and depth, and equal flange and web thicknesses. The "M" stands for miscellaneous, and it is a shape that does not fit exactly into any of the W, HP, or S categories. Both the M- and HP-shapes are designated in the same manner as the W-shape: for example, M14 × 18 and HP14 × 117.

Other frequently used cross-sectional shapes are shown in Figure 1.7. Bars can have circular, square, or rectangular cross sections. If the width of a rectangular shape is more than 8 inches, it is classified as a *plate* instead of a *bar*. Bars and plates are formed by hot-rolling. Hollow shapes can be produced by either bending plate material into the desired shape and welding the seam or by hot-working to produce a seamless shape. Hollow shapes include circular sections (called *pipe*) and tubes, both square and rectangular.

Other shapes are produced by hot-rolling, but those described above are the ones most frequently used. In most cases, one of these standard shapes will satisfy the design requirements. If the requirements are especially severe, then a built-up section, such as one of those shown in Figure 1.8, may be needed. Sometimes a standard shape is augmented by additional cross-sectional elements, as when a cover plate is welded to one or both flanges of a W-shape. This is an effective way of strengthening an existing structure that is being

FIGURE 1.7 ≡

Bars

Plate

Pipe

Tubes

rehabilitated or modified for some use other than the one for which it was designed. Sometimes a built-up shape must be used because none of the standard rolled shapes are large enough; that is, the cross section does not have enough area or moment of inertia. In such cases, plate girders can be used. These can be I-shaped sections, with two flanges and a web; or box sections, with two flanges and two webs. The components are welded together and can be designed to have exactly the properties needed. Built-up shapes can also be created by attaching two or more standard rolled shapes together. A widely used combination is a pair of angles placed back-to-back and connected at intervals along their length. This is called a *double-angle shape.* There are many other possibilities, and some will be illustrated throughout this book.

Another category of steel products for structural applications is cold-formed steel. Structural shapes of this type are created by bending thin material such as sheet steel or plate into the desired shape without heating. Typical cross sections are shown in Figure 1.9. Only relatively thin material can be used, and the resulting shapes are suitable only for light applications. An advantage of this product is its versatility, since almost any conceivable cross-sectional shape

FIGURE 1.8 ≡

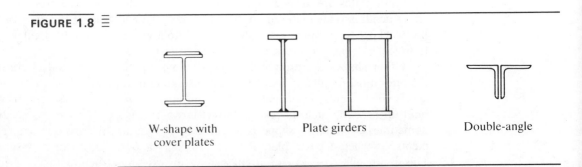

W-shape with
cover plates

Plate girders

Double-angle

FIGURE 1.9 ≡

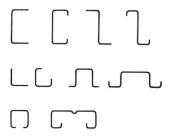

can easily be formed. In addition, cold-working will increase the yield point of the steel, and under certain conditions this may be accounted for in design.[8] This increase comes at the expense of reduced ductility, however. Because of the thinness of the cross-sectional elements, the problem of instability (to be discussed in Chapters 4 and 5) is a particularly important factor in the design of cold-formed steel structures.

2 CONCEPTS IN STRUCTURAL STEEL DESIGN

2.1 ≡ DESIGN PHILOSOPHIES

As discussed earlier, the design of a structural member entails the selection of a cross section that will safely and economically resist the applied loads. Economy usually means minimum weight — that is, the minimum amount of steel. This corresponds to the cross section with the smallest weight per foot, which is the one with the smallest cross-sectional area. Although other considerations, such as ease of construction, may ultimately affect the choice of member size, the process begins with the selection of the lightest cross-sectional shape that will do the job. Having established this objective, we must decide how to do it safely, and this is where there are different approaches to design. There are essentially three different philosophies, and we will discuss them here in a general way, saving specifics for later chapters.

In *allowable stress design*, a member is selected that has cross-sectional properties such as area and moment of inertia large enough to prevent the maximum stress from exceeding an allowable, or permissible, stress. This allowable stress will be in the elastic range of the material and will be less than the yield stress F_y (see Figure 1.4). A typical value might be $0.60F_y$. The allowable stress is obtained by dividing either the yield stress F_y or the ultimate tensile strength F_u by a factor of safety. This approach to design is also called *elastic design* or *working stress design*. Working stresses are those resulting from the working loads, which are the applied loads. A properly designed member will be stressed to no more than the allowable stress when subjected to working loads.

Plastic design is based on a consideration of failure conditions rather than working load conditions. A member is selected by using the criterion that the structure will fail at a load substantially higher than the working load. Failure in this context means either collapse or extremely large deformations. The term *plastic* is used because, at failure, parts of the member will be subjected to very large strains, large enough to put the member into the plastic range (see Figure 1.3b). When the entire cross section becomes plastic at enough locations, "plastic hinges" will form at those locations, creating a *collapse mechanism*. Since the actual loads will be less than the failure loads by a factor of safety known as the *load factor*, members designed this way are not unsafe, despite being designed based on what happens at failure. The design procedure is roughly as follows:

1. Multiply the working loads by the load factor to obtain the failure loads.

2. Determine the cross-sectional properties needed to resist failure under these loads. (A member with these properties is said to have sufficient strength and would be at the verge of failure when subjected to the factored loads.)

3. Select the lightest cross-sectional shape that has these properties.

Members designed by plastic theory would reach the point of failure under the factored loads but are safe under actual working loads.

Load and resistance factor design (LRFD) is similar to plastic design in that strength, or the failure condition, is considered. Load factors are applied to the loads, and a member is selected that will have enough strength to resist the factored loads. In addition, the theoretical strength of the member is reduced by the application of a resistance factor. The criterion that must be satisfied in the selection of a member is

$$\text{factored load} \leq \text{factored strength} \tag{2.1}$$

In this expression, the factored load is actually the sum of all working loads to be resisted by the member, each multiplied by its own load factor. For example, dead loads will have load factors that are different from those for live loads. The factored strength is the theoretical strength multiplied by a resistance factor. Equation 2.1 can therefore be written as

$$\Sigma(\text{loads} \times \text{load factors}) \leq \text{resistance} \times \text{resistance factor} \tag{2.2}$$

Since the factored load is a failure load greater than the total actual working load, the load factors are usually greater than unity. On the other hand, the factored strength is a reduced, usable strength, and the resistance factor is usually less than unity. The factored loads are the loads that bring the structure or member to its limit. In terms of safety, this *limit state* can be fracture, yielding, or buckling, and the factored resistance is the useful strength of the member, reduced from the theoretical value by the resistance factor. The limit state can also be one of serviceability, such as a maximum acceptable deflection.

2.2 ≡ AMERICAN INSTITUTE OF STEEL CONSTRUCTION SPECIFICATION

Because the emphasis of this book is on the design of structural steel building members and their connections, the specification of the American Institute of Steel Construction is the design specification of most importance here. It is written and kept current by an AISC committee comprising structural engineering practitioners, educators, steel producers, and fabricators. New editions are published periodically, and supplements are issued when interim revisions are needed. Allowable stress design has been the primary method used for structural steel buildings since the first AISC Specification was issued in 1923, although recent editions have incorporated provisions for plastic design. In 1986, AISC issued the first specification for load and resistance factor design of structural steel buildings and a companion *Manual of Steel Construction*. The purpose of these two documents was to provide an alternative to allowable stress design, much as plastic design is an alternative. The second edition of the *Manual*,[13] published in 1994, incorporates the 1993 Specification. The provisions of the LRFD Specification are based on research reported in eight papers published in 1978 in the structural journal of the American Society of Civil Engineers (ASCE).[14–21] Unless otherwise stated, references to the AISC Specification and the *Manual of Steel Construction* will be references to the LRFD versions.

Load and resistance factor design is not a recent concept; since 1974 it has been used in Canada, where it is known as *limit states design*. It is also the basis of most European building codes. In this country, LRFD has been an accepted method of design for reinforced concrete for years and is the primary method authorized in the American Concrete Institute's Building Code, where it is known as *strength design*.[22] The AASHTO Specification permits either service load design (allowable stress design) or strength design, also called *load factor design*.[6] The AASHTO provisions for strength design incorporate both load and resistance factors for reinforced concrete, but only load factors for structural steel.

The AISC Specification is published as a stand-alone document but is also part of the *Manual of Steel Construction*, which will be discussed in the next section. Except for specialized steel products such as cold-formed steel, which is covered by a different specification,[8] the AISC Specification is the standard by which virtually all structural steel buildings in this country are designed and constructed. Because of this, it is imperative that the student of structural steel design have ready access to this document. The details of the Specification will be covered in the chapters that follow, but the overall organization will be discussed here.

The Specification consists of four parts: the main body, the appendixes, the Numerical Values section, and the Commentary. The body is alphabetically organized into Chapters A through M. Within each chapter, major headings are labeled with the chapter designation followed by a number. Further subdivisions are numerically labeled. For example, the types of structural steel authorized are listed in Chapter A, "General Provisions," under Section A3. Material, and, under this, Section 1. Structural Steel. The main body of the Specification is followed by appendixes to selected chapters. The appendixes are labeled B, E, F, G, H, J, and K to correspond to the chapters that are referenced. This section is followed by the Numerical Values section, which contains tables of numerical values for some of the Specification requirements. The Numerical Values section is followed by the Commentary, which gives background and elaboration on many of the provisions of the Specification. It uses the same organizational scheme as the Specification, so material applicable to a particular section can be easily located. The appendixes, the Numerical Values section, and the Commentary are considered to be official parts of the Specification and carry the same authority as material in the main body.

2.3 ≡ LOAD AND RESISTANCE FACTORS USED IN THE AISC SPECIFICATION

Equation 2.2 can be written more precisely as

$$\Sigma \gamma_i Q_i \le \phi R_n \tag{2.3}$$

where

Q_i = a load effect (a force or a moment)

γ_i = a load factor

R_n = the nominal resistance, or strength, of the component under consideration

ϕ = a resistance factor

The factored resistance ϕR_n is called the *design strength*. The summation on the left side of Equation 2.3 is over the total number of load effects (including, but not limited to, dead load and live load), where each load effect can be associated with a different load factor. Not only can each load effect have a different load factor, but also the value of the load factor for a particular load effect will depend on the combination of loads under consideration. Load combinations to be considered are given in Chapter A, "General Provisions," of the Specification as

$1.4D$	(A4-1)
$1.2D + 1.6L + 0.5(L_r \text{ or } S \text{ or } R)$	(A4-2)
$1.2D + 1.6(L_r \text{ or } S \text{ or } R) + (0.5L \text{ or } 0.8W)$	(A4-3)
$1.2D + 1.3W + 0.5L + 0.5(L_r \text{ or } S \text{ or } R)$	(A4-4)
$1.2D \pm 1.0E + 0.5L + 0.2S$	(A4-5)
$0.9D \pm (1.3W \text{ or } 1.0E)$	(A4-6)

where

D = dead load

L = live load due to equipment and occupancy

L_r = roof live load

S = snow load

R = rain or ice load*

W = wind load

E = earthquake load

---

* This does not include *ponding*, a phenomenon that will be discussed in Chapter 5 of this book.

The identifier to the right of each of the above load combinations is the numbering scheme used by AISC for equations and expressions, with the letter representing the chapter, the first number the section, and the second number the sequence within that section.

As previously mentioned, the load factor for a given load effect is not the same in all load combinations. For example, in A4-2, γ for the live load L is 1.6, whereas in A4–3, it is 0.5. This is because the live load is being taken as the dominant effect in A4-2 and one of the three effects, L_r, S, or R, will be dominant in A4-3. In each combination, one of the effects is considered to be at its "lifetime maximum" value and the others at their "arbitrary point in time" values. These load factors and load combinations are the ones recommended in *Minimum Design Loads*[4] and are based on extensive statistical studies.

The resistance factor ϕ for each type of resistance is given by AISC in the Specification chapter dealing with that resistance. These factors range in value from 0.75 to 1.0.

Before considering the theoretical basis of the load and resistance factors, we will illustrate their application with an axially loaded compression member. The resistance factor for this type of member is $\phi_c = 0.85$. For such a member subjected to dead and live loads only, the only load combinations that could possibly control are A4-1: $1.4P_D$ and A4-2: $1.2P_D + 1.6P_L$, where P_D is the dead load and P_L is the live load. The larger value from these two combinations must be less than $\phi R_n = \phi_c P_n = 0.85P_n$, the reduced strength, or design strength, of the compression member. If, for example, the second load combination were larger, it would govern, and the requirement of Equation 2.3 would be

$$1.2P_D + 1.6P_L \le 0.85P_n$$

2.4 ≡ PROBABILISTIC BASIS OF LOAD AND RESISTANCE FACTORS

Both the load and the resistance factors specified by AISC are based on probabilistic concepts. The resistance factors account for uncertainties in material properties, design theory, and fabrication and construction practices. Although a complete treatment of probability theory is beyond the scope of this book, a brief summary of the basic concepts will be given here.

Experimental data can be represented in the form of a histogram, or bar graph, as shown in Figure 2.1, with the abscissa representing sample values, or events, and the ordinate representing either the number of samples having a certain value or the frequency of occurrence of a certain value. Each bar can represent a single sample value or a range of values. If the ordinate is the percentage of values rather than the actual number of values, the graph is referred to as a *relative* frequency distribution. In such a case the sum of the ordinates will be 100%. If the abscissa values are random events, and enough samples are used, each ordinate can be interpreted as the probability, expressed as a percentage, of that sample value or event occurring. The relative frequency can also be expressed in decimal form, with values between 0 and 1.0. Thus, the sum of the ordinates will be unity, and if each bar has a unit width, the total area of the diagram will also be unity. This implies a probability of 1.0 that an event will fall within the boundaries of the diagram. Furthermore, the probability that a certain value or something smaller will occur is equal to the area of the diagram to the left of that value. The probability of an event having a value falling between *a* and *b* in Figure 2.1 is equal to the area of the diagram between *a* and *b*.

Before proceeding, some definitions are in order. The *mean* $\bar{x}$ of a set of sample values, or *population*, is the arithmetic average:

$$\bar{x} = \frac{1}{n} \sum_{i=1}^{n} x_i$$

where x_i is a sample value and n is the number of values. The *median* is the midmost value of x, and the *mode* is the most frequently occurring value. The *variance* v is a measure of the overall variation of the data from the mean and is defined as

$$v = \frac{1}{n} \sum_{i=1}^{n} (x_i - \bar{x})^2$$

The *standard deviation* s is the square root of the variance:

$$s = \sqrt{\frac{1}{n} \sum_{i=1}^{n} (x_i - \bar{x})^2}$$

Like the variance, the standard deviation is a measure of the overall variation, but it has the same units and the same order of magnitude as the data. The *coefficient of variation* V is the standard deviation divided by the mean:

$$V = \frac{s}{\bar{x}}$$

FIGURE 2.1

FIGURE 2.2 ≡

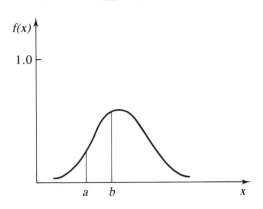

If the actual frequency distribution is replaced by a theoretical continuous function that closely approximates the data, it is called a *probability density function*. Such a function is illustrated in Figure 2.2. Probability functions are designed such that the total area under the curve is unity. That is, for a function $f(x)$,

$$\int_{-\infty}^{+\infty} f(x)\, dx = 1.0$$

which means that the probability that one of the sample values or events will occur is equal to 1.0. The probability of one of the events between a and b in Figure 2.2 is equal to the area under the curve between a and b, or

$$\int_{a}^{b} f(x)\, dx$$

When a theoretical probability density function is used, the following notation is conventional:

μ = mean

σ = standard deviation

The probabilistic basis of the load and resistance factors used by AISC is presented in the ASCE structural journal and will be summarized here.[14] Load effects Q and resistances R are random variables and depend on many factors. Loads can be estimated or obtained from measurements and inventories of actual structures, and resistances can be computed or determined experimentally. Discrete values of Q and R from observations can be plotted as frequency distribution histograms or represented by theoretical probability density functions. This latter representation will be used in the material that follows.

If the probability density functions for load effects Q and resistances R are plotted on the same graph, as in Figure 2.3, the region corresponding to $Q > R$ represents failure, and $Q < R$ represents survival. If the distributions of Q and R

FIGURE 2.3 ≡

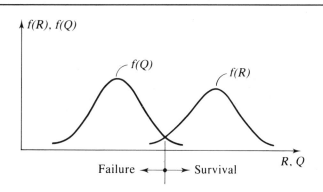

are combined into one function, $R - Q$, positive values of $R - Q$ correspond to survival. Equivalently, if a probability density function of R/Q, the factor of safety, is used, survival is represented by values of R/Q greater than 1.0. The corresponding probability of failure is the probability that R/Q is less than 1; that is,

$$P_F = P\left[\left(\frac{R}{Q}\right) < 1\right]$$

Taking the natural logarithm of both sides of the inequality,

$$P_F = P\left[\ln\left(\frac{R}{Q}\right) < \ln 1\right] = P\left[\ln\left(\frac{R}{Q}\right) < 0\right]$$

The frequency distribution curve of $\ln(R/Q)$ is shown in Figure 2.4.

FIGURE 2.4 ≡

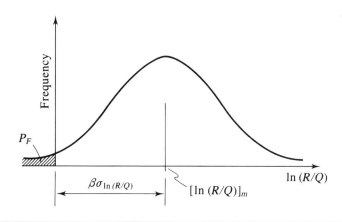

The *standardized* form of the variable $\ln(R/Q)$ can be defined as

$$U = \frac{\ln\left(\dfrac{R}{Q}\right) - \left[\ln\left(\dfrac{R}{Q}\right)\right]_m}{\sigma_{\ln(R/Q)}}$$

where

$$\left[\ln\left(\frac{R}{Q}\right)\right]_m = \text{the mean value of } \ln\left(\frac{R}{Q}\right)$$

$$\sigma_{\ln(R/Q)} = \text{standard deviation of } \ln\left(\frac{R}{Q}\right)$$

This transformation converts the abscissa U to multiples of standard deviations and places the mean of U at $U = 0$. The probability of failure can then be written as

$$P_F = P\left[\ln\left(\frac{R}{Q}\right) < 0\right] = P\left(\left\{U\sigma_{\ln(R/Q)} + \left[\ln\left(\frac{R}{Q}\right)\right]_m\right\} < 0\right)$$

$$= P\left\{U < -\frac{\left[\ln\left(\dfrac{R}{Q}\right)\right]_m}{\sigma_{\ln(R/Q)}}\right\} = F_u\left\{-\frac{\left[\ln\left(\dfrac{R}{Q}\right)\right]_m}{\sigma_{\ln(R/Q)}}\right\}$$

where F_u is the *cumulative distribution function* of U, or the probability that U will not exceed the argument of the function. If we let

$$\beta = \frac{\left[\ln\left(\dfrac{R}{Q}\right)\right]_m}{\sigma_{\ln(R/Q)}} \tag{2.4}$$

then

$$\left[\ln\left(\frac{R}{Q}\right)\right]_m = \beta\sigma_{\ln(R/Q)}$$

The variable β can be interpreted as the number of standard deviations from the origin that the mean value of $\ln(R/Q)$ is. For safety, the mean value *must* be more than zero, and as a consequence, β is called the *safety index* or *reliability index*. The larger this value, the larger the margin of safety. This means that the probability of failure, represented by the shaded area in Figure 2.4 labeled P_F, will be smaller. The reliability index is a function of both the load effect Q and the resistance R. By using the same reliability index for each type of member when subjected to the same type of loading, there is relative uniformity of strength from member to member. Based on the recommendations given in the ASCE structural journal,[14] the "target" values of β shown in Table 2.1 were selected and used in computing both load and resistance factors.

TABLE 2.1 ≡ Target Values of β

Type of Component	Loading Condition		
	$D + (L \text{ or } S)$	$D + L + W$	$D + L + E$
Members	3.0	2.5	1.75
Connections	4.5	4.5	4.5

The ASCE structural journal[14] shows that

$$\phi = \frac{R_m}{R_n} e^{-0.55\beta V_R} \tag{2.5}$$

where

R_m = mean value of the resistance R

R_n = nominal or theoretical resistance

V_R = coefficient of variation of R

Equation 2.5 is the expression for the resistance factor ϕ as given in the Commentary to the Specification.

2.5 ≡ MANUAL OF STEEL CONSTRUCTION

Anyone engaged in structural steel design in the United States must have access to AISC's *Manual of Steel Construction*.[13] This publication contains the AISC Specification and numerous design aids in the form of tables and graphs as well as a "catalog" of the most widely available structural shapes. This textbook is written under the assumption that the reader will have the *Manual* at hand at all times. To encourage the manual's use, tables and graphs will not be reproduced in this book, so that reference to the *Manual* is necessary.

The *Manual* consists of two volumes. Volume I, subtitled "Structural Members, Specifications and Codes," contains Parts 1–7 and deals primarily with member design. Volume II, subtitled "Connections," contains Parts 8–12 and is devoted to connection design. The primary focus of this textbook will be on Volume I, which is divided into seven parts as follows:

Part 1. Dimensions and Properties: This part contains details on standard rolled shapes, pipe, and structural tubing, including all necessary cross-sectional dimensions and properties such as area and moment of inertia. Information is also given on their availability in various strengths. The steels covered are the ones authorized by the AISC Specification for use in building construction and include the following:

ASTM A36: All-purpose carbon grade steel

ASTM A529: Structural carbon steel

ASTM A572: High-strength, low-alloy structural steel

ASTM A242: Atmospheric corrosion-resistant, high-strength, low-alloy structural steel

ASTM A588: Atmospheric corrosion-resistant, high-strength, low-alloy structural steel

ASTM A852: Quenched and tempered alloy structural steel plate

ASTM A514: Quenched and tempered alloy structural steel plate

Part 2. Essentials of LRFD: This is a condensed introduction to the basics of load and resistance factor design of steel structures. Numerical examples are included.

Part 3. Column Design: This part contains numerous tables to facilitate the design of axially loaded compression members and beam-columns. Most of these tables are specialized for steels with yield stresses of 36 ksi and 50 ksi. In addition, design examples illustrating the use of the tables are given.

Part 4. Beam and Girder Design: As with Part 3, many design aids are furnished, including graphs as well as tables. While many of these deal with the requirements of the AISC Specification, some — such as the Beam Diagrams and Formulas — pertain to structural analysis. This part also contains discussion of beam and girder design procedures as well as design examples.

Part 5. Composite Design: Composite members, usually beams or columns, are structural components composed of two materials — structural steel and reinforced concrete. Composite beams are ordinarily used when a system of parallel beams supports a reinforced concrete floor slab. In this application, elements welded to the top flange are embedded in the concrete, forming the connection between the two materials. Composite columns consist of structural steel shapes encased in reinforced concrete. Part 5 contains background information, design aids, and examples.

Part 6. Specifications and Codes: This part contains the AISC Specification and the Commentary, a specification for high-strength bolts,[23] and other documents.

Part 7. Miscellaneous Data and Mathematical Tables: This part deals with wire, sheet steel, and various properties of steel and other building materials. Also included are mathematical formulas and conversion factors for different systems of units.

Volume II, comprising Parts 8–13, contains tables to assist in the design of bolted and welded connections as well as tables that provide details on "standard" connections. (Part 13 is a list of construction industry organizations.)

The AISC Specification is only a small part of the *Manual*. Many of the terms and constants used in other parts of the *Manual* are for the purpose of facilitating the design process and are not necessarily part of the Specification. In some instances, recommendations may be given that are not requirements of the Specification but only "rules of thumb" based on common practice

Although such information is not in conflict with the Specification, it is important to recognize what is a *requirement* (when adopted by a building code) and what is not.

2.6 ≡ DESIGN COMPUTATIONS AND PRECISION

The computations required in engineering design and analysis are done with either an electronic calculator or a digital computer, which may be a microcomputer, minicomputer, or mainframe computer. When one is doing manual computations with the aid of an electronic calculator, a decision must be made regarding the degree of precision needed. The problem of how many significant figures to use in engineering computations has no simple solution. Recording too many significant digits is misleading and can imply an unrealistic degree of precision. On the other hand, recording too few figures can lead to chaos and meaningless results. The question of precision was mostly academic before the early 1970s, when the chief calculating tool was the slide rule. The guiding principle at that time was to read and record numbers as accurately as possible, which meant three to four significant figures.

There are many inherent inaccuracies and uncertainties in structural design, including variations in material properties and loads; load estimates sometimes border on educated guesses. It hardly makes sense to perform computations with 12 significant figures and record the answer to that degree of precision when the yield stress is rounded to the nearest 1000 psi, or two significant figures. To avoid results that are even less precise, however, it is reasonable to assume that the given parameters of a problem, such as the yield stress, are correct and then decide on the degree of precision required in subsequent calculations.

A further complication arises when electronic calculators are used. If all of the computations in a given problem are done in one continuous series of operations on a calculator, the number of significant figures used is whatever the calculator uses, perhaps 10 or 12. But if intermediate values are rounded off, recorded, and used in subsequent computations, then a consistent number of significant figures will not have been used. Furthermore, the manner in which the computations are grouped will have an influence on the final result. In general, the result will be no more accurate than the least accurate number used in the computation — and sometimes less because of round-off error. For example, consider a number calculated on a 12-digit calculator and recorded to five significant figures. If this number is multiplied by a similarly obtained and rounded number, the product will be precise to five significant figures at most, regardless of the number of digits displayed on the calculator. It is not reasonable to record this number to more than five significant figures.

It is also unreasonable to record the results of every calculator multiplication or division to a predetermined number of significant figures in order to have a consistent degree of precision throughout. A reasonable approach is to perform operations on the calculator in any convenient manner and record intermediate values to whatever degree of precision is deemed adequate (without clearing the intermediate value from the calculator if it can be used in the next computation). The final results should then be expressed to a precision consistent with this procedure. It is difficult to determine what this precision should be for the typical structural steel design problem. More precision than three or four significant figures is probably unrealistic in most cases, and results based on less than three may be too approximate to be of any value. The practice in this book will be to record intermediate values to three or four digits (usually four), depending on the circumstances, and to record final results to three digits.

Although this is not a high degree of precision, it should not be made worse by sloppy numerical practices. When numbers are rounded, they should be rounded properly and not truncated. Thus, 1.666666 rounded to three significant figures is 1.67, not 1.66. Care should also be taken to differentiate between number of significant figures and number of decimal places; the two are not always the same. For example, 0.00323 is expressed to three significant figures, while 0.003 is expressed to three decimal places but only one significant figure.

Much routine structural analysis and design can be done with the aid of a digital computer and the appropriate computer programs. To use this powerful tool in a responsible manner, however, the engineer must have a working knowledge of the methods used by a program so that the correct input data will be used and the results properly interpreted. The ability to write computer programs is also a useful skill. Even though a structural engineer may never need to write an original program, the need for modifying an existing program is common. Most of the chapters in this book include programming assignments. The purpose of these is twofold: to reinforce the concepts and procedures involved in a particular application, and to sharpen programming skills, which become dull if not used.

3

TENSION MEMBERS

INTRODUCTION

Tension members are defined as structural elements that are subjected to axial tensile forces. They are found in a variety of structures and include truss members, cables in suspension and cable-stayed bridges, bracing for buildings and bridges, and cables in suspended roof systems. Any cross-sectional configuration may be used, since for a given material, the only determinant of strength is the cross-sectional area. Circular rods and rolled angle shapes are commonly used. Shapes built up from rolled shapes or a combination of rolled shapes and plates are sometimes used when large loads must be resisted. The most common built-up configuration is probably the double-angle section, shown in Figure 3.1 along with other typical cross sections. Because the use of this section is so widespread, tables of properties of various combinations of angles are included in the AISC *Manual of Steel Construction*.

The stress in an axially loaded tension member is given by

$$f = \frac{P}{A}$$

where P is the magnitude of the load, and A is the cross-sectional area normal to the load. The stress as given by this equation is exact, provided the cross section under consideration is not adjacent to the point of application of the load, where the distribution of stress is not uniform.

FIGURE 3.1 ≡

FIGURE 3.2 ≡

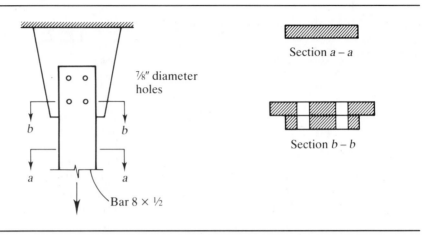

Section *a – a*

Section *b – b*

⅞″ diameter holes

b   *b*

a   *a*

Bar 8 × ½

If the cross-sectional area of a tension member varies along its length, the stress is a function of the particular section under consideration. The presence of holes in a member also influences the stress at a cross section through the hole or holes. At such a location, the cross-sectional area will be reduced by an amount equal to the area removed by the holes. Tension members are frequently connected at their ends with bolts, as shown in Figure 3.2. The tension member, a bar $8 \times \frac{1}{2}$ in., is connected to a *gusset plate*, a connection element whose purpose is to transfer the load from the member. While the area of the bar at section *a–a* is $(\frac{1}{2})(8) = 4$ in.^{2}, the area at section *b–b* is $4 - (2)(\frac{1}{2})(\frac{7}{8}) = 3.12$ in.^{2} and will be more highly stressed. This reduced area is referred to as the *net area*, or *net section*, and the unreduced area is the *gross area*.

The typical design problem is to select a member with sufficient cross-sectional area so that the factored load does not exceed the design strength (the nominal strength multiplied by the resistance factor). A closely related problem is that of analysis, or review, of a given member, wherein the design strength is computed and compared with the factored load. In general, analysis is a direct procedure, while design is an iterative process and may require some trial and error.

Tension members are covered in Chapter D of the Specification. Requirements that are common with other types of members are covered in Chapter B, "Design Requirements."

3.2 ≡ **DESIGN STRENGTH**

A tension member can fail by reaching one of two limit states: yielding or fracture. To prevent yielding and the accompanying excessive deformation, the load on the gross section must be small enough so that the stress on the gross section is less than the yield stress F_y. To prevent fracture, the stress on the net section must be less than the tensile strength F_u. In each case, the stress P/A must be less than the limiting stress F. This means that the load P must be less than F times A:

$$P < FA$$

The left side of this expression is the applied factored load, and the right side is the strength. The *nominal* strength in yielding is

$$P_n = F_y A_g$$

and the nominal strength in fracture is

$$P_n = F_u A_e$$

where A_e is the *effective* net area, which may be equal to either the net area or, in some cases, a smaller area. This will be discussed in the next section.

Although yielding will first occur on the net cross section, the deformation within the length of the connection will generally be smaller than the deformation in the remainder of the tension member. This is because the net section exists over a relatively small length of the member, and the total elongation is a product of the length and the strain (a function of the stress). Since most of the member will have an unreduced cross section, attainment of the yield stress on the gross area will result in larger total elongation.

The resistance factor $\phi = \phi_t$ is smaller for fracture than for yielding, reflecting the more serious nature of reaching the limit state of fracture.

For yielding, $\phi_t = 0.90$

For fracture, $\phi_t = 0.75$

Equation 2.3,

$$\sum \gamma_i Q_i \le \phi R_n$$

can be written for tension members as

$$\sum \gamma_i Q_i \le \phi_t P_n$$

or

$$P_u \le \phi_t P_n$$

where P_u is the governing combination of factored loads. Because there are two limit states, both of the following conditions must be satisfied:

$$P_u \le 0.90 F_y A_g$$
$$P_u \le 0.75 F_u A_e$$

The smaller of these is the design strength of the member.

Values of F_y, F_u, and $0.75F_u$ can be found in Table 2 in the Numerical Values section of the AISC Specification, but some of the high-strength steels are grouped in such a way that it is not always possible to associate a given material with a particular yield stress and tensile strength by inspection. This information can be obtained from Tables 1-1 and 1-2 in Part 1 of the AISC *Manual of Steel Construction* (hereafter referred to as the *Manual*). For example, to determine the properties of a W33 × 221 of ASTM A242 steel, first refer to Table 1-2 and determine the shape group to which this section belongs — in this case, Group 3. In Table 1-1, all the steels available for Group 3 shapes are represented by shaded areas under the Group 3 heading. For A242 steel, the only alloy available for Group 3 shapes has a yield stress F_y of 46 ksi and a tensile strength F_u of 67 ksi.

The exact amount of area to be deducted from the gross area to account for the presence of bolt holes depends on the fabrication procedure. The usual practice is to drill or punch standard holes (i.e., not oversized) with a diameter $\frac{1}{16}$ in. larger than the fastener diameter. To account for possible roughness around the edges of the hole, Section B2 of the AISC Specification (henceforth, references to the Specification will be in the form AISC B2) requires the addition of $\frac{1}{16}$ in. to the actual hole diameter. This amounts to using an effective hole diameter $\frac{1}{8}$ in. larger than the fastener diameter. In the case of slotted holes, $\frac{1}{16}$ in. should be added to the actual *width* of the hole. Details related to standard, oversized, and slotted holes can be found in AISC J3.2, "Size and Use of Holes" (in Chapter J, "Connections, Joints, and Fasteners").

☰ EXAMPLE 3.1

A bar 5 × $\frac{1}{2}$ of A36 steel is used as a tension member. It is connected to a gusset plate with four $\frac{5}{8}$-in. diameter bolts as shown in Figure 3.3. Assume that the effective net area A_e is equal to the actual net area and compute the design strength.

SOLUTION

For the limit state of yielding,

$$A_g = 5(\tfrac{1}{2}) = 2.5 \text{ in.}^2$$

The *nominal* strength is

$$P_n = F_y A_g = 36(2.5) = 90 \text{ kips}$$

and the *design* strength is

$$\phi_t P_n = 0.90(90) = 81 \text{ kips}$$

For fracture of the net section,

$$
\begin{aligned}
A_n &= A_g - A_{\text{holes}} \\
&= 2.5 - (\tfrac{1}{2})(\tfrac{3}{4}) \times 2 \text{ holes} \\
&= 2.5 - 0.75 = 1.75 \text{ in.}^2 \\
A_e &= A_n = 1.75 \text{ in.}^2 \quad \text{(for this example)}
\end{aligned}
$$

FIGURE 3.3 ≡

The *nominal* strength is

$$P_n = F_u A_e = 58(1.75) = 101.5 \text{ kips}$$

and the *design* strength is

$$\phi_t P_n = 0.75(101.5) = 76.1 \text{ kips}$$

The smaller value controls.

Answer   Design strength = 76.1 kips

It would appear that the effects of stress concentrations at holes have been overlooked. In reality, stresses at holes can be as high as three times the average stress on the net section, and at fillets of rolled shapes they can be more than twice the average.[24] Because of the ductile nature of structural steel, the usual design practice is to neglect such localized overstress. After yielding begins at a point of stress concentration, additional stress is transferred to adjacent areas of the cross section. It is this stress redistribution that is responsible for the "forgiving" nature of structural steel. Its ductility permits the initially yielded zone to deform without fracture as the stress on the remainder of the cross section continues to increase. Under certain conditions, however, steel may lose its ductility and stress concentrations can precipitate brittle fracture. These situations include fatigue loading and extremely low temperature.

≡ **EXAMPLE 3.2**

A single angle tension member, an $L3\frac{1}{2} \times 3\frac{1}{2} \times \frac{3}{8}$, is connected to a gusset plate with $\frac{7}{8}$-in. diameter bolts as shown in Figure 3.4. A36 steel is used. The service loads are 35 kips dead load and 15 kips live load. Investigate this member for compliance with the AISC Specification. Assume that the effective net area is 85% of the computed net area (computation of effective net area will be covered in Section 3.3).

FIGURE 3.4 ☰

Section

SOLUTION   Load combinations:

(A4-1): $1.4D = 1.4(35) = 49$ kips

(A4-2): $1.2D + 1.6L = 1.2(35) + 1.6(15) = 66$ kips

The second combination controls; $P_u = 66$ kips.

Design strength:

Gross section: $A_g = 2.48$ in.^{2} (from Part 1 of the *Manual*)

$$\phi_t P_n = \phi_t F_y A_g = 0.90(36)(2.48) = 80.4 \text{ kips}$$

Net section: $A_n = 2.48 - \frac{3}{8}(\frac{7}{8} + \frac{1}{8}) = 2.105$ in.^{2}

$$A_e = 0.85(2.105) = 1.789 \text{ in.}^2 \text{ (in this example)}$$

$$\phi_t P_n = \phi_t F_u A_e = 0.75(58)(1.789) = 77.8 \text{ kips (controls)}$$

Answer   Since $P_u < \phi_t P_n$ (66 kips < 77.8 kips), the member is satisfactory.

In the preceding example, load combination A4-2 controls. When only dead load and live loads are present, this combination will always control when the dead load is less than eight times the live load. In future examples, 1.4D (condition A4-1) will not be checked when it obviously does not control.

3.3 ☰ EFFECTIVE NET AREA

Of the several factors influencing the performance of a tension member, the manner in which it is connected is the most important. A connection almost always weakens the member, and the measure of its influence is called the *joint efficiency*. This factor is a function of the ductility of the material, fastener spacing, stress concentrations at holes, fabrication procedure, and a phenomenon known as *shear lag*. All of these contribute to reducing the effectiveness of the member, but shear lag is the most important.

FIGURE 3.5 ≡

Section

Shear lag occurs when all elements of the cross section are not connected. This happens, for example, when only one leg of an angle is bolted to a gusset plate (Figure 3.5). The consequence of this partial connection is that the connected element becomes overloaded and the unconnected part is not fully stressed. Lengthening the connected region relative to the member length will reduce this effect. Research reported in the ASCE structural journal[25] suggests that shear lag be accounted for by using a reduced, or effective, net area obtained by multiplying the actual net area by the reduction factor, U:

$$U = 1 - \left(\frac{\bar{x}}{L}\right) \leq 0.9 \qquad \text{(AISC Equation B3-2)}$$

In this expression, $\bar{x}$ is the distance from the centroid of the connected area to the shear plane of the connection, and L is the length of the connection. In the event a member has two symmetrically located planes of connection, $\bar{x}$ is measured from the centroid of the nearest one-half of the area. Figure 3.6 illustrates $\bar{x}$ for various cross-sectional configurations.

FIGURE 3.6 ≡

For bolted connections, the effective net area is

$$A_e = UA_n$$

For welded connections,

$$A_e = UA_g$$

If all elements of the cross section are connected, as in Example 3.1, then there is no reduction, and $A_e = A_n$.

FIGURE 3.7 ≡

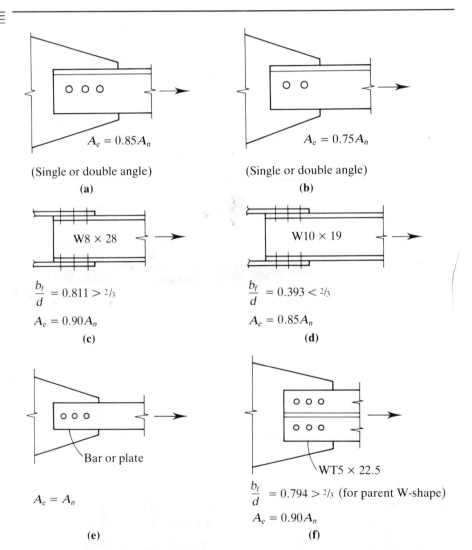

$A_e = 0.85A_n$

(Single or double angle)

(a)

$A_e = 0.75A_n$

(Single or double angle)

(b)

W8 × 28

$\dfrac{b_f}{d} = 0.811 > \frac{2}{3}$

$A_e = 0.90A_n$

(c)

W10 × 19

$\dfrac{b_f}{d} = 0.393 < \frac{2}{3}$

$A_e = 0.85A_n$

(d)

Bar or plate

$A_e = A_n$

(e)

WT5 × 22.5

$\dfrac{b_f}{d} = 0.794 > \frac{2}{3}$ (for parent W-shape)

$A_e = 0.90A_n$

(f)

Based on average values of $\bar{x}/L$ for various types of bolted tension member connections, the Commentary to AISC B3 gives values of the reduction factor U that may be used in lieu of the computed value of $1 - (\bar{x}/L)$. These average values of U for bolted connections are based on two broad categories of connections: those with two fasteners per line in the direction of the applied load and those with three or more per line. Only three different values are given; they correspond to the following conditions.

1. For W-, M-, and S-shapes that have a flange width-to-web depth ratio of at least $\frac{2}{3}$ (and tee shapes cut from these) and are connected through the flanges with at least three fasteners per line in the direction of applied load,

   $U = 0.90$

2. For all other shapes (including built-up shapes) with at least three fasteners per line,

   $U = 0.85$

3. For all members with only two fasteners per line,

   $U = 0.75$

Application of the above rules is illustrated in Figure 3.7.

It should be emphasized that this reduction factor is applied only when some elements of the cross section are not connected. For tension members such as single plates and bars (as in Example 3.1), the effective area is taken as the full computed net area.

EXAMPLE 3.3

Determine the effective net area for the tension member shown in Figure 3.8.

SOLUTION

$$A_n = A_g - A_{\text{holes}}$$
$$= 5.75 - \tfrac{1}{2}(\tfrac{5}{8} + \tfrac{1}{8})(2) = 5.00 \text{ in.}^2$$

Since only one element (one leg) of the cross section is connected, the net area must be reduced. From the properties tables in Part 1 of the *Manual*, the distance from the centroid to the outside face of the leg of an $L6 \times 6 \times \frac{1}{2}$ is

$$\bar{x} = 1.68 \text{ in.}$$

The length of the connection is

$$L = 3 + 3 = 6 \text{ in.}$$

$$\therefore \quad U = 1 - \left(\frac{\bar{x}}{L}\right) = 1 - \left(\frac{1.68}{6}\right) = 0.720 < 0.9 \quad \text{(OK)}$$

$$A_e = U A_n = 0.720(5.00) = 3.60 \text{ in.}^2$$

FIGURE 3.8 ≡

$L6 \times 6 \times \frac{1}{2}$

⅝ in. bolts

3" 3"

Section

The average value of U from the Commentary could also be used. Since this shape is not a W, M, S, or tee and has more than two bolts in the direction of the load, the reduction factor U can be taken as 0.85, and

$$A_e = UA_n = 0.85(5.00) = 4.25 \text{ in.}^2$$

Either U value is acceptable, but the value obtained from AISC Equation B3-2 is more accurate. However, the average values of U are useful during preliminary design, when actual section properties are not known.

AISC B3 gives two special cases for welded connections:

1. For members connected by transverse welds alone,

A_e = area of connected element

Figure 3.9 illustrates the difference between transverse and longitudinal welds.

2. For plates or bars connected by longitudinal welds at their ends as in Figure 3.10,

$$U = 1.0 \quad \text{for } \ell \geq 2w$$
$$= 0.87 \quad \text{for } 1.5w \leq \ell < 2w$$
$$= 0.75 \quad \text{for } w \leq \ell < 1.5w$$

FIGURE 3.9 ≡

Longitudinal

Transverse

FIGURE 3.10 ≣

where
$$\ell = \text{length of the pair of welds} \geq w$$
$$w = \text{distance between the welds}$$

≣ **EXAMPLE 3.4**

If the tension member of Example 3.3 is welded as shown in Figure 3.11, determine the effective net area.

SOLUTION

As in the previous example, only part of the cross section is connected, and a reduced effective net area must be used. The connection is made with a combination of longitudinal and transverse welds, so it is not one of the special cases for welded members.

$$U = 1 - \left(\frac{\bar{x}}{L}\right) = 1 - \left(\frac{1.68}{5.5}\right) = 0.695 < 0.9$$

Answer $A_e = UA_g = 0.695(5.75) = 4.00 \text{ in.}^2$

FIGURE 3.11 ≣

$5\frac{1}{2}''$ $L6 \times 6 \times \frac{1}{2}$

3.4 ≡ STAGGERED FASTENERS

If a tension member connection is made with bolts, the net area will be maximized if the fasteners are placed in a single line. Sometimes space limitations, such as a limit on dimension a in Figure 3.12a, necessitate using more than one line. If this must be done, the reduction in cross-sectional area is minimized if the fasteners are arranged in a staggered pattern as shown. Sometimes staggered fasteners are required by the geometry of a connection like the one in Figure 3.12b. In either case, any given cross section will pass through fewer holes than if the fasteners are not staggered.

If the amount of stagger is small enough, the influence of an offset hole may be felt by a nearby cross section, and fracture along an inclined path such as $abcd$ in Figure 3.12c is possible. In such a case, the relationship $f = P/A$ does not apply, and stresses on the inclined portion b–c are a combination of tensile and shearing stresses. Several approximate methods have been proposed for accounting for the effects of staggered holes. Cochran[26] suggests the use of a net area equal to the product of the plate thickness and a net width. This is computed as follows: Decide on a possible failure line to investigate and let the net width be equal to the gross width minus the quantity

$$d' = d - \frac{s^2}{4g} \tag{3.1}$$

FIGURE 3.12 ≡

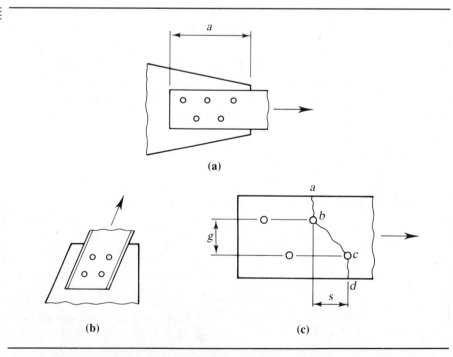

(a)

(b)

(c)

for each staggered hole in the chain, minus d for each unstaggered hole. In this expression, d is the hole diameter, s (pitch) is the spacing of two adjacent holes in the direction parallel to the load, and g (gage) is the transverse spacing. The AISC Specification uses this same approximation, but in a slightly different form. Section B2 requires that the net width be computed by subtracting the sum of the hole diameters from the gross width and adding, for each inclined line in the chain, the quantity $s^2/4g$. That is,

$$w_n = w_g - \sum d + \sum \frac{s^2}{4g}$$

In those instances where more than one failure pattern is conceivable, all possibilities should be investigated, and the one corresponding to the smallest load capacity should be used. Note that this method will not accommodate failure patterns with lines parallel to the applied load.

EXAMPLE 3.5

Compute the net area for the plate shown in Figure 3.13. The holes are for 1-in. diameter bolts.

SOLUTION

The effective hole diameter is $1 + \frac{1}{8} = 1\frac{1}{8}$ in.

Line $abde$:

$$w_n = 16 - 2(1.125) = 13.75 \text{ in.}$$

Line $abcde$:

$$w_n = 16 - 3(1.125) + \frac{2(3)^2}{4(5)} = 13.52 \text{ in.}$$

The second condition governs:

Answer $A_n = tw_n = 0.75(13.52) = 10.14 \text{ in.}^2$

FIGURE 3.13 ≡

FIGURE 3.14 ≡

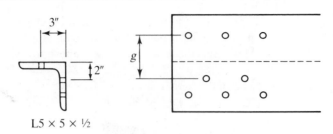

L5 × 5 × ½

Since each fastener resists an equal share of the load (an assumption used in the design of simple connections; see Chapter 7), different potential failure lines may be subjected to different loads. For example, line *abcde* in Figure 3.13 must resist the full load, whereas *ijfh* will be subjected to $\frac{8}{11}$ of the applied load. This is because $\frac{3}{11}$ of the load will have been transferred from the member before *ijfh* receives any load.

When lines of fasteners are present in both legs of an angle, and the fasteners in these lines are staggered with respect to one another, the net area is found by first "unfolding" the angle to obtain an equivalent plate. This plate is then analyzed like any other. The unfolding is done along the middle surface, giving a gross width of plate equal to the sum of the leg lengths minus the angle thickness. AISC B2 specifies that any gage line crossing the heel of the angle be reduced by an amount equal to the angle thickness. Thus, the distance *g* in Figure 3.14, to be used in the $s^2/4g$ term, would be $3 + 2 - \frac{1}{2} = 4\frac{1}{2}$ inches.

≡ **EXAMPLE 3.6**

Find the design tensile strength of the angle shown in Figure 3.15. A36 steel is used, and holes are for $\frac{7}{8}$-in. diameter bolts.

SOLUTION

Compute the net width:

$$w_g = 8 + 6 - \tfrac{1}{2} = 13.5 \text{ in.}^2$$

Effective hole diameter $= \frac{7}{8} + \frac{1}{8} = 1$ in.

Line *abdf*:

$$w_n = 13.5 - 2(1) = 11.5 \text{ in.}$$

Line *abceg*:

$$w_n = 13.5 - 3(1) + \frac{(1.5)^2}{4(2.5)} = 10.73 \text{ in.}$$

Because $\frac{1}{10}$ of the load has been transferred from the member by the fastener at *d*, this potential failure line must resist only $\frac{9}{10}$ of the load. Therefore, the net

FIGURE 3.15 ≡

L8 × 6 × ½

width of 10.73 in. should be multiplied by $\frac{10}{9}$ to obtain a net width that can be compared with those lines that resist the full load. Use $w_n = 10.73(\frac{10}{9}) = 11.92$ in.

Line *abcdeg*:

$$g_{cd} = 3 + 2.25 - 0.5 = 4.75 \text{ in.}$$

$$w_n = 13.5 - 4(1) + \frac{(1.5)^2}{4(2.5)} + \frac{(1.5)^2}{4(4.75)} + \frac{(1.5)^2}{4(3)} = 10.03 \text{ in.}$$

The last case controls:

$$A_n = t(w_n) = 0.5(10.03) = 5.015 \text{ in.}^2$$

Since both legs of the angle are connected,

$$A_e = A_n = 5.015 \text{ in.}^2$$

Design strength based on fracture:

$$\phi_t P_n = 0.75 F_u A_e = 0.75(58)(5.015) = 218 \text{ kips}$$

Design strength based on yielding:

$$\phi_t P_n = 0.90 F_y A_g = 0.90(36)(6.75) = 219 \text{ kips}$$

Answer   Fracture controls; design strength = 218 kips.

The product of the calculated gross width and angle thickness is the exact gross area of the angle, just as if the legs of the angle were rectangular. This is because the rectangular shape can be obtained by removing area at the heel fillet and adding it at the toes.

The AISC Specification furnishes no guidance when staggered rows of fasteners are placed in rolled shapes other than angles. The equivalent-plate concept is complicated by the presence of elements of different thicknesses in shapes such as channels and wide flanges. In such cases, the use of areas (rather

FIGURE 3.16 ≡

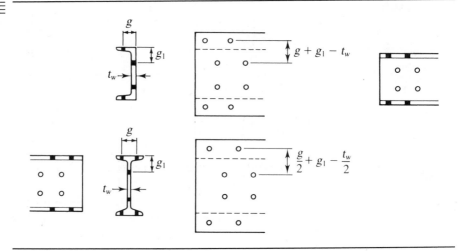

than widths) and reduced hole diameters, given by Equation 3.1, is recommended. To determine gage distances for use in this expression, however, one needs to visualize the shape as an unfolded plate. This is illustrated in Figure 3.16.

≡ **EXAMPLE 3.7**

Find the design strength of the S-shape shown in Figure 3.17. The holes are for $\frac{3}{4}$-in.-diameter bolts. Use A36 steel.

SOLUTION

Compute the net area:

$$A_n = A_g - \sum (t \times \text{hole diameter})$$

$$\text{Effective hole diameter} = \frac{3}{4} + \frac{1}{8} = \frac{7}{8}$$

Line ad:

$$A_n = 14.7 - 4\left(\frac{7}{8}\right)(0.622) = 12.52 \text{ in.}^2$$

Line $abcd$:

Gage distance for use in the $s^2/4g$ term:

$$\frac{g}{2} + g_1 - \frac{t_w}{2} = \frac{3.5}{2} + 2.75 - \frac{0.550}{2} = 4.225 \text{ in.}$$

Treat the holes at b and c as the staggered holes:

$$A_n = 14.7 - 4(0.622)\frac{7}{8} - 2(0.550)\left[\frac{7}{8} - \frac{(1.5)^2}{4(4.225)}\right] = 11.71 \text{ in.}^2$$

FIGURE 3.17 ≡

Line *abcd* controls. Since all elements of the cross section are connected,

$$A_e = A_n = 11.71 \text{ in.}^2$$

For the net section,

$$\phi_t P_n = 0.75 F_u A_e = 0.75(58)(11.71) = 509 \text{ kips}$$

For the gross section,

$$\phi_t P_n = 0.90 F_y A_g = 0.90(36)(14.7) = 476 \text{ kips}$$

Yielding of the gross section controls.

Answer   Design strength = 476 kips.

3.5 ≡ BLOCK SHEAR

For certain connection configurations, a segment or "block" of material at the end of the member can tear out. For example, the connection of the single angle tension member shown in Figure 3.18 is susceptible to this phenomenon, called

FIGURE 3.18 ≡

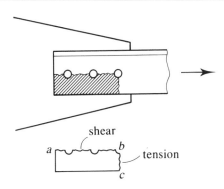

block shear. For the case illustrated, the shaded block would tend to fail by shear along the longitudinal section *ab* and by tension on the transverse section *bc*. While this topic is not covered explicitly in AISC Chapter D ("Tension Members"), the introductory paragraph directs the reader to Chapter J ("Connections, Joints, and Fasteners"), Section J4.3 ("Block Shear Rupture Strength").

The procedure is based on the assumption that one of the two failure surfaces fractures and the other yields. That is, fracture on the shear surface is accompanied by yielding on the tension surface or fracture on the tension surface with yielding on the shear surface. Both surfaces contribute to the total strength, and the resistance to block shear will be the sum of the strengths of the two surfaces.

The nominal strength in tension is $F_u A_{nt}$ for fracture and $F_y A_{gt}$ for yielding, where A_{nt} and A_{gt} are the net and gross areas along the tension surface (*bc* in Figure 3.18). Taking the shear yield stress and ultimate stress as 60% of the values for tension, the nominal strength for shear fracture is $0.6F_u A_{nv}$, and the strength for shear yielding is $0.6F_y A_{gv}$, where A_{nv} and A_{gv} are the net and gross areas along the shear surface (*ab* in Figure 3.18).

There are two possible failure modes. For shear yield and tension fracture, the design strength is

$$\phi R_n = \phi[0.6F_y A_{gv} + F_u A_{nt}]$$
(AISC Equation J4-3a)

For shear fracture and tension yield,

$$\phi R_n = \phi[0.6F_u A_{nv} + F_y A_{gt}]$$
(AISC Equation J4-3b)

In both cases, $\phi = 0.75$. Since the limit state is fracture, the controlling equation will be the one that has the larger fracture term.

≡ EXAMPLE 3.8

Check the block-shear design strength of the tension member shown in Figure 3.19. The holes are for $\frac{7}{8}$-in. diameter bolts, and A36 steel is used.

SOLUTION   Shear areas:

$$A_{gv} = \frac{3}{8}(7.5) = 2.812 \text{ in.}^2$$

and since there are 2.5 hole diameters,

$$A_{nv} = \frac{3}{8}[7.5 - 2.5(1)] = 1.875 \text{ in.}^2$$

FIGURE 3.19 ≡

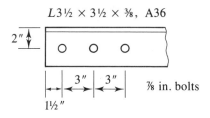

Tension areas:

$$A_{gt} = \frac{3}{8}(1.5) = 0.5625 \text{ in.}^2$$

$$A_{nt} = \frac{3}{8}(1.5 - 0.5) = 0.375 \text{ in.}^2$$

AISC Equation J4-3a:

$$\phi R_n = \phi[0.6F_y A_{gv} + F_u A_{nt}]$$
$$= 0.75[0.6(36)(2.812) + 58(0.375)]$$
$$= 0.75[60.74 + 21.75] = 61.9 \text{ kips}$$

AISC Equation J4-3b:

$$\phi R_n = \phi[0.6F_u A_{nv} + F_y A_{gt}]$$
$$= 0.75[0.6(58)(1.875) + 36(0.5625)]$$
$$= 0.75[65.25 + 20.25] = 64.1 \text{ kips}$$

The second equation has the larger fracture term (the one involving F_u); therefore, the second equation governs.

Answer   Design strength for block shear = 64.1 kips.

3.6 ≡ DESIGN OF TENSION MEMBERS

The design of a tension member involves finding a member with adequate gross and net areas. If the member has a bolted connection, the selection of a suitable cross section requires an accounting for the area lost because of holes. For a member with a rectangular cross section, the calculations are relatively straightforward. If a rolled shape is to be used, however, the area to be deduced cannot be predicted in advance since the thickness at the location of the holes is not known.

A secondary consideration in the design of tension members is slenderness. If a structural member has a small cross section in relation to its length, it is said to be slender. A more precise measure of the slenderness is the slenderness ratio L/r, where L is the member length and r is the minimum radius of gyration of the cross-sectional area. The minimum radius of gyration is the one corresponding to the minor principal axis of the cross section. This value is tabulated for all rolled shapes in the properties Tables in Part 1 of the *Manual*.

Although slenderness is critical to the strength of a compression member, it is inconsequential for a tension member. In many situations, however, it is good practice to limit the slenderness of tension members. If the axial load in a slender tension member is removed and small transverse loads are applied, undesirable vibrations or deflections might occur. This could happen, for example, in a slack bracing rod subjected to wind loads. For this reason, AISC B7 suggests a maximum slenderness ratio of 300. This is only a recommended value since slenderness has no structural significance for tension members, and the limit may be exceeded when special circumstances warrant it. This limit does not apply to cables, and the Specification explicitly excludes rods.

The central problem of all member design, including tension member design, is to find a cross section such that the sum of the factored loads does not exceed the member strength; that is,

$$\sum \gamma Q \le \phi R_n$$

For tension members, this takes the form

$$P_u \le \phi_t P_n \quad \text{or} \quad \phi_t P_n \ge P_u$$

where P_u is the sum of the factored loads. To avoid reaching the limit state of yielding,

$$0.90 F_y A_g \ge P_u \quad \text{or} \quad A_g \ge \frac{P_u}{0.90 F_y}$$

To avoid fracture,

$$0.75 F_u A_e \ge P_u \quad \text{or} \quad A_e \ge \frac{P_u}{0.75 F_u}$$

If $A_e = U A_n$, then

$$A_n \ge \frac{P_u}{0.75 F_u U}$$

The slenderness ratio limitation will be satisfied if

$$r \ge \frac{L}{300}$$

where r is the minimum radius of gyration of the cross section and L is the member length.

≡ **EXAMPLE 3.9**

A tension member with a length of 5 feet 9 inches must resist a service dead load of 18 kips and a service live load of 52 kips. Select a member with a rectangular cross section. Use A36 steel and assume a connection with one line of $\frac{7}{8}$-in. diameter bolts.

SOLUTION

$$P_u = 1.2D + 1.6L = 1.2(18) + 1.6(52) = 104.8 \text{ kips}$$

$$\text{required } A_g = \frac{P_u}{0.90F_y} = \frac{104.8}{0.90(36)} = 3.235 \text{ in.}^2$$

$$\text{required } A_e = \frac{P_u}{0.75F_u} = \frac{104.8}{0.75(58)} = 2.409 \text{ in.}^2$$

Since $A_e = A_n$ for this member, the gross area corresponding to the required net area is

$$A_g = A_n + A_{\text{hole}}$$

$$= 2.409 + \left(\frac{7}{8} + \frac{1}{8}\right)t = 2.409 + t$$

Try $t = 1$ in.

$$A_g = 2.409 + 1(1) = 3.409 \text{ in.}^2$$

Since $3.409 > 3.235$, the required gross area is 3.409 in.^{2}, and

$$w_g = \frac{A_g}{t} = \frac{3.409}{1} = 3.409 \text{ in.}$$

Rounding to the nearest $\frac{1}{8}$ in., try a $1 \times 3\frac{1}{2}$ cross section. Check the slenderness ratio:

$$I_{\text{min}} = \frac{3.5(1)^3}{12} = 0.2917 \text{ in.}^4$$

$$A = 1(3.5) = 3.5 \text{ in.}^2$$

Since $I = Ar^2$,

$$r_{\text{min}} = \sqrt{\frac{I_{\text{min}}}{A}} = \sqrt{\frac{0.2917}{3.5}} = 0.2887 \text{ in.}$$

$$\text{maximum } \frac{L}{r} = \frac{5.75(12)}{0.2887} = 239 < 300 \quad \text{(OK)}$$

Answer   Use a bar $3\frac{1}{2} \times 1$ in.

Since this member is less than 8 in. wide, it is classified as a bar rather than a plate. Bars should be specified to the nearest $\frac{1}{4}$ inch in width and to the nearest $\frac{1}{8}$ inch in thickness (the precise classification system is given in Part 1 of the *Manual* under the heading "Bars and Plates").

≣ **EXAMPLE 3.10**

Design an unequal-leg angle tension member 15 feet long to resist a service dead load of 35 kips and a service live load of 70 kips. Use A36 steel. The connection is as shown in Figure 3.20.

SOLUTION   Factored loads: $P_u = 1.2D + 1.6L = 1.2(35) + 1.6(70) = 154$ kips

$$\text{required } A_g = \frac{P_u}{\phi_t F_y} = \frac{154}{0.90(36)} = 4.753 \text{ in.}^2$$

$$\text{required } A_e = \frac{P_u}{\phi_t F_u} = \frac{154}{0.75(58)} = 3.540 \text{ in.}^2$$

If the average value of U from the Commentary is used,

$$\text{required } A_n = \frac{\text{required } A_e}{U} = \frac{3.540}{0.85} = 4.165 \text{ in.}^2$$

The radius of gyration should not exceed

$$\frac{L}{300} = \frac{15(12)}{300} = 0.600 \text{ in.}$$

The net area requirement can be reduced to a gross area requirement as follows. Since

$$A_n = A_g - 2\left(\frac{7}{8}\right)t = A_g - 1.75t$$

the corresponding required gross area is

$$\text{required } A_g = \text{required } A_n + 1.75t = 4.165 + 1.75t$$

The minimum acceptable gross area is either $4.165 + 1.75t$ or 4.753, whichever is larger.

FIGURE 3.20 ≣

¾ in. diameter bolts

Since only a finite number of angle leg thicknesses are available, the lightest member of each thickness (the one with the smallest area) will be selected. A table, such as the one that follows, is useful for keeping track of all the requirements.

t (in.)	Required A_g (in.²)	Shape	Furnished A_g (in.²)	r_{min} (in.)
$\frac{1}{4}$	4.75	—		
$\frac{5}{16}$	4.75	—		
$\frac{3}{8}$	4.82	—		
$\frac{7}{16}$	4.93	L8 × 6 × $\frac{7}{16}$	5.93	1.31
$\frac{1}{2}$	5.04	L7 × 4 × $\frac{1}{2}$	5.25	0.872
$\frac{9}{16}$	5.15	L6 × 4 × $\frac{9}{16}$	5.31	0.866
$\frac{5}{8}$	5.26	L6 × 4 × $\frac{5}{8}$	5.86	0.864

Answer   Use an L7 × 4 × $\frac{1}{2}$.

No selection is made for the $\frac{1}{4}$-inch, $\frac{5}{16}$-inch, and $\frac{3}{8}$-inch thicknesses because there are no unequal-leg angle shapes that meet the area requirements. The search is stopped after the $\frac{5}{8}$-inch selection because succeeding required areas will be larger than the ones already found.

Although the table used in this example allows a systematic selection of the lightest member, it is not necessary. A trial or "guess" procedure works just as well, provided only a few trials are needed. The only requirement is that the lightest section be found, and any method that has this result is correct. Intelligent trial selections in any structural design problem require good engineering judgment, which usually comes with experience.

When structural shapes or plates are connected to form a built-up shape, they must be connected not only at the ends of the member, but also at intervals along its length. A continuous connection is not required. This type of connection is called *stitching*, and the fasteners used are termed *stitch bolts*. The usual practice is to locate the points of stitching so that L/r for any component part does not exceed L/r for the built-up member. AISC D2 recommends that built-up shapes whose component parts are separated by intermittent fillers be connected at the fillers at intervals such that the maximum L/r for any component does not exceed 300. Built-up shapes consisting of plates or a combination of plates and shapes are addressed in AISC Section J3.5 of Chapter J ("Connections, Joints, and Fasteners"). In general, the spacing of fasteners or welds should not exceed 24 times the thickness of the thinner plate, or 12 inches. If the member is of "weathering" steel subject to atmospheric corrosion, the maximum spacing is 14 times the thickness, or 7 inches.

3.7 ≡ THREADED RODS AND CABLES

In those cases where slenderness is not a consideration, rods with circular cross sections and cables are often used as tension members. The distinction between the two is that rods are solid and cables are made from individual strands wound together in ropelike fashion. Rods and cables are frequently used in suspended roof systems and as hangers or suspension members in bridges. Rods are also used in bracing systems; in some cases, they are pretensioned to prevent them from going slack when external loads are removed. Figure 3.21 illustrates typical rod and cable connection methods.

When the end of a rod is to be threaded, an upset end is sometimes used. This is an enlargement of the end in which the threads are to be cut. Threads reduce the cross-sectional area, and upsetting the end produces a larger gross area to start with. Standard upset ends with threads will actually have more net area in the threaded portion than in the unthreaded part. Upset ends are relatively expensive, however, and in most cases unnecessary.

The effective cross-sectional area in the threaded portion of a rod is called the *stress area* and is a function of the unthreaded diameter and the number of threads per inch. The ratio of stress area to nominal area varies but has a lower bound of approximately 0.75. The nominal tensile strength of the threaded rod can therefore be written as

$$P_n = A_s F_u$$
$$= 0.75 A_b F_u$$

FIGURE 3.21 ≡

where

A_s = stress area

A_b = nominal (unthreaded) area

This is the nominal strength given in Table J3.2 in Section J3.6 of the AISC Specification. The resistance factor for this case is $\phi_t = 0.75$.

If upset ends are used, the tensile capacity at the major thread diameter must be greater than F_y times the unthreaded body area (AISC Table J3.2, footnote c).

EXAMPLE 3.11

A threaded rod is to be used as a bracing member that must resist a service tensile load of 2 kips dead load and 6 kips live load. What size rod is required if A36 steel is used?

SOLUTION   Factored load:

$$P_u = 1.2(2) + 1.6(6) = 12 \text{ kips}$$

Since $\phi_t P_n \geq P_u$,

$$\phi_t(0.75F_u)A_g \geq P_u$$

$$\text{required } A_g = \frac{P_u}{\phi_t(0.75)F_u} = \frac{12}{0.75(0.75)(58)} = 0.3678 \text{ in.}^2$$

From $A_g = \dfrac{\pi d^2}{4}$,

$$\text{required } d = \sqrt{\frac{4(0.3678)}{\pi}} = 0.684 \text{ in.}$$

Answer   Use a $\frac{3}{4}$-in. diameter threaded rod ($A_g = 0.442$ in.^{2})

To prevent damage during construction, rods should not be too slender. Although there is no specification requirement, a common practice is to use a minimum diameter of $\frac{5}{8}$ inch.

Flexible cables, in the form of strands or wire rope, are used in applications where high strength is required and rigidity is unimportant. In addition to their use in bridges and cable roof systems, they are also used in hoists and derricks, as guy lines for towers, and as longitudinal bracing in metal building systems. The difference between strand and wire rope is illustrated in Figure 3.22. A strand consists of individual wires wound helically around a central core, and a wire rope is made of several strands laid helically around a core.

FIGURE 3.22 ≡

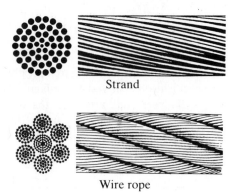

Strand

Wire rope

Selection of the correct cable for a given loading is usually based on both strength and deformation considerations. In addition to ordinary elastic elongation, there is an initial stretching caused by seating or shifting of the individual wires, which results in a permanent stretch. For this reason, cables are often prestretched. Wire rope and strand are made from steels of much higher strength than structural steels and are not covered by the AISC Specification. The breaking strengths of various cables, as well as details of available fixtures for connections, can be obtained from manufacturers' literature. *Cable Roof Structures* contains useful data of this type.[27]

3.8 ≡ TENSION MEMBERS IN ROOF TRUSSES

Many of the tension members that structural engineers design are components of trusses. For this reason, some general discussion of roof trusses is in order. For a more comprehensive treatment of the subject, see Lothars.[28]

When trusses are used in buildings, they usually function as the main supporting elements of roof systems where long spans are required. They are used in those situations where the cost and weight of a beam would be prohibitive. (A truss may be thought of as a deep beam with much of the web removed.) Roof trusses are often used in industrial or mill buildings, although construction of this type has largely given way to rigid frames. Typical roof construction with trusses supported by load-bearing walls is illustrated in Figure 3.23. In this type of construction, one end of the connection of the truss to the walls usually can be considered as pinned and the other as roller-supported. Thus the truss can be analyzed as an externally statically determinate structure. The supporting walls can be reinforced concrete, concrete block, brick, or a combination of these materials.

Roof trusses normally are spaced uniformly along the length of the building and are tied together by longitudinal beams called *purlins* and by x-bracing. The

FIGURE 3.23 ≡

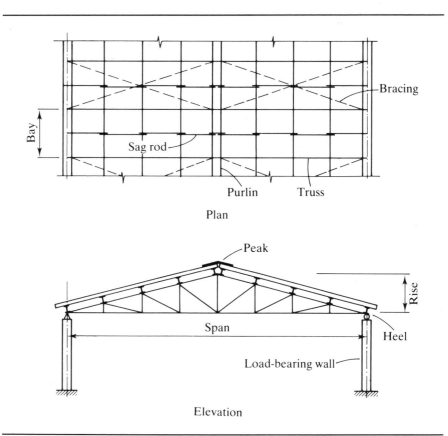

Plan

Elevation

primary function of the purlins is to transfer loads to the top chord of the truss, but they can also act as part of the bracing system. Bracing is usually provided in the planes of both the top and bottom chords, but it is not required in every bay since lateral forces can be transferred from one braced bay to the other through the purlins.

Ideally, purlins are located at the truss joints so that the truss can be treated as a pin-connected structure loaded only at the joints. Sometimes, however, the roof deck cannot span the distance between joints, and intermediate purlins may be needed. In such cases the top chord will be subjected to significant bending as well as axial compression and must be designed as a beam-column (Chapter 6).

Sag rods are tension members used to provide lateral support for the purlins. Since most of the loads applied to the purlins are vertical, there will be a component parallel to a sloping roof, and this will cause the purlin to bend (sag) in that direction (Figure 3.24).

Sag rods can be located at the midpoint, the third points, or at more frequent intervals along the purlins, depending on the amount of support needed.

FIGURE 3.24 ≡

This will be a function of the truss spacing, the slope of the top chord, the resistance of the purlin to this type of bending (most shapes used for purlins are very weak in this respect), and the amount of support furnished by the roofing. If a metal deck is used, it will usually be rigidly attached to the purlins, and sag rods may not be needed at all. Sometimes, however, the weight of the purlin itself is enough to cause problems, and sag rods may be needed to provide support during construction before the deck is in place.

If sag rods are used, they are designed to support the component of roof loads parallel to the roof. Each segment between purlins is assumed to support everything below it; thus the top rod is designed for the load on the roof area tributary to the rod, from the heel of the truss to the peak (Figure 3.25). Although the force will be different in each segment of rod, the usual practice is to use one size throughout. The extra amount of material in question is insignificant, and the use of the same size for each segment eliminates the possibility of a mix-up during construction.

A possible treatment at the peak or ridge is shown in Figure 3.26a. The tie rod beween ridge purlins must resist the load from all of the sag rods on either side. The tensile force in this horizontal member has as one of its components the force in the upper sag-rod segment. A free-body diagram of one ridge purlin illustrates this effect (Figure 3.26b).

FIGURE 3.25 ≡

FIGURE 3.26 ≡

(a)   (b)

≡ **EXAMPLE 3.12**

Fink trusses spaced at 20 feet on centers support $W6 \times 12$ purlins as shown in Figure 3.27a. The purlins are supported at their midpoints by sag rods. Use A36 steel and design the sag rods and the tie rod at the ridge for the following service loads:

Metal deck: 2 psf

Built-up roof: 5 psf

Snow: 18 psf of horizontal projection of the roof surface

Purlin weight: 12 pounds per foot (lb/ft) of length

FIGURE 3.27 ≡

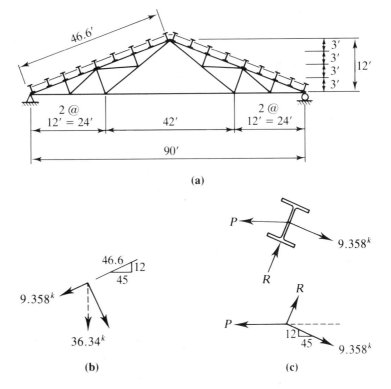

(a)

(b)   (c)

SOLUTION **Loads:**

Tributary area for deck and built-up roof = 20(46.6) = 932 ft^2

The total load from the deck and roof is

(2 + 5)(932) = 6520 lb

tributary area for snow load = 20(45) = 900 ft^2

total snow load = 18(900) = 16,200 lb

purlins: 12(20)(9) = 2160 lb

total dead load = 6520 + 2160 = 8680 lb

Check load combinations:

(A4-2): 1.2D + 0.5S = 1.2(8680) + 0.5(16,200) = 18,520 lb

(A4-3): 1.2D + 1.6S = 1.2(8680) + 1.6(16,200) = 36,340 lb

Combination A4-3 controls. (By inspection, A4-1, A4-4, and A4-5 will not govern.)

Component parallel to the roof (Figure 3.27b):

$$T = (36.34)\frac{12}{46.6} = 9.358 \text{ kips}$$

$$\text{required } A_g = \frac{T}{\phi_t(0.75F_u)} = \frac{9.358}{0.75(0.75)(58)} = 0.2868 \text{ in.}^2$$

Answer Use a $\frac{5}{8}$-in. diameter threaded rod (A_g = 0.3068 in.^{2}).

Tie rod at the ridge (Figure 3.27c):

$$P = (9.358)\frac{46.6}{45} = 9.691 \text{ kips}$$

$$\text{required } A_g = \frac{9.691}{0.75(0.75)(58)} = 0.2970 \text{ in.}^2$$

Answer Use a $\frac{5}{8}$-in. diameter threaded rod (A_g = 0.3068 in.^{2}).

For the usual truss geometry and loading, the bottom chord and some of the web members will be in tension. When wind effects are included and consideration is given to different wind directions, the force in some web members may alternate between tension and compression. If this occurs, the affected member must be designed to function as both a tension member and a compression member.

In bolted trusses, double-angle sections are frequently used for both chord and web members. This facilitates the connection of members meeting at a joint

FIGURE 3.28 ≡

by permitting the use of a single gusset plate as illustrated in Figure 3.28. When structural tee-shapes are used as chord members in welded trusses, the web angles can usually be welded to the stem of the tee. If the force in a web member is small, single angles can be used, although this eliminates the plane of symmetry from the truss and causes the web member to be eccentrically loaded. Chord members are usually fabricated as continuous pieces and spliced if necessary.

The fact that chord members are continuous and joints are bolted or welded would seem to invalidate the usual assumption that the truss is pin-connected. It is true that this joint rigidity does introduce some bending moment into the members, but it is usually small and considered to be a secondary effect. The usual practice is to ignore it. Bending caused by loads directly applied to members between the joints, however, must be taken into account.

Properly detailed trusses have the *working lines* of their members intersect at the *working point* at each joint. For a bolted truss, the bolt lines are the working lines, and in welded trusses the centroidal axes of the welds are the working lines. The usual truss-analysis procedure assumes member lengths to be measured from working point to working point.

≡ **EXAMPLE 3.13**

Select a structural tee for the bottom chord of the Warren roof truss shown in Figure 3.29a. The trusses are welded and spaced at 20 feet. Assume that the bottom chord connection is made with 9-in.-long longitudinal fillet welds at the flange. Use A36 steel and the following load data (wind will not be considered in this example):

Purlins: M8 × 6.5

Snow: 20 psf of horizontal projection

Metal deck: 2 psf

Roofing: 4 psf

Insulation   3 psf

FIGURE 3.29 ☰

(a)

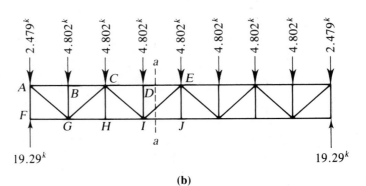

(b)

(c)

SOLUTION   Loads:

Snow: 20(40)(20) = 16,000 lb

Dead load (exclusive of purlins): Deck: 2 psf

Roof: 4

Insulation: 3

Total: 9 psf

Total dead load = 9(40)(20) = 7200 lb

Total purlin weight = 6.5(20)(9) = 1170 lb

Estimate the truss weight as 10% of the other loads:

0.10(16,000 + 7200 + 1170) = 2437 lb

Loads at an interior joint:

$$D = \frac{7200}{8} + \frac{2437}{8} + 6.5(20) = 1335 \text{ lb}$$

$$S = \frac{16,000}{8} = 2000 \text{ lb}$$

At an exterior joint, the tributary roof area is half of that at an interior joint. The corresponding loads are

$$D = \frac{7200}{2(8)} + \frac{2437}{2(8)} + 6.5(20) = 732.3 \text{ lb}$$

$$S = \frac{16,000}{2(8)} = 1000 \text{ lb}$$

Load combination A4-3 will control:

$$P_u = 1.2D + 1.6S$$

At an interior joint,

$$P_u = 1.2(1.335) + 1.6(2.0) = 4.802 \text{ kips}$$

At an exterior joint,

$$P_u = 1.2(0.7323) + 1.6(1.0) = 2.479 \text{ kips}$$

The loaded truss is shown in Figure 3.29b.

The bottom chord is designed by determining the force in each member of the bottom chord and selecting a cross section to resist the largest force. In this problem, the force in member IJ will control. For the free body left of section a–a (Figure 3.29c),

$$\sum M_E = 19.29(20) - 2.479(20) - 4.802(15 + 10 + 5) - 4F_{IJ} = 0$$
$$F_{IJ} = 48.02 \text{ kips}$$

Gross section:

$$\text{Required } A_g = \frac{F_{IJ}}{0.90F_y} = \frac{48.02}{0.90(36)} = 1.48 \text{ in.}^2$$

Net section:

$$\text{Required } A_e = \frac{F_{IJ}}{0.75F_u} = \frac{48.02}{0.75(58)} = 1.10 \text{ in.}^2$$

Try a WT5 × 6:

$$A_g = 1.77 \text{ in.}^2 > 1.48 \text{ in.}^2 \quad \text{(OK)}$$

Since the connection is made with longitudinal welds, this will not be one of the special cases for welded members subject to shear lag.

$$U = 1 - \left(\frac{\bar{x}}{L}\right) = 1 - \left(\frac{1.36}{9}\right) = 0.849 < 0.9$$

$$A_e = UA_g = 0.849(1.77) = 1.50 \text{ in.}^2 > 1.10 \text{ in.}^2 \quad (\text{OK})$$

If we assume that the bottom chord is braced at the panel points,

$$\frac{L}{r} = \frac{5(12)}{0.785} = 76.43 < 300 \quad (\text{OK})$$

Answer   Use a WT5 × 6.

3.9 ≡ PIN-CONNECTED MEMBERS

When a member is to be pin-connected, a hole is made in both the member and the parts to which it is connected, and a pin is placed through the holes. This provides a connection that is as moment-free as can be fabricated. Tension members connected in this manner are subjected to several failure modes, which are covered in AISC D3 and will be discussed in the following paragraphs.

The eyebar is a special type of pin-connected member in which the end containing the pin hole is enlarged (Figure 3.30). The design strength is based on yielding of the gross section. Detailed rules for proportioning eyebars are given in AISC D3 and will not be repeated here. These requirements are based on experience and test programs for forged eyebars, but they are conservative when applied to the ones thermally cut from plates (the present fabrication method). Eyebars were widely used in the past as single tension members in bridge trusses or were linked together in chain-like fashion in suspension bridges. They are rarely used today.

Pin-connected members should be designed for the following limit states (see Figure 3.31):

1. **Tension** on the net effective area (Figure 3.31a):

$$\phi_t = 0.75, \qquad P_n = 2tb_{eff}F_u \qquad\qquad \text{(AISC Equation D3-1)}$$

2. **Shear** on the effective area (Figure 3.31b):

$$\phi_{sf} = 0.75, \qquad P_n = 0.6A_{sf}F_u \qquad\qquad \text{(AISC Equation D3-2)}$$

3. **Bearing.** This requirement is given in Chapter J ("Connections, Joints, and Fasteners"). Referring to Figure 3.31c:

$$\phi = 0.75, \qquad P_n = 1.8F_yA_{pb} \qquad\qquad \text{(AISC Equation J8-1)}$$

4. **Tension** on the gross section:

$$\phi = 0.90, \qquad P_n = F_yA_g \qquad\qquad \text{(AISC Equation D1-1)}$$

FIGURE 3.30 ≡

where

t = thickness of connected part

$b_{eff} = 2t + 0.63 \leq b$

b = distance from edge of pin hole to edge of member, perpendicular to direction of force

$A_{sf} = 2t(a + d/2)$

a = distance from edge of pin hole to edge of member, parallel to direction of force

d = pin diameter

A_{pb} = projected bearing area = dt

FIGURE 3.31 ≡

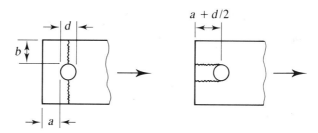

(a) Fracture of net section   **(b) Longitudinal shear**

(c) Bearing

Additional requirements for the relative proportions of the pin and the member are covered in AISC D3.

≡ PROBLEMS ≡

NOTES   1. Unless otherwise indicated, all given loads are *service* loads.

2. In *design* problems with bolted connections, use average U values from the Commentary.

3.1   Compute the design strength for a bar $6 \times \frac{1}{2}$ tension member with $\frac{7}{8}$-in.-diameter bolts in a single line. A36 steel is used.

FIGURE P3.1

3.2   Compute the design strength for a bar $5\frac{1}{2} \times \frac{3}{8}$ tension member with $\frac{3}{4}$-in.-diameter bolts in two lines. A242 steel is used ($F_y = 50$ ksi, $F_u = 70$ ksi).

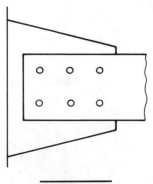

FIGURE P3.2

3.3   A bar $7 \times \frac{1}{2}$ is subjected to a dead load of 35 kips and a live load of 35 kips. It is connected with two lines of $\frac{3}{4}$-in. bolts. If A36 steel is used, is the member adequate?

3.4   A $6 \times 6 \times \frac{3}{4}$ angle is connected to a gusset plate with a single line of nine $\frac{3}{4}$-in.-diameter bolts. How much service dead and live load can be resisted in tension if a live load-to-dead load ratio of 0.5 is used? Use A36 steel and an average value of U from the Commentary.

3.5   A W14 × 176 tension member is connected with two lines of $\frac{7}{8}$-in. diameter bolts through each flange. Each line contains four bolts. Obtain U from AISC equation B3-2 and determine the design tensile strength for (a) A36 steel and (b) A242 steel.

1.5"   3 @ 3" = 9"

FIGURE P3.5

3.6   A WT9 × 25 tension member is connected by welding the flange to a gusset plate with 11-in.-long longitudinal welds. Determine the design strength for A36 steel.

3.7   For the tension member of Problem 3.5, compare the net area reduction factor U as obtained from AISC Equation B3-2 with the average value given in the Commentary. Which would you use? Why?

3.8   A bar $5 \times \frac{3}{8}$ is connected at its ends with 5-in.-long welds along each side. Compute the design strength for A36 steel.

3.9   Determine the smallest net area for the plate shown in Figure P3.9.

5"   5"

2"
3"
2"

2½"   5"

(thickness = ⅜";
¾" diameter bolts)

FIGURE P3.9

3.10   Compute the smallest net width of the flat bar shown. The holes are for $\frac{7}{8}$-in.-diameter bolts.

FIGURE 3.10

3.11   Determine the smallest net width for the plate in Figure P3.11. The holes are for $\frac{7}{8}$-in.-diameter bolts.

FIGURE P3.11

3.12   Two C10 × 20 shapes are used for a built-up tension member as shown. The holes are for $\frac{1}{2}$-in.-diameter bolts. Determine the design strength if A36 steel is used.

FIGURE P3.12

3.13 What is the smallest net width of the plate shown in the accompanying figure? The holes are for $1\frac{1}{2}$-in. bolts.

FIGURE P3.13

3.14 A C12 × 30 tension member is connected with $1\frac{1}{8}$-in.-diameter bolts as shown. Determine the design tensile strength for (a) A36 steel and (b) A572 Grade 50 steel.

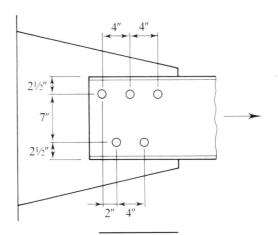

FIGURE P3.14

3.15 A PL $\frac{1}{2}$ × 10 tension member of A36 steel is connected with $1\frac{1}{8}$-in.-diameter bolts as shown. What is the design strength?

FIGURE P3.15

3.16   The tension member shown is a C10 × 30 of A36 steel, and the holes are for $\frac{7}{8}$-in.-diameter bolts. Determine the design strength.

FIGURE P3.16

3.17   The tension member shown in Figure P3.17 contains holes of diameter d. At what spacing s will the net area for a section through one hole be the same as that for a failure line passing through two holes?

FIGURE P3.17

3.18   An 8 × 4 × $\frac{1}{2}$ angle of A36 steel is used as a tension member. Holes for $\frac{5}{8}$-in.-diameter bolts are arranged in the pattern shown. What is the design strength?

FIGURE P3.18

3.19   Determine the smallest net area of the tension member shown. The holes are for $\frac{7}{8}$-in.-diameter bolts.

FIGURE P3.19

3.20   A bar $5\frac{1}{2} \times \frac{3}{8}$ is connected to a $\frac{3}{8}$-in.-thick gusset plate with $\frac{3}{4}$-in.-diameter bolts as shown. A242 steel is used for the tension member, and A36 is used for the gusset plate. Investigate the gusset plate for block shear (i.e., determine the design strength for block shear).

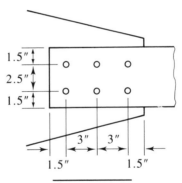

FIGURE P3.20

3.21   Select an equal-leg angle to resist a dead load of 25 kips and a live load of 60 kips. The connection will be made with one row of $\frac{3}{4}$-in.-diameter bolts, using more than two bolts. The length of the member is 10 ft. Use A36 steel.

3.22   Select a double-angle selection for a 15-ft tension member subjected to a dead load of 45 kips and a live load of 105 kips. Use unequal-leg angles with the long legs back-to-back. Assume one line of three or more 1-in.-diameter bolts and a $\frac{3}{8}$-in.-thick gusset plate. Use A572 Grade 50 steel.

3.23   Select a single angle to resist a tensile dead load of 55 kips and a live load of 145 kips. The member is 30 ft long and is to be connected with longitudinal welds. Use A36 steel and an estimated U value of 0.85.

3.24   Select the most economical double-angle shape to resist a tensile dead load of 65 kips and a live load of 65 kips. The member length is 20 ft. Assume A36 steel and a connection with longitudinal welds. Use an estimated U value of 0.85.

3.25   Select a double-angle tension member to resist a dead load of 12 kips and a live load of 36 kips. The member is 16 ft long and is connected to a $\frac{5}{8}$-in.-thick gusset plate with one line of four $\frac{3}{4}$-in.-diameter bolts. Use A36 steel.

3.26   Select the lightest American Standard Channel that will safely support a *factored* tensile load of 260 kips. The member is 18 ft long, and there is one line of more than two 1-in.-diameter bolts in each flange at the connection. Use A36 steel.

3.27   Select a pair of American Standard Channels for the tension member shown in Figure P3.27. Design for a service dead load of 45 kips and a service live load of 135 kips. The length is 20 ft. Use A36 steel and 1-in.-diameter bolts.

FIGURE P3.27

3.28   Select a threaded rod for a factored tensile load of 180 kips. Use A36 steel.

3.29   What size threaded rod is required for member AB in Figure P3.29? The given load is a service live load. (Neglect the weight of member CB.) Use A36 steel.

FIGURE P3.29

3.30   A pipe is supported at 12-foot intervals by a bent, threaded rod as shown. If an 8-in.-diameter standard weight steel pipe full of water is used, what size rod is required? Use A36 steel.

FIGURE P3.30

3.31   Use A36 steel and select a structural tee for the top chord of the welded roof truss in Figure P3.31. All connections are made with longitudinal plus transverse welds. Assume $U = 0.85$. The spacing of trusses in the roof system is 15 feet. Design for the following loads:

Snow: 18 psf of horizontal projection

Roofing: 15 psf

MC8 × 8.5 purlins

Truss weight: 1000 lb (estimated)

FIGURE P3.31

3.32   Select single-angle shapes for the web tension members of the truss loaded as shown in Figure P3.32. The given loads are *factored loads*. All connections are with longitudinal welds. Use A36 steel and assume $U = 0.85$.

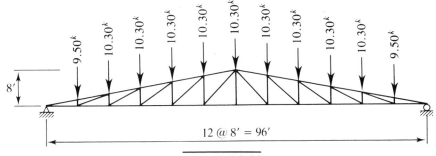

FIGURE P3.32

3.33   Compute the factored joint loads for the truss of Problem 3.32 for the following conditions:

Trusses spaced at 18 ft.

Weight of roofing = 8 psf.

Snow load = 20 psf of horizontal projection.

W10 × 33 purlins located only at the joints.

Total estimated truss weight = 5000 lb.

3.34   Design the tension members of the roof truss shown in Figure P3.34. Use double-angle shapes throughout and assume $\frac{3}{8}$-in.-thick gusset plates and welded connections (either longitudinal or longitudinal plus transverse; assume $U = 0.85$). The trusses are spaced at 30 feet. Use A36 steel and design for the following loads:

Metal deck: 4 psf of roof surface

Built-up roof: 12 psf of roof surface

Purlins: 3 psf of roof surface (estimated)

Snow: 20 psf of horizontal projection

Truss weight: 5 psf of horizontal projection (estimated)

FIGURE P3.34

3.35   Design sag rods for the truss of Problem 3.34. Assume that once the metal deck is attached, it will provide lateral support for the purlins; therefore, the sag rods need to be designed for the purlin weight only. A36 steel.

3.36   Write a computer program to compute design tensile strengths for given rolled shapes. Let the area deducted for bolt holes vary from 10% to 50% of the gross area in increments of 0.01 square inches.

Input: Name of shape (W10 × 112, S5 × 10, etc.), gross area, yield stress, and ultimate tensile strength.

Output: Same as input plus a table of values giving the area deducted, the net area, and the design strengths corresponding to U values of 1.0, 0.90, 0.85, and 0.75.

4 COMPRESSION MEMBERS

4.1 ≡ DEFINITION

Compression members are those structural elements that are subjected only to axial compressive forces; that is, the loads are applied along a longitudinal axis through the centroid of the member cross section, and the stress can be taken as $f_a = P/A$, where f_a is considered to be uniform over the entire cross section. This ideal state is never achieved in reality, however, and some eccentricity of the load is inevitable. This will result in bending, but it can usually be regarded as secondary and can be neglected if the theoretical loading condition is closely approximated. This cannot always be done if there is a *computed* bending moment, and situations of this type will be considered in Chapter 6, "Beam-Columns."

The most common type of compression member occurring in buildings and bridges is the *column*, a vertical member whose primary function is to support vertical loads. In many instances these members are also called upon to resist bending, and in these cases the member is a *beam-column*. Compression members can also be found in trusses and as components of bracing systems. Smaller compression members not classified as columns are sometimes referred to as *struts*.

4.2 ≡ COLUMN THEORY

Consider the long, slender compression member shown in Figure 4.1a. If the axial load P is slowly applied, it will ultimately reach a value large enough to cause the member to become unstable and assume the shape indicated by the dashed line. The member is said to have buckled, and the corresponding load is called the *critical buckling load*. If the member is more stocky, as the one in Figure 4.1b, a larger load will be required to bring the member to the point of instability. For extremely stocky members, failure may be by compressive yielding rather than buckling. For these stocky members and for more slender columns before they buckle, the compressive stress P/A is uniform over the cross section at any point along the length. As we shall see, the load at which buckling occurs is a function of slenderness, and for very slender members this load could be quite small.

If the member is so slender (a precise definition of slenderness will be given

FIGURE 4.1 ≡

shortly) that the stress just before buckling is below the proportional limit —
that is, the member is still elastic — the critical buckling load is given by

$$P_{cr} = \frac{\pi^2 EI}{L^2} \tag{4.1}$$

where E is the modulus of elasticity of the material, I is the moment of inertia of
the cross-sectional area with respect to the minor principal axis, and L is the
length of the member between points of support. For Equation 4.1 to be valid,
the member must be elastic, and its ends must be free to rotate but not translate
laterally. This end condition is satisfied by hinges or pins, as shown in Figure
4.2. This remarkable relationship was first formulated by Swiss mathematician

FIGURE 4.2 ≡

Leonhard Euler and published in 1759. The critical load is sometimes referred to as the *Euler load* or the *Euler buckling load*. The validity of Equation 4.1 has been demonstrated convincingly by numerous tests. Its derivation has become somewhat of a standard exercise in textbooks on differential equations, but it will be given here to illustrate the importance of the end conditions.

For convenience, the member will be oriented with its longitudinal axis along the x-axis of the coordinate system given in Figure 4.3. The roller support is to be interpreted as restraining the member from translating either up or down. An axial compressive load is applied and gradually increased. If a temporary transverse load is applied so as to deflect the member into the shape indicated by the dashed line, the member will return to its original position when this temporary load is removed if the axial load is less than the critical buckling load. The critical buckling load P_{cr} is defined as the load that is just large enough to maintain the deflected shape when the temporary transverse load is removed.

It can be shown that the differential equation giving the deflected shape of an elastic member subjected to bending is

$$\frac{d^2 y}{dx^2} = -\frac{M}{EI} \tag{4.2}$$

where x locates a point along the longitudinal axis of the member, y is the deflection of the axis at that point, and M is the bending moment at the point. E and I were previously defined, but it should be emphasized that the moment of inertia I is with respect to the axis of bending (buckling). This equation was derived by Jacob Bernoulli and independently by Euler, who specialized it for the column buckling problem.[29] If we begin at the point of buckling, then from Figure 4.3 the bending moment is $P_{cr} y$. Equation 4.2 can then be written as

$$y'' + \frac{P_{cr}}{EI} y = 0$$

FIGURE 4.3 ≡

where the prime denotes differentiation with respect to x. This is a second-order, linear, ordinary differential equation with constant coefficients and has the solution

$$y = A \cos(cx) + B \sin(cx)$$

where

$$c = \sqrt{\frac{P_{cr}}{EI}}$$

and A and B are constants. These constants are evaluated by applying the following boundary conditions:

At $x = 0$, $y = 0$:　　$0 = A \cos(0) + B \sin(0)$　　$A = 0$

At $x = L$, $y = 0$:　　$0 = B \sin(cL)$

This last condition requires that $\sin(cL)$ be zero if B is not to be zero (the trivial solution, corresponding to $P = 0$). For $\sin(cL) = 0$,

$$cL = 0, \pi, 2\pi, 3\pi, \ldots = n\pi, \quad n = 0, 1, 2, 3, \ldots$$

$$cL = \sqrt{\frac{P_{cr}}{EI}}\, L = n\pi, \qquad L^2 \frac{P_{cr}}{EI} = n^2\pi^2$$

and

$$P_{cr} = \frac{n^2\pi^2 EI}{L^2}$$

The various values of n correspond to different buckling modes; $n = 1$ represents the first mode, $n = 2$ the second, and so forth. A value of zero gives the trivial case of no load. These buckling modes are illustrated in Figure 4.4.

FIGURE 4.4 ≡

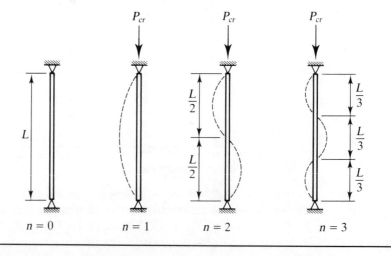

Values of n larger than 1 are not possible unless the compression member is physically restrained from deflecting at the points where the reversal of curvature would occur.

The solution to the differential equation is therefore

$$y = B \sin \left(\frac{n\pi x}{L} \right)$$

and the coefficient B is indeterminate. This state of affairs is a consequence of approximations made in formulating the differential equation; a linear representation of a nonlinear phenomenon was used.

For the usual case of a compression member with no supports between its ends, $n = 1$ and the Euler equation is written as

$$P_{cr} = \frac{\pi^2 EI}{L^2} \tag{4.3}$$

It will be convenient to rewrite this as follows:

$$P_{cr} = \frac{\pi^2 EI}{L^2} = \frac{\pi^2 E A r^2}{L^2} = \frac{\pi^2 EA}{(L/r)^2}$$

where A is the cross-sectional area and r is the radius of gyration with respect to the axis of buckling. The ratio L/r is the slenderness ratio and is the measure of a compression member's slenderness, with large values corresponding to slender members.

If the critical load is divided by the cross-sectional area, the critical buckling stress is obtained:

$$F_{cr} = \frac{P_{cr}}{A} = \frac{\pi^2 E}{(L/r)^2} \tag{4.4}$$

This is the compressive stress at which buckling will occur about the axis corresponding to r. Since buckling will take place as soon as the load reaches the value given by Equation 4.3, the column will become unstable about the principal axis corresponding to the largest slenderness ratio. This usually means the axis with the smaller moment of inertia (we will examine exceptions to this later). Thus, the minimum moment of inertia and radius of gyration of the cross section should be used in Equations 4.3 and 4.4.

≡ **EXAMPLE 4.1**

A W12 × 50 is used as a column to support an axial compressive load of 145 kips. The length is 20 feet, and the ends are pinned. Without regard to load or resistance factors, investigate this member for stability. (The grade of steel need not be known: The critical buckling load is a function of the modulus of elasticity, not the yield stress or ultimate tensile strength.)

SOLUTION   For a W12 × 50,

$$\text{minimum } r = r_y = 1.96 \text{ in.}$$

$$\text{maximum } \frac{L}{r} = \frac{20(12)}{1.96} = 122.4$$

$$P_{cr} = \frac{\pi^2 EA}{(L/r)^2} = \frac{\pi^2(29,000)(14.7)}{(122.4)^2} = 280.8 \text{ kips}$$

Answer   Since the applied load of 145 kips is less than P_{cr}, the column remains stable and has an overall factor of safety against buckling of $280.8/145 = 1.94$.

Early researchers soon found that Euler's equation did not give reliable results for stocky, or less slender, compression members. This is because of the small slenderness ratio for members of this type, which results in a large buckling stress (from Equation 4.4). If the stress at which buckling occurs is greater than the proportional limit of the material, the relation between stress and strain is not linear, and the modulus of elasticity E can no longer be used. (In the previous example, the stress at buckling is $P_{cr}/A = 280.8/14.7 = 19.10$ ksi, which is well below the proportional limit for any grade of structural steel.) This difficulty was initially resolved by Friedrich Engesser, who proposed in 1889 the use of a variable tangent modulus E_t in Equation 4.3. For a material with a stress–strain curve like the one in Figure 4.5, E is not a constant for stresses greater than the proportional limit F_{pl}. The tangent modulus E_t is defined as the slope of the tangent to the stress–strain curve for values of f between F_{pl} and F_y. If the compressive stress at buckling, P_{cr}/A, is in this region, it can be shown that

$$P_{cr} = \frac{\pi^2 E_t I}{L^2} \tag{4.5}$$

This is identical to the Euler equation, except that E_t is substituted for E.

The stress–strain curve in Figure 4.5 is different from the ones shown earlier for ductile steel (in Figures 1.3 and 1.4) because it has a pronounced region of nonlinearity. This curve is typical of a compression test of a short length of W-shape called a *stub column*, rather than the result of testing a tensile specimen. The nonlinearity is primarily the result of the presence of residual stresses in the W-shape. When a hot-rolled shape cools after rolling, all elements of the cross section do not cool at the same rate. The tips of the flanges, for example, cool at a faster rate than the junction of the flange and the web. This uneven cooling induces stresses that remain permanently. Other factors, such as welding and cold-bending to create curvature in a beam, can contribute to the residual stress, but the cooling process is the chief source.

Note that E_t is smaller than E, and for the same L/r corresponds to a smaller critical load P_{cr}. Because of the variability of E_t, computation of P_{cr} in the inelastic range by the use of Equation 4.5 is difficult. In general, a trial-and-

FIGURE 4.5 ≡

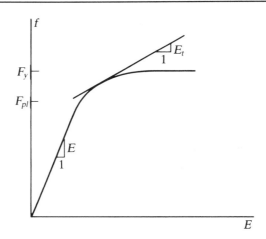

error approach must be used, and a compressive stress–strain curve such as the one in Figure 4.5 must be used to determine E_t for trial values of P_{cr}. For this reason, most design specifications, including AISC, use empirical formulas for inelastic columns.

Engesser's tangent modulus theory had its detractors, who pointed out several inconsistencies. Engesser was convinced by their arguments, and in 1895 he refined his theory to incorporate a reduced modulus, which has a value between E and E_t. Test results, however, always agreed more closely with the tangent modulus theory. In 1947, F. R. Shanley[30] resolved the apparent inconsistencies in the original theory, and today the tangent modulus formula, Equation 4.5, is accepted as the correct one for inelastic buckling. Although the load predicted by this equation is actually a lower bound on the true value of the critical load, the difference is very slight.[31]

For a given material, the critical buckling stress can be plotted as a function of slenderness as in Figure 4.6. The tangent modulus curve is tangent to the Euler curve at the point corresponding to the proportional limit of the material. The composite curve, called a *column strength curve*, completely describes the stability of any column of a given material. Other than E and E_t, which are properties of the material, the strength is a function only of the slenderness ratio.

Effective Length

Both the Euler and tangent modulus equations are based on the following assumptions:

1. The column is perfectly straight, with no initial crookedness.

2. The load is axial, with no eccentricity.

3. The column is pinned at both ends.

FIGURE 4.6 ≡

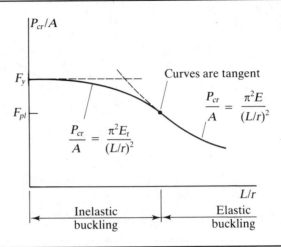

The first two conditions mean that there is no bending moment in the member before buckling. As mentioned previously, some accidental moment will be present, but in most cases it can be neglected. The requirement for pinned ends, however, is a serious limitation, and provisions must be made for other support conditions. The pinned-end condition is one that requires that the member be restrained from lateral translation, but not rotation, at the ends. Since it is virtually impossible to construct a frictionless pin connection, even this support condition can only be closely approximated at best. Obviously, all columns must be free to deform axially.

Other end conditions can be accounted for in the derivation of Equation 4.3. In general, the bending moment will be a function of x, and this results in a nonhomogeneous differential equation. The boundary conditions will be different from those in the original derivation, but the overall procedure will be the same. The form of the resulting equation for P_{cr} will also be the same. For example, consider a compression member pinned at one end and fixed against rotation and translation at the other as in Figure 4.7. The Euler equation for this case, derived in the same manner as Equation 4.3, is

$$P_{cr} = \frac{2.05\pi^2 EI}{L^2}$$

or

$$P_{cr} = \frac{2.05\pi^2 EA}{(L/r)^2} = \frac{\pi^2 EA}{(0.70L/r)^2}$$

Thus, this compression member has the same load capacity as a column that is pinned at both ends and is only 70% as long as the given column. Similar expressions can be found for columns with other end conditions.

The column-buckling problem can also be formulated in terms of a fourth-

FIGURE 4.7 ≡

order differential equation instead of Equation 4.2. This proves to be convenient when dealing with boundary conditions other than pinned ends.

For convenience, the equations for critical buckling load will be written as

$$P_{cr} = \frac{\pi^2 E A}{(KL/r)^2} \quad \text{or} \quad P_{cr} = \frac{\pi^2 E_t A}{(KL/r)^2} \tag{4.6a/4.6b}$$

where KL is the effective length, and K is called the *effective length factor*. The effective length factor for the end conditions specified above is 0.70. For the most favorable condition of both ends fixed against rotation and translation, $K = 0.5$. Values of K for these and other cases can be determined with the aid of Table C-C2.1 in the Commentary to the AISC Specification. The three conditions mentioned thus far are included, as well as some where end translation is possible. Two values of K are given: a theoretical value and a recommended design value to be used when the ideal end condition is approximated. This means that unless a "fixed" end is perfectly fixed, use the more conservative design values. Only under the most extraordinary circumstances would the use of the theoretical values be justified. Note, however, that the theoretical and recommended design values are the same for conditions (d) and (f). This is because any deviation from a perfectly frictionless hinge or pin introduces rotational restraint and tends to reduce K. Hence, use of the theoretical values in these two cases is conservative.

The use of the effective length KL in place of the actual length L in no way alters any of the relationships discussed thus far. The column strength curve given in Figure 4.6 is unchanged except for renaming the abscissa KL. The critical buckling stress corresponding to a given length, actual or effective, remains the same.

4.3 ≡ AISC REQUIREMENTS

The basic requirements for compression members are covered in Chapter E of the AISC Specification. The relationship between loads and strength (Equation 2.3) takes the form

$$P_u \leq \phi_c P_n$$

where

P_u = sum of factored loads

P_n = nominal compressive strength = $A_g F_{cr}$

F_{cr} = critical buckling stress

ϕ_c = resistance factor for compression members = 0.85

Instead of expressing the critical buckling stress F_{cr} as a function of the slenderness ratio KL/r, the Specification uses the slenderness parameter

$$\lambda_c = \frac{KL}{r\pi} \sqrt{\frac{F_y}{E}} \qquad \text{(AISC Equation E2-4)}$$

which incorporates the material properties but is nondimensional. For elastic columns, Equation 4.4 can be written as

$$F_{cr} = \frac{\pi^2 E}{(KL/r)^2} = \frac{1}{\lambda_c^2} F_y$$

To account for the effects of initial crookedness, this value is reduced as follows:

$$F_{cr} = \frac{0.877}{\lambda_c^2} F_y$$

For inelastic columns, the tangent modulus equation, Equation 4.6b, is replaced by

$$F_{cr} = (0.658^{\lambda_c^2})F_y$$

which also accounts for initial crookedness. Thus a direct solution can be obtained, avoiding the trial-and-error approach inherent in the use of the tangent modulus equation. If the boundary between elastic and inelastic columns is taken as $\lambda_c = 1.5$, the AISC equations for critical buckling stress can be summarized as follows:

For $\lambda_c \leq 1.5$,

$$F_{cr} = (0.658^{\lambda_c^2})F_y \qquad \text{(AISC Equation E2-2)}$$

For $\lambda_c > 1.5$,

$$F_{cr} = \frac{0.877}{\lambda_c^2} F_y \qquad \text{(AISC Equation E2-3)}$$

These requirements are represented graphically in Figure 4.8.

FIGURE 4.8 ≡

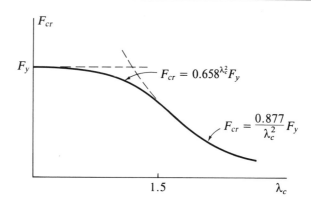

AISC Equations E2-2 and E2-3 are a condensed version of five equations[32] that cover five ranges of λ_c. These equations are based on experimental and theoretical studies that account for the effects of residual stresses and an initial out-of-straightness of $L/1500$, where L is the member length.

AISC B7 recommends a maximum slenderness ratio KL/r of 200 for compression members. Even though this is only a suggested limit, it is a practical upper limit since columns more slender than this will have little strength and not be economical.

EXAMPLE 4.2

Compute the design compressive strength of a W14 × 74 with a length of 20 feet and pinned ends. A36 steel is used.

SOLUTION   Slenderness ratio:

$$\text{Maximum } \frac{KL}{r} = \frac{1.0(20 \times 12)}{2.48} = 96.77 < 200 \quad \text{(OK)}$$

$$\lambda_c = \frac{KL}{r\pi}\sqrt{\frac{F_y}{E}} = \frac{96.77}{\pi}\sqrt{\frac{36}{29{,}000}} = 1.085$$

$$\lambda_c < 1.5 \quad \therefore \quad F_{cr} = (0.658)^{\lambda_c^2}F_y$$
$$= (0.658)^{(1.085)^2}(36) = 21.99 \text{ ksi}$$

$$P_n = A_g F_{cr} = 21.8(21.99) = 479.5 \text{ kips}$$

$$\phi_c P_n = 0.85(479.5) = 408 \text{ kips}$$

Answer   Design compressive strength = 408 kips.

The mode of failure considered thus far is referred to as *flexural* buckling, since the member is subjected to flexure, or bending, when it becomes unstable. For some cross-sectional configurations, the member will fail by twisting (torsional buckling) or by a combination of twisting and bending (flexural-torsional buckling). These infrequent cases will be considered in Section 4.6.

Local Stability

The strength corresponding to any buckling mode cannot be developed, however, if the elements of the cross section are so thin that *local* buckling occurs. This type of instability is a localized buckling or wrinkling at an isolated location. If this takes place, the cross section is no longer fully effective, and the member has failed. I- and H-shaped cross sections with thin flanges or webs are susceptible to this phenomenon, and their use should be avoided when possible. Otherwise, the compressive strength given by AISC Equations E2-2 and E2-3 must be reduced. The measure of this susceptibility is the width-thickness ratio of each cross-sectional element. Two types of elements must be considered: unstiffened elements, which are free of support along one edge parallel to the direction of load, and stiffened elements, which are supported along both edges.

Limiting values of width-thickness ratios are given in AISC B5, "Local Buckling," where cross-sectional shapes are classified as *compact*, *noncompact*, or *slender*, according to the values of the ratios. For uniformly compressed elements, as in an axially loaded compression member, the strength must be reduced if the shape has any slender elements. The width-thickness ratio is given the generic name of λ. Depending on the particular cross-sectional element, λ is either the ratio b/t or h/t_w, both to be defined presently. If λ is greater than the specified limit, denoted by λ_r, the shape is slender, and we must account for the potential for local buckling. (A discussion of the compact and noncompact categories will be postponed until Chapter 5, "Beams.") For I- or H-shapes, the projecting flange is considered to be an unstiffened element, and its width can be taken as half the full nominal width. Using AISC notation,

$$\lambda = \frac{b}{t} = \frac{b_f/2}{t_f} = \frac{b_f}{2t_f}$$

where b_f and t_f are the width and thickness of the flange. The upper limit is

$$\lambda_r = \frac{95}{\sqrt{F_y}}$$

The webs of I- and H-shapes are stiffened elements, and the stiffened width is to be taken as the distance between the roots of the flanges. The width-thickness parameter is

$$\lambda = \frac{h}{t_w}$$

where h is the distance between the roots of the flanges, and t_w is the web thickness. The upper limit is

$$\lambda_r = \frac{253}{\sqrt{F_y}}$$

Values of the ratios $b_f/2t_f$ and h/t_w are tabulated in the dimensions and properties tables in Part 1 of the *Manual*.

Stiffened and unstiffened elements of various cross-sectional shapes are illustrated in Figure 4.9. The appropriate compression member limits from AISC B5 are also given.

FIGURE 4.9 ≡

$b/t \le 95/\sqrt{F_y}$

$h/t_w \le 253/\sqrt{F_y}$

$b/t \le 95/\sqrt{F_y}$

$b/t \le 76/\sqrt{F_y}$

$d/t \le 127/\sqrt{F_y}$

$b/t \le 95/\sqrt{F_y}$
$h/t_w \le 253/\sqrt{F_y}$

$b/t \le 238/\sqrt{F_y}$
$h/t_w \le 253/\sqrt{F_y}$

EXAMPLE 4.3

Investigate the column of Example 4.2 for local stability.

SOLUTION For a W14 × 74, $b_f = 10.07$ in., $t_f = 0.785$ in., and

$$\frac{b_f}{2t_f} = \frac{10.07}{2(0.785)} = 6.4$$

Values of $b_f/2t_f$ are also tabulated in the properties tables.

$$\frac{95}{\sqrt{F_y}} = \frac{95}{\sqrt{36}} = 15.8 > 6.4 \quad (OK)$$

$$\frac{h}{t_w} = 25.3 \quad \text{(from properties tables)}$$

$$\frac{253}{\sqrt{F_y}} = \frac{253}{\sqrt{36}} = 42.2 > 25.3 \quad (OK)$$

Answer Local instability is not a problem.

It is permissible to use a cross-sectional shape that does not satisfy the width-thickness ratio requirements, but such a member will not be permitted to carry as large a load as one that does. In other words, the design strength must be reduced. This is accomplished by modifying F_{cr} by factors obtained from Appendix B of the Specification. In most cases, however, a rolled shape that satisfies the width-thickness requirements can be found, and this procedure will not be necessary.

Tables for Compression Members

The *Manual* contains many useful tables for analysis and design. For compression members whose strength is governed by flexural buckling (the kind considered thus far), Tables 3-36, 3-50, and 4 in the Numerical Values section of the Specification and the column load tables in Part 3 of the *Manual*, "Column Design," are the most useful. Table 3-36 gives values of $\phi_c F_{cr}$ as a function of KL/r for $F_y = 36$ ksi. Table 3-50 does the same for $F_y = 50$ ksi, and Table 4 gives $\phi_c F_{cr}/F_y$ as a function of λ_c. (All *Manual* tables for $F_y = 50$ ksi, such as Table 3-50, are distinguished from tables for $F_y = 36$ ksi by gray shading.) The Column Load tables give the design strength of selected shapes for various values of the effective length. Tables 3-36 and 3-50 stop at the recommended upper limit of $KL/r = 200$, and the Column Load tables include values of KL up to those corresponding to $KL/r = 200$. The use of each of these tables will be illustrated in the following example.

EXAMPLE 4.4

Compute the design strength of the compression member of Example 4.2 with the aid of (a) Table 3-36, (b) Table 4, and (c) the Column Load tables.

SOLUTION

a. From Example 4.2, $KL/r = 96.77$. Since $F_y = 36$ ksi, Table 3-36 is used. Values of $\phi_c F_{cr}$ are only given for integer values of KL/r; for decimal values, KL/r may be rounded *up*, or linear interpolation may be used. For uniformity, interpolation will be used in this book for all tables unless otherwise indicated.

For $KL/r = 96.77$,

$$\phi_c F_{cr} = 18.69 \text{ ksi}$$
$$\phi_c P_n = \phi_c A_g F_{cr} = A_g(\phi_c F_{cr}) = 21.8(18.69) = 407 \text{ kips}$$

b. From Example 4.2, $\lambda_c = 1.085$.

From Table 4, for $\lambda_c = 1.085$,

$$\phi_c \frac{F_{cr}}{F_y} = 0.519$$

$$\phi_c P_n = A_g\left(\phi_c \frac{F_{cr}}{F_y} \right)F_y = 21.8(0.519)(36) = 407 \text{ kips}$$

c. The column load tables in Part 3 of the *Manual* give the design strength $\phi_c P_n$ for selected W, HP, pipe, tube, double-angle, WT, and single-angle shapes. The tabular values for the symmetrical shapes (W, HP, pipe, and tube) were calculated by using the minimum radius of gyration for each shape.

From Example 4.2, $K = 1.0$

$$KL = 1.0(20) = 20 \text{ ft}$$

For a W14 × 74, A36 steel, and $KL = 20$ ft,

$$\phi_c P_n = 407 \text{ kips}$$

The values from Tables 3-36, 3-50, and 4 are based on flexural buckling and AISC Equations E2-2 and E2-3. This means that local stability is assumed; therefore, width-thickness ratio limits must not be exceeded. The design strengths given in the Column Load Tables account for the reduction required when the width-thickness limits are exceeded.

From a practical standpoint, if a compression member to be analyzed can be found in the Column Load Tables, then these tables should be used. Otherwise, Tables 3-36 and 3-50 offer a computational advantage over Table 4. If the yield strength is other than 36 ksi or 50 ksi, Table 4 is the only one that can be used.

4.4 ≡ DESIGN

The selection of an economical rolled shape to resist a given compressive load is simple with the aid of the Column Load tables. One enters the table with the effective length and moves horizontally until the desired design strength (or something slightly larger) is found. In some cases, the search must be continued in order to be certain the lightest shape has been found. Usually the category of shape (W, WT, etc.) will have been decided upon in advance. Often the overall nominal dimensions will also be known because of architectural or other requirements. As pointed out earlier, all tabulated values correspond to a slenderness ratio of 200 or less. The tabulated unsymmetrical shapes — the structural tees and the single and double-angles — require special consideration and will be covered in Section 4.6.

≡ EXAMPLE 4.5

A compression member is subjected to service loads of 165 kips dead load and 535 kips live load. The member is 26-ft long and pinned at each end. Use A36 steel and select a W14 shape.

SOLUTION   Factored loads:

$$P_u = 1.2D + 1.6L = 1.2(165) + 1.6(535) = 1054 \text{ kips}$$

∴ required design strength $\phi_c P_n = 1054$ kips

From the Column Load Tables for $KL = 26$ ft, a W14 × 176 has a design strength of 1150 kips.

Answer   Use a W14 × 176.

≡ EXAMPLE 4.6

Select the lightest W-shape to resist a factored compressive load P_u of 190 kips. The effective length is 24 feet. Use ASTM A572 Grade 50 steel.

SOLUTION   The appropriate strategy here is to find the lightest shape for each nominal size and then choose the lightest overall. The choices are as follows:

W4, W5, and W6:	None of the tabulated shapes will work.
W8:	W8 × 58, $\phi_c P_n = 194$ kips
W10:	W10 × 49, $\phi_c P_n = 239$ kips
W12:	W12 × 53, $\phi_c P_n = 247$ kips
W14:	W14 × 61, $\phi_c P_n = 276$ kips

Note that the load capacity is not proportional to the weight (or cross-sectional area). Although the W8 × 58 has the smallest design strength of the four choices, it is the second heaviest.

Answer   Use a W10 × 49.

For shapes not in the Column Load Tables, a trial-and-error approach must be used. The general procedure is to assume a shape and then compute its design strength. If the strength is too small (unsafe) or too large (uneconomical), another trial must be made. A systematic approach to making the trial selection is as follows:

1. Assume a value for the critical buckling stress F_{cr}.

2. From the requirement that $\phi_c P_n \geq P_u$, let

$$\phi_c A_g F_{cr} \geq P_u$$

and

$$A_g \geq \frac{P_u}{\phi_c F_{cr}}$$

3. Select a shape that satisfies this area requirement.

4. Compute F_{cr} and $\phi_c P_n$ for the trial shape.

5. Revise if necessary. If the design strength is very close to the required value, the next tabulated size can be tried. Otherwise, repeat the entire procedure using the value of F_{cr} found for the current trial shape as a value for Step 1.

6. Check local stability (check width-thickness ratios). Revise if necessary.

Instead of assuming values of F_{cr}, the term $\phi_c F_{cr}$ can be assumed. This will be convenient if the yield stress is either 36 ksi or 50 ksi. In these cases, Table 3-36 or 3-50 in the Numerical Values section of the Specification can be used as an aid. Since the tabulated values of $\phi_c F_{cr}$ are for $1 \leq KL/r \leq 200$, the tables will define the upper and lower limits for $\phi_c F_{cr}$.

EXAMPLE 4.7

Select a W18 shape of A36 steel to resist a factored load of 1054 kips. The effective length, KL, is 26 feet.

SOLUTION   From Table 3-36, $\phi_c F_{cr}$ is seen to lie between 5.33 ksi and 30.60 ksi.

Try $\phi_c F_{cr} = 20$ ksi.

$$\text{required } A_g = \frac{P_u}{\phi_c F_{cr}} = \frac{1054}{20} = 52.7 \text{ in.}^2$$

Try W18 × 192:

$$A_g = 56.4 \text{ in.}^2 > 52.7 \text{ in.}^2$$

$$\frac{KL}{r_{min}} = \frac{26(12)}{2.79} = 111.8$$

From Table 3-36, $\phi_c F_{cr} = 15.85$ ksi. Although this is less than the assumed value, the area furnished is greater than that required, so the design strength should be computed before rejecting this shape.

$$\phi_c P_n = \phi_c F_{cr} A_g = 15.85(56.4) = 894 \text{ kips} < 1054 \text{ kips} \quad \text{(N.G.)*}$$

Try $\phi_c F_{cr} = 15.85$ ksi:

$$\text{required } A_g = \frac{1054}{15.85} = 66.5 \text{ in.}^2$$

Try W18 × 234:

$$A_g = 68.8 \text{ in.}^2 > 66.5 \text{ in.}^2$$

$$\frac{KL}{r_{min}} = \frac{26(12)}{2.85} = 109.5$$

$$\phi_c F_{cr} = 16.28 \text{ ksi}$$

$$\phi_c P_n = \phi_c F_{cr} A_g = 16.28(68.8) = 1120 \text{ kips} > 1054 \text{ kips} \quad \text{(OK)}$$

Since this shape is not listed in the column load tables, the width-thickness ratios must be checked:

$$\frac{b_f}{2t_f} = 2.8 < \frac{95}{\sqrt{36}} = 15.8 \quad \text{(OK)}$$

$$\frac{h}{t_w} = 13.8 < \frac{253}{\sqrt{36}} = 42.2 \quad \text{(OK)}$$

Answer Use a W18 × 234.

4.5 ≡ MORE ON EFFECTIVE LENGTH

The concept of effective length was introduced in Section 4.2, "Column Theory." All compression members are treated as pin-ended regardless of the actual end conditions, but with an effective length KL that may differ from the actual length. With this modification, the load capacity of compression

---

* The notation "N.G." means "no good."

members is a function of only the slenderness parameter λ_c. For a given material, it is a function of the slenderness ratio KL/r.

If a compression member is supported differently with respect to each of its principal axes, the effective length will be different for the two directions. In Figure 4.10, a W-shape is used as a column and is braced by horizontal members in two perpendicular directions at the top. These members prevent translation of the column in all directions, but the connections, the details of which are not shown, permit small rotations to take place. Under these conditions, the member can be treated as pin-connected at the top. For the same reasons, the connection to the support at the bottom may also be treated as a pin connection. Generally speaking, a rigid, or fixed, condition is very difficult to achieve, and unless some special provisions are made, ordinary connections will usually closely approximate a hinge or pin connection. At midheight, the column is braced, but only in one direction. Again, the connection is such that translation is prevented, but no restraint against rotation is furnished. This brace prevents translation in a direction perpendicular to the weak axis of the

FIGURE 4.10 ≡

$KL = 5'$

(a) Minor Axis Buckling

$KL = 10'$

(b) Major Axis Buckling

cross section but provides no restraint perpendicular to the strong axis. As shown schematically in Figure 4.10, if the member were to buckle about the x-axis, the effective length would be 10 feet, whereas buckling about the weak axis would have to be in the second buckling mode, corresponding to an effective length of 5 feet. Since the strength is inversely proportional to the square of the slenderness ratio, a column will buckle in the direction corresponding to the largest slenderness ratio, so $K_x L/r_x$ must be compared with $K_y L/r_y$. In Figure 4.10, the ratio $10(12)/r_x$ must be compared with $5(12)/r_y$ (where r_x and r_y are in inches), and the larger ratio would be used for the determination of the nominal axial compressive strength P_n.

EXAMPLE 4.8

A W12 × 65, 24-feet long, is pinned at both ends and braced in the weak direction at the third points (Figure 4.11). A36 steel is used. Determine the design compressive strength.

SOLUTION

$$\frac{K_x L}{r_x} = \frac{24(12)}{5.28} = 54.55$$

$$\frac{K_y L}{r_y} = \frac{8(12)}{3.02} = 31.79$$

$K_x L/r_x$, the larger value, controls. From Table 3-36 with $KL/r = 54.55$,

$$\phi_c F_{cr} = 26.17 \text{ ksi}$$
$$\phi_c P_n = A_g(\phi_c F_{cr}) = 19.1(26.17) = 500 \text{ kips}$$

Answer   Design strength = 500 kips.

FIGURE 4.11 ≡

x-direction   y-direction

The design strengths given in the column load tables are based on the slenderness ratio with respect to the y-axis. A procedure for using the tables with

$K_x L$, however, can be developed by examining the procedure used to determine the tabular values. For a given rolled shape and F_y, each design compressive strength can be computed as follows:

1. Specify a value of KL.
2. Compute KL/r_y.
3. Compute λ_c.
4. Compute F_{cr}.
5. Compute $\phi_c P_n = \phi_c A_g F_{cr}$.

This is based on the assumption that $KL = K_y L$. If the capacity with respect to x-axis buckling is desired, the table can be entered with $KL = K_x L/(r_x/r_y)$, and the tabulated load will be based on

$$\frac{KL}{r_y} = \frac{K_x L/(r_x/r_y)}{r_y} = \frac{K_x L}{r_x}$$

The ratio r_x/r_y is given in the column load tables for each shape listed.

EXAMPLE 4.9

The compression member shown in Figure 4.12 is pinned at both ends and supported in the weak direction at midheight. A service load of 400 kips, with equal parts of dead and live load, must be supported. Use A36 steel and select the lightest W-shape.

SOLUTION

Factored load $= P_u = 1.2(200) + 1.6(200) = 560$ kips

Assume that the weak direction controls and enter the column load tables with $KL = 9$ feet. Beginning with the smallest shapes, the first one found that will work is a W10 × 77 with a design strength of 632 kips.

FIGURE 4.12 ≡

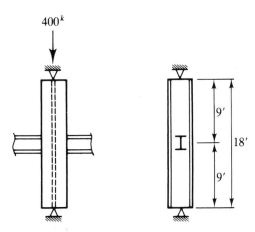

Check the strong axis:

$$\frac{K_x L}{r_x/r_y} = \frac{18}{1.73} = 10.40 \text{ ft} > 9 \text{ ft} \quad \therefore \; K_x L \text{ controls for this shape.}$$

Enter the tables with $KL = 10.4$ feet. A W10 × 77 is still the lightest W10, with a design strength of 612 kips (interpolated).

Continuing the search, we investigate a W12 × 72:

$$\frac{K_x L}{r_x/r_y} = \frac{18}{1.75} = 10.3 \text{ ft} > 9 \text{ ft}$$

$K_x L$ again controls, and the design strength is 592 kips.

Determine the lightest W14. The lightest one with a possibility of working is a W14 × 74. Since this is heavier than the lightest one found so far, it will not be considered.

Answer   Use a W12 × 72.

Whenever possible, the designer should provide extra support for the weak direction of a column. Otherwise, the member is inefficient: It has an excess of strength in one direction. When $K_x L$ and $K_y L$ are different, $K_y L$ will control unless r_x/r_y is smaller than $K_x L/K_y L$. When the two ratios are equal, the column has equal strength in both directions. Whether this is possible in any given situation can be ascertained by examining the range of values of r_x/r_y. For the W-shapes in the column load tables, r_x/r_y ranges between 1.6 and 1.8 except for some of the lighter shapes.

≡ **EXAMPLE 4.10**

The column shown in Figure 4.13 is subjected to a factored axial load of 840 kips. Use A36 steel and select a W-shape.

SOLUTION $K_x L = 20$ ft, maximum $K_y L = 8$ ft:

$$\frac{K_x L}{K_y L} = \frac{20}{8} = 2.5$$

This is an indication that $K_x L$ probably controls. Try $r_x/r_y = 1.7$:

$$\frac{K_x L}{r_x/r_y} = \frac{20}{1.7} = 11.76 > K_y L$$

FIGURE 4.13 ≡

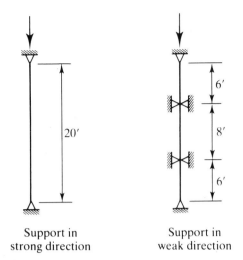

Support in
strong direction

Support in
weak direction

Rounding this to $KL = 12$ ft and entering the column load tables, try a
W10 × 112 ($\phi_c P_n = 865$ kips):

$$\text{actual } \frac{K_x L}{r_x/r_y} = \frac{20}{1.74} = 11.5 \text{ ft} < 12 \text{ ft} \quad \therefore \phi_c P_n > \text{required } 840 \text{ kips}$$

(By interpolation, $\phi_c P_n = 876$ kips.)

Check a W12 × 106:

$$\frac{K_x L}{r_x/r_y} = \frac{20}{1.76} = 11.4 \text{ ft}$$

For $KL = 12$ ft, $\phi_c P_n = 853$ kips > 840 kips   (OK)

Investigate W14 shapes. For $r_x/r_y = 1.7$ (this is the approximate ratio for all
likely possibilities),

$$\frac{K_x L}{r_x/r_y} = \frac{20}{1.7} = 11.76 \text{ ft} > K_y L = 8 \text{ ft}$$

For $KL = 12$ ft, a W14 × 109, with a capacity of 905 kips, is the lightest W14.

Since 12 feet is a conservative approximation of the actual effective length, this
shape is satisfactory.

Answer   Use a W12 × 106 (lightest of the three possibilities).

For isolated columns that are not part of a continuous frame, Table C-C2.1
in the Commentary to the Specification will usually suffice. Consider, however,
the rigid frame in Figure 4.14. The columns in this frame are not independent
members, but part of a continuous structure. Except for those in the lower

FIGURE 4.14 ≣

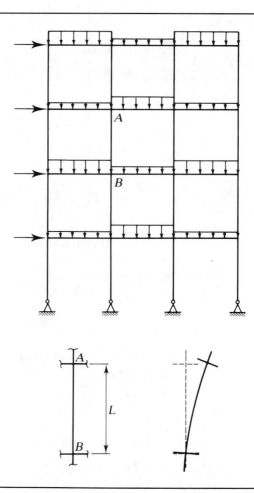

story, the columns are restrained at both ends by their connection to beams and other columns. This frame is also unbraced, meaning that horizontal displacements of the frame are possible, and all columns are subject to sidesway. If Table C-C2.1 is used for this frame, the lower story columns are best approximated by condition (*f*), and a value of *K* = 2 might be used. For a column such as *AB*, a value of *K* = 1.2, corresponding to condition (c), could be selected. A more rational procedure, however, will account for the degree of restraint provided by connecting members.

The rotational restraint provided by the beams, or girders, at the end of a column is a function of the rotational stiffness of the members intersecting at the joint. Rotational stiffness is defined as the moment, when applied to one end of a member, that is needed to produce a unit rotation at that end if the other end is fixed. For a homogeneous member with a uniform cross section, this is given by $4EI/L$, where I is the moment of inertia with respect to the axis of bending. The rotational restraint at the end of a column will be proportional to the ratio of column stiffnesses to girder stiffnesses at that end, given by

$$G = \frac{\text{total column stiffness}}{\text{total girder stiffness}} = \frac{\sum 4EI/L \text{ for columns}}{\sum 4EI/L \text{ for girders}} = \frac{\sum I_c/L_c}{\sum I_g/L_g} \qquad (4.7)$$

where E is a constant for structural steel. If a very slender column is connected to girders with large cross sections, the girders will effectively prevent rotation of the column. The ends of the column are approximately fixed, and K is relatively small. This corresponds to small values of G given by Equation 4.7. On the other hand, the ends of the stiff columns connected to flexible beams can more freely rotate and approach the pinned condition, giving relatively large values of G and K. The relationship between G and K has been quantified in the Jackson–Mooreland Alignment Charts,[33] which are reproduced in Figure C-C2.2 in the Commentary. To obtain a value of K from one of the nomograms, first calculate the value of G at each end of the column, letting one value be G_A and the other be G_B. Connect G_A and G_B with a straight line, and read the value of K on the middle scale. The effective length factor obtained in this manner is with respect to the axis of bending, which is the axis perpendicular to the plane of the frame. A separate analysis must be made for buckling about the other axis. Normally the beam-to-column connections in this direction will not transmit moment, sidesway is prevented by bracing, and K can be taken as 1.0.

≡ **EXAMPLE 4.11**

The rigid frame shown in Figure 4.15 is unbraced. Each member is oriented so that its web is in the plane of the frame. Determine the effective length factor $K_x L$ for columns AB and BC.

FIGURE 4.15 ≡

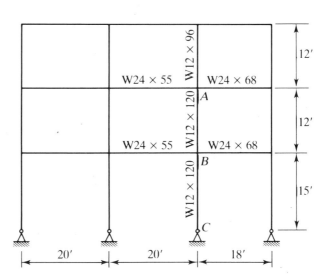

SOLUTION   Column *AB*:

Joint A:

$$G = \frac{\sum I_c/L_c}{\sum I_g/L_g} = \frac{833/12 + 1070/12}{1350/20 + 1830/18} = \frac{158.6}{169.2} = 0.94$$

Joint B:

$$G = \frac{\sum I_c/L_c}{\sum I_g/L_g} = \frac{1070/12 + 1070/15}{169.2} = \frac{160.5}{169.2} = 0.95$$

Answer   From the alignment chart for sidesway uninhibited, with $G_A = 0.94$ and $G_B = 0.95$, $K = 1.3$ for column AB.

Column *BC*:

Joint B: As before,

$G = 0.95.$

Joint C: At a pin connection, the situation is analogous to that of a very stiff column attached to infinitely flexible girders — that is, girders of zero stiffness. The ratio of column stiffness to girder stiffness would therefore be infinite for a perfectly frictionless hinge. Since this end condition can only be approximated in practice, the discussion accompanying the alignment chart recommends that G be taken as 10.0.

Answer   From the alignment chart with $G_A = 0.95$ and $G_B = 10.0$, $K = 1.85$ for column *BC*.

Unbraced frames are able to support lateral loads because of their moment-resisting joints. Often the frame is augmented by a bracing system of some sort; such frames are called *braced frames*. The additional resistance to lateral loads can take the form of diagonal bracing, as illustrated in Figure 4.16, or rigid shear walls. In either case, the tendency for columns to sway is blocked within a given panel, or bay, for the full height of the frame. This forms a cantilever structure that is resistant to horizontal displacements and also provides horizontal support for the other bays. Depending on the size of the structure, more than one bay may require bracing. Columns that are members of braced frames are prevented from sidesway and have some degree of rotational restraint at their ends. Thus, they are in a category that lies somewhere between cases (a) and (d) in Table C-C2.1 of the Commentary, and K is between 0.5 and 1.0. A value of 1.0 is therefore always conservative for members of braced frames, and this is the value prescribed by AISC C2.1 unless an analysis is made. Such an analysis can be made with the alignment chart for braced frames. Use of this

FIGURE 4.16 ≡

(a) Diagonal bracing

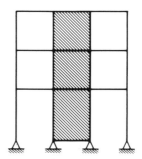

(b) Shear Walls
(masonry, reinforced concrete,
or steel plate)

nomogram will result in an effective length factor somewhat less than 1.0, and some savings could be realized.

As with any design aid, the alignment charts should be used only under the conditions for which they were derived. These are discussed in Section C2 of the Commentary to the Specification and will not be enumerated here. Most of the conditions will usually be approximately satisfied; if they are not, the deviation will be on the conservative side. One condition that usually is not satisfied is the requirement that all behavior be elastic. If the slenderness parameter λ_c is less than 1.5, the column will buckle inelastically, and the effective length factor obtained from the alignment chart will be overly conservative. A large number of columns are in this category. A convenient procedure for determining K for inelastic columns allows the alignment charts to be used.[34, 35] To demonstrate the procedure, we begin with the critical buckling load for an inelastic column given by Equation 4.6b. If this is divided by the cross-sectional area, the buckling stress is obtained:

$$F_{cr} = \frac{\pi^2 E_t}{(KL/r)^2}$$

The rotational stiffness of a column in this state would be $4E_t I/L$, and the appropriate value of G for use in the alignment chart is

$$G_{\text{inelastic}} = \frac{\sum E_t I_c/L_c}{\sum EI_g/L_g} = \frac{E_t}{E} G_{\text{elastic}}$$

Since E_t is less than E, $G_{\text{inelastic}}$ is less than G_{elastic}, and the effective length factor K will be reduced, resulting in a more economical design. To evaluate E_t/E, called the *stiffness reduction factor*, consider the following relationship for a column with pinned ends:

$$\frac{F_{cr(\text{inelastic})}}{F_{cr(\text{elastic})}} = \frac{\pi^2 E_t/(L/r)^2}{\pi^2 E/(L/r)^2} = \frac{E_t}{E} \tag{4.8}$$

Since AISC uses an approximation for the inelastic portion of the column strength curve, Equation 4.8 is an approximation when AISC Equations E2-2 and E2-3 are used for F_{cr}. If we let

$$F_{cr} = \frac{P_{cr}}{A} \approx \frac{P_u/\phi_c}{A}$$

then $F_{cr(\text{inelastic})}/F_{cr(\text{elastic})}$ will be a function of $P_u/(\phi_c A)$. For example, for $P_u/(\phi_c A) = 26$ ksi and $F_y = 36$ ksi,

$$F_{cr(\text{inelastic})} \approx 26 = 0.658^{\lambda_c^2} F_y = 0.658^{\lambda_c^2}(36)$$
$$\lambda_c^2 = 0.7776$$
$$F_{cr(\text{elastic})} = \frac{0.877}{\lambda_c^2} F_y = \frac{0.877}{0.7776}(36) = 40.7$$

The stiffness reduction factor is therefore

$$\frac{F_{cr(\text{inelastic})}}{F_{cr(\text{elastic})}} = \frac{26}{40.7} = 0.639$$

Since ϕ_c is a constant, the stiffness reduction factor can also be expressed as a function of P_u/A.

Values of the stiffness reduction factor as a function of P_u/A are given in Table 3-1 in Part 3 of the *Manual*.

≡ EXAMPLE 4.12

A rigid unbraced frame is shown in Figure 4.17. All members are oriented so that bending is about the strong axis. Lateral support is provided at each joint by simply connected bracing in the direction perpendicular to the frame. Determine the effective length factors with respect to each axis for member *AB*. The factored axial load on this member is 180 kips, and A36 steel is used.

FIGURE 4.17 ≡

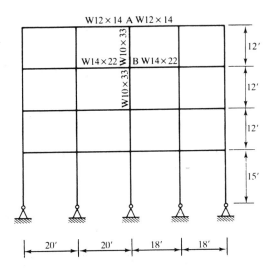

SOLUTION   Compute elastic G factors:

Joint A:

$$\frac{\sum (I_c/L_c)}{\sum (I_g/L_g)} = \frac{170/12}{88.6/20 + 88.6/18} = \frac{14.17}{9.35} = 1.52$$

Joint B:

$$\frac{\sum (I_c/L_c)}{\sum (I_g/L_g)} = \frac{2(170/12)}{199/20 + 199/18} = \frac{28.3}{21.0} = 1.35$$

From the alignment chart for unbraced frames, $K_x = 1.43$, based on elastic behavior.

$$\lambda_c = \frac{K_x L}{r_x \pi} \sqrt{\frac{F_y}{E}} = \frac{1.43(12)(12)}{4.19\pi} \sqrt{\frac{36}{29,000}} = 0.5512$$

Since λ_c is less than 1.5, the inelastic K factor should be used.

$$\frac{P_u}{A} = \frac{180}{9.71} = 18.5 \text{ ksi}$$

From Table 3-1 in Part 3 of the *Manual*, the stiffness reduction factor = 0.83.

Joint A:

$$G_{\text{inelastic}} = \text{stiffness reduction factor} \times G_{\text{elastic}} = 0.83(1.52) = 1.26$$

Joint B:

$$G_{\text{inelastic}} = 0.827(1.35) = 1.12$$

Answer   From the alignment chart, $K_x = 1.37$. Because of the support conditions normal to the frame, K_y can be taken as 1.0.

4.6 ≣ TORSIONAL AND FLEXURAL-TORSIONAL BUCKLING

When an axially loaded compression member becomes unstable in an overall sense (that is, not locally unstable), it can buckle in one of three ways (see Figure 4.18):

1. **Flexural buckling.** This is the type of buckling that has been considered up to now and is a deflection due to bending, or flexure, about the axis corresponding to the largest slenderness ratio (Figure 4.18a). This is usually the minor principal axis — the one with the smallest radius of gyration. Compression members with any type of cross-sectional configuration can fail in this way.

2. **Torsional buckling.** This is failure by twisting about the longitudinal axis of the member. It can occur only with doubly symmetric cross sections with very slender cross-sectional elements (Figure 4.18b). Standard hot-rolled shapes are not susceptible to torsional buckling, but members built up from thin plate elements should be investigated.

3. **Flexural-torsional buckling.** This is a combination of flexural buckling and torsional buckling; the member bends and twists simultaneously (Figure 4.18c). This failure mode can occur only with unsymmetrical cross sections, both those with one axis of symmetry, such as structural tees, double-angle shapes, and equal-leg single angles, and those with no axis of symmetry, such as unequal-leg single angles.

The AISC Specification requires an analysis of torsional or flexural-torsional buckling when appropriate. Section E3 of the Specification covers double-angle and tee-shaped members, and Appendix E3 provides a more general approach that can be used for other unsymmetric shapes.

The procedure prescribed in Appendix E3 will be discussed first. It is based on the use of a slenderness parameter λ_e in place of λ_c. This can be derived as follows:

From the Euler buckling stress,

$$F_e = \frac{\pi^2 E}{(KL/r)^2}$$

FIGURE 4.18 ≡

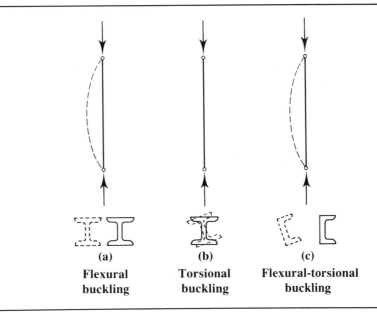

(a)
Flexural
buckling

(b)
Torsional
buckling

(c)
Flexural-torsional
buckling

the slenderness ratio can be written as

$$\frac{KL}{r} = \sqrt{\frac{\pi^2 E}{F_e}}$$

If F_e is defined as the elastic buckling stress corresponding to the controlling mode of failure, whether flexural, torsional, or flexural-torsional, then the corresponding slenderness ratio is

$$\left(\frac{KL}{r}\right)_e = \sqrt{\frac{\pi^2 E}{F_e}}$$

and the corresponding slenderness parameter is

$$\lambda_e = \frac{(KL/r)_e}{\pi} \sqrt{\frac{F_y}{E}} = \sqrt{\frac{F_y(KL/r)_e^2}{\pi^2 E}} = \sqrt{\frac{F_y}{F_e}}$$

An outline of the overall procedure is as follows:

1. Determine F_e for torsional elastic buckling or flexural-torsional elastic buckling from the equations given in Appendix E3.

2. Compute the effective slenderness parameter λ_e.

3. Compute the critical stress F_{cr} from the usual equations (AISC Equations E2-2 and E2-3), but use λ_e instead of λ_c. The design strength is then

$$\phi_c P_n = \phi_c A_g F_{cr}$$

where ϕ_c is 0.85, the same as it is for flexural buckling.

The equations for F_e given in AISC Appendix E3 are based on the well-established theory given in *Theory of Elastic Stability*.[36] Except for some changes in notation, they are the same equations as those given in that work, with no simplifications. For doubly symmetric shapes (torsional buckling),

$$F_e = \left[\frac{\pi^2 E C_w}{(K_z L)^2} + GJ \right] \frac{1}{I_x + I_y} \qquad \text{(AISC Equation A-E3-5)}$$

For singly symmetric shapes (flexural-torsional buckling),

$$F_e = \frac{F_{ey} + F_{ez}}{2H} \left(1 - \sqrt{1 - \frac{4 F_{ey} F_{ez} H}{(F_{ey} + F_{ez})^2}} \right) \qquad \text{(AISC Equation A-E3-6)}$$

For shapes with *no* axis of symmetry (flexural-torsional buckling),

$$(F_e - F_{ex})(F_e - F_{ey})(F_e - F_{ez}) - F_e^2 (F_e - F_{ey})(x_0/\bar{r}_0)^2$$
$$- F_e^2 (F_e - F_{ex})(y_0/\bar{r}_0)^2 = 0 \qquad \text{(AISC Equation A-E3-7)}$$

This last equation is a cubic; F_e is the smallest root. Fortunately, there will be little need for solving this equation since completely unsymmetric shapes are rarely used as compression members. The previously undefined terms used in these three equations are defined as follows:

C_w = warping constant (in.)

K_z = effective length factor for *torsional* buckling. This is based on the amount of end restraint against twisting about the longitudinal axis.

G = shear modulus (ksi)

J = torsional constant (equal to the polar moment of inertia only for circular cross sections) (in.^{4})

$$F_{ex} = \frac{\pi^2 E}{(K_x L/r_x)^2} \qquad \text{AISC Equation (A-E3-10)}$$

$$F_{ey} = \frac{\pi^2 E}{(K_y L/r_y)^2} \qquad \text{(AISC Equation A-E3-11)}$$

where y is the axis of symmetry for singly symmetric shapes.

$$F_{ez} = \left[\frac{\pi^2 E C_w}{(K_z L)^2} + GJ \right] \frac{1}{A \bar{r}_0^2} \qquad \text{(AISC Equation A-E3-12)}$$

$$H = 1 - \left(\frac{x_0^2 + y_0^2}{\bar{r}_0^2} \right) \qquad \text{(AISC Equation A-E3-9)}$$

x_0, y_0 = coordinates of the shear center of the cross section with respect to the centroid (in inches). The shear center is the point on the cross section through which a transverse load on a beam must pass if the member is to bend without twisting. The shear center will be covered in more detail in Chapter 5, " Beams."

$$\bar{r}_0^2 = x_0^2 + y_0^2 + \frac{I_x + I_y}{A} \qquad \text{(AISC Equation A-E3-8)}$$

Values of the constants used in the three equations for F_e can be found in the tables of Torsion Properties and Flexural-Torsional Properties in Part 1 of the *Manual*. For W, M, S, and HP shapes, J and C_w are given. The values of J, C_w, $\bar{r}_0$, and H are given for channels, single angles, and structural tees. The tables for double angles give values of $\bar{r}_0$ and H (J and C_w are double the values given for single angles).

As previously pointed out, the need for a torsional buckling analysis of a doubly symmetric shape will be rare. Similarly, shapes with no axis of symmetry are rarely used for compression members, and flexural-torsional buckling analysis of these types of members will seldom, if ever, need to be done. For these reasons, we will limit further consideration to flexural-torsional buckling of shapes with one axis of symmetry. Furthermore, the most commonly used of these shapes, the double angle, is a built-up shape, and a consideration of it will be postponed until Section 4.7.

For singly symmetric shapes, the flexural-torsional elastic buckling stress F_e is found from AISC Equation A-E3-6. In this equation, y is the axis of symmetry, and flexural-torsional buckling will only take place about this axis. The x-axis is subject only to flexural buckling. When the slenderness ratio for the y-axis, $K_y L/r_y$, is greater than the slenderness ratio for the x-axis, $K_x L/r_x$, the strength will be governed by y-axis buckling. When $K_x L/r_x$ is greater than $K_y L/r_y$, the strength can be governed by flexural buckling about the x-axis, but flexural-torsional buckling about the y-axis should also be checked, since it could still control.

EXAMPLE 4.13

Compute the design compressive strength of a WT13.5 × 80.5. The effective length with respect to the x-axis is 25 ft 6 in., the effective length with respect to the y-axis is 20 ft, and the effective length with respect to the z-axis is 20 ft. A36 steel is used.

SOLUTION   Compare slenderness ratios:

$$\frac{K_x L}{r_x} = \frac{25.5(12)}{3.96} = 77.27$$

$$\frac{K_y L}{r_y} = \frac{20(12)}{3.24} = 74.07$$

Since $K_x L/r_x$ is larger, there is a possibility that flexural buckling about the x-axis could control. From Table 3-36,

$$\phi_c F_{cr} = 22.35 \text{ ksi}$$

and

$$\phi_c P_n = \phi_c A_g F_{cr} = A_g(\phi_c F_{cr}) = 23.7(22.35) = 530 \text{ kips}$$

Flexural-torsional buckling about y-axis:

$$F_{ey} = \frac{\pi^2 E}{(K_y L/r_y)^2} = \frac{\pi^2(29,000)}{(74.07)^2} = 52.17 \text{ ksi}$$

$$F_{ez} = \left[\frac{\pi^2 E C_w}{(K_z L)^2} + GJ\right]\frac{1}{A\bar{r}_0^2}$$

$$= \left[\frac{\pi^2(29,000)(42.7)}{(20 \times 12)^2} + 11,200(7.31)\right]\frac{1}{23.7(5.67)^2} = 107.7 \text{ ksi}$$

$$F_{ey} + F_{ez} = 52.17 + 107.7 = 159.9 \text{ ksi}$$

$$F_e = \frac{F_{ey} + F_{ez}}{2H}\left[1 - \sqrt{1 - \frac{4F_{ey}F_{ez}H}{(F_{ey} + F_{ez})^2}}\right]$$

$$= \frac{159.9}{2(0.813)}\left[1 - \sqrt{1 - \frac{4(52.17)(107.7)(0.813)}{(159.9)^2}}\right] = 45.81 \text{ ksi}$$

$$\lambda_e = \sqrt{\frac{F_y}{F_e}} = \sqrt{\frac{36}{45.81}} = 0.8865$$

Since this is less than 1.5, use AISC Equation E2-2 with λ_e in place of λ_c:

$$F_{cr} = (0.658^{\lambda_e^2})F_y = (0.658)^{(0.8865)^2}(36) = 25.91 \text{ ksi}$$

$$\phi_c P_n = \phi_c A_g F_{cr} = 0.85(23.7)(25.91) = 522 \text{ kips} \quad \text{(controls)}$$

Answer   Design strength = 522 kips.

The procedure for flexural-torsional buckling analysis of double-angles and tees given in AISC Section E3 is a modification of the procedure given in AISC Appendix E3. There is some notational change as well: F_e becomes F_{crft}, F_{ey} becomes F_{cry}, and F_{ez} becomes F_{crz}. The stress F_{cry} is found from AISC E2 and is based on flexural buckling about the y-axis.

To obtain F_{crz}, we can drop the first term of AISC Equation A-E3-12 to give

$$F_{crz} = \frac{GJ}{A\bar{r}_0^2}$$

This is permissible, since for double-angles and tees, the first term is negligible compared to the second term.

The flexural buckling stress F_{cry} is computed with the usual equations of AISC Chapter E, using KL/r corresponding to the y-axis (the axis of symmetry).

The nominal strength can then be computed as

$$P_n = A_g F_{crft},$$

where

$$F_{crft} = \left(\frac{F_{cry} + F_{crz}}{2H}\right)\left[1 - \sqrt{1 - \frac{4F_{cry}F_{crz}H}{(F_{cry} + F_{crz})^2}}\right] \qquad \text{(AISC Equation E3-1)}$$

All other terms from Appendix E3 remain unchanged. This procedure, to be used with double-angles and tees only, is more accurate than the procedure given in Appendix E3.

≡ **EXAMPLE 4.14**

Compute the design strength of the shape in Example 4.13 by using the equations of AISC E3.

SOLUTION

Compute F_{cry}. From Example 4.13, $K_y L/r_y = 74.07$. From AISC Equation E2-4, the slenderness parameter is

$$\lambda_c = \frac{KL}{r\pi}\sqrt{\frac{F_y}{E}} = \frac{74.07}{\pi}\sqrt{\frac{36}{29,000}} = 0.8307 < 1.5$$

From AISC Equation E2-2,

$$F_{cr} = F_{cry} = (0.658^{\lambda_c^2})F_y = (0.658)^{(0.8307)^2}(36) = 26.97 \text{ ksi}$$

From AISC E3,

$$F_{crz} = \frac{GJ}{A\bar{r}_0^2} = \frac{11,200(7.31)}{23.7(5.67)^2} = 107.5 \text{ ksi}$$

$$F_{cry} + F_{crz} = 26.97 + 107.5 = 134.5 \text{ ksi}$$

$$F_{crft} = \left(\frac{F_{cry} + F_{crz}}{2H}\right)\left[1 - \sqrt{1 - \frac{4F_{cry}F_{crz}H}{(F_{cry} + F_{crz})^2}}\right]$$

$$= \frac{134.5}{2(0.813)}\left[1 - \sqrt{1 - \frac{4(26.97)(107.5)(0.813)}{(134.5)^2}}\right] = 25.48 \text{ ksi}$$

$$\phi_c P_n = \phi_c A_g F_{crft} = 0.85(23.7)(25.48) = 513 \text{ kips}$$

Answer   Design strength = 513 kips.

The results of these two examples show that the error in using Appendix E3 for this shape is on the unconservative side. The procedure of Example 4.14 (which uses AISC Specification E3) should always be used for double angles and tees. In practice, however, most double angles and tees can be found in the column load tables. These tables are based on the recommended procedure of AISC E3 and can be used to verify the result of Example 4.14. The tables give two values of the design strength, one based on flexural buckling about the x-axis and one based on either flexural buckling about the y-axis or flexural-torsional buckling about the y-axis, whichever gives the smaller strength.

Tables are also provided for single-angle compression members. The design strengths given are not based on flexural-torsional buckling theory, but on an approximation given in the separate specification for single-angle members in Part 6 of the *Manual, Specifications and Codes.*

4.7 ≡ BUILT-UP MEMBERS

If the cross-sectional properties of a built-up compression member are known, its analysis is the same as for any other compression member, provided the component parts of the cross section are properly connected. AISC E4 contains many details related to this connection, with separate requirements for members composed of two or more rolled shapes and for members composed of plates or a combination of plates and shapes. Before considering the connection problem, we will review the computation of cross-sectional properties of built-up shapes.

The design strength of a built-up compression member is a function of the slenderness parameter λ_c. This means that the principal axes and the corresponding radii of gyration about these axes must be determined. For homogeneous cross sections, the principal axes coincide with the centroidal axes. The procedure will be illustrated in the following example. It will be assumed that the components of the cross section are properly connected.

≡ **EXAMPLE 4.15**

The column shown in Figure 4.19 is fabricated by welding a $\frac{3}{8}$-in. × 4-in. cover plate to the flange of a W18 × 35. A36 steel is used for both components. The effective length is 15 feet with respect to both axes. Assume that the components are connected in such a way that the member is fully effective, and compute the design strength based on flexural buckling.

SOLUTION

With the addition of the cover plate, the shape is slightly unsymmetric, but the flexural-torsional effects will be negligible.

The vertical axis of symmetry is one of the principal axes, and its location need not be computed. The horizontal principal axis will be found by application of the *principle of moments*: The sum of moments of component areas about any axis (in this example, a horizontal axis along the top of the plate will be used) must be equal to the moment of the total area. We will use Table 4.1 to keep track of the computations.

$$\bar{y} = \frac{\sum Ay}{\sum A} = \frac{95.30}{11.8} = 8.076 \text{ in.}$$

FIGURE 4.19 ≡

W18 × 35

TABLE 4.1 ≡

Component	A	y	Ay
Plate	1.500	0.1875	0.2812
W	10.3	9.225	95.02
$\sum$	11.8		95.30

With the location of the horizontal centroidal axis known, the moment of inertia with respect to this axis can be found by using the *parallel-axis theorem*:

$$I = \bar{I} + Ad^2$$

where

$\bar{I}$ = moment of inertia about the centroidal axis of a component area

A = area of the component

I = moment of inertia about an axis parallel to the centroidal axis

d = perpendicular distance between the two axes

The contributions from each component area are computed and summed to obtain the moment of inertia of the composite area. Table 4.2, an expanded version of Table 4.1, includes these computations. For the vertical axis,

$$I_y = \frac{1}{12}(3/8)(4)^3 + 15.3 = 17.30 \text{ in.}^4 \quad \text{(controls)}$$

$$r_y = \sqrt{\frac{I_y}{A}} = \sqrt{\frac{17.30}{11.8}} = 1.211 \text{ in.}$$

$$\lambda_c = \frac{KL}{r\pi}\sqrt{\frac{F_y}{E}} = \frac{15(12)}{1.211\pi}\sqrt{\frac{36}{29,000}} = 1.667 > 1.5$$

$$F_{cr} = \left[\frac{0.877}{\lambda_c^2}\right]F_y = \left[\frac{0.877}{(1.667)^2}\right](36) = 11.36 \text{ ksi}$$

$$\phi_c P_n = \phi_c A_g F_{cr} = 0.85(11.8)(11.36) = 114 \text{ kips}$$

Answer   Design strength = 114 kips.

TABLE 4.2 ≡

Component	A	y	Ay	$\bar{I}$	d	$\bar{I} + Ad^2$
Plate	1.500	0.1875	0.2812	0.01758	7.889	93.37
W	10.3	9.225	95.02	510	1.149	523.6
$\sum$	11.8		95.30			$617.0 = I_x$

Connection Requirements for Built-Up Members Composed of Rolled Shapes

The most common built-up shape is one that is composed of rolled shapes, namely, the double-angle shape. This type of member will be used to illustrate the requirements for this category of built-up members. Figure 4.20 shows a truss tension member connected to gusset plates at each end. To maintain the back-to-back separation of the angles along the length, we place fillers, or spacers, of the same thickness as the gusset plate between the angles at equal intervals. The intervals must be small enough so that the member functions as a unit. If the member buckles about the x-axis (flexural buckling) the connectors are not subjected to any calculated load, and the connection problem is simply one of maintaining the relative positions of the two components. To ensure that the built-up member acts as a unit, AISC requires that the slenderness of an individual component is no greater than three-fourths of the slenderness of the built-up member; that is,

$$\frac{a}{r_i} \le \frac{3}{4}\frac{KL}{r}$$

where

a = spacing of the connectors

r_i = minimum radius of gyration of the component

KL/r = maximum slenderness ratio of the built-up member.

If the member buckles about the axis of symmetry — that is, it is subjected to flexural or flexural-torsional buckling about the y-axis — the connectors are

FIGURE 4.20 ≡

FIGURE 4.21 ≡

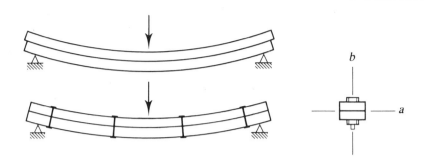

subjected to shearing forces. This can be visualized by considering the two planks used as a beam in Figure 4.21. If the planks are unconnected, they will slip along the surface of contact when loaded and will function as two separate beams. When connected by bolts (or any other fasteners, such as nails), the two planks will behave as a unit, and the resistance to slip will be provided by shear in the bolts. This is the behavior taking place in the double-angle shape when bending about its y-axis. If the plank beam is oriented so that bending takes place about its other axis (the b-axis), then both planks bend in exactly the same manner, and there is no slippage and hence no shear. This is analogous to bending about the x-axis of the double-angle shape. When the fasteners are subjected to shear, a modified slenderness ratio larger than the actual value may be required.

AISC E4 considers two categories of connectors: (1) snug-tight bolts and (2) welds or fully tightened bolts. These connection methods will be covered in detail in Chapter 7, "Simple Connections." The Column Load Tables for double-angles are based on the use of welds or fully tightened bolts. For this category,

$$\left(\frac{KL}{r}\right)_m = \sqrt{\left(\frac{KL}{r}\right)_0^2 + 0.82 \frac{\alpha^2}{(1 + \alpha^2)} \left(\frac{a}{r_{ib}}\right)^2} \qquad \text{(AISC Equation E4-2)}$$

where

$(KL/r)_0$ = original unmodified slenderness ratio
$(KL/r)_m$ = modified slenderness ratio
r_{ib} = radius of gyration of component about axis parallel to member axis of buckling
α = separation ratio = $\dfrac{h}{2r_{ib}}$
h = distance between component centroids (perpendicular to member axis of buckling)

When the connectors are snug-tight bolts,

$$\left(\frac{KL}{r}\right)_m = \sqrt{\left(\frac{KL}{r}\right)_0^2 + \left(\frac{a}{r_i}\right)^2} \qquad \text{(AISC Equation E4-1)}$$

≡ EXAMPLE 4.16

Determine the design strength of the compression member shown in Figure 4.22. Two angles, $5 \times 3 \times \frac{1}{2}$, are oriented with the long legs back-to-back and separated by $\frac{3}{8}$ inch. The effective length KL is 16 feet and there are three fully tightened intermediate connectors. Compute the design strength for A36 steel.

SOLUTION   Check slenderness ratios:

$$\frac{KL}{r_x} = \frac{16(12)}{1.59} = 120.8$$

$$\frac{KL}{r_y} = \frac{16(12)}{1.25} = 153.6$$

Since KL/r_y is greater than KL/r_x, buckling about the y-axis will control. Although the design strength can be determined directly from the Column Load Tables, it will be computed here to illustrate the use of the modified slenderness ratio $(KL/r)_m$. The spacing of the connectors is

$$a = \frac{16(12)}{4 \text{ spaces}} = 48 \text{ in.}$$

$$\frac{a}{r_i} = \frac{a}{r_z} = \frac{48}{0.648} = 74.07 < (0.75)(153.6) = 115.2 \quad \text{(OK)}$$

$$r_{ib} = r_y = 0.829 \text{ in.}$$

$$h = 2(0.75) + \frac{3}{8} = 1.875 \text{ in.}$$

$$\alpha = \frac{h}{2r_{ib}} = \frac{1.875}{2 \times 0.829} = 1.131$$

From AISC Equation E4-2,

$$\left(\frac{KL}{r}\right)_m = \sqrt{\left(\frac{KL}{r}\right)_0^2 + 0.82 \frac{\alpha^2}{(1 + \alpha^2)} \left(\frac{a}{r_{ib}}\right)^2}$$

$$= \sqrt{(153.6)^2 + 0.82 \frac{(1.131)^2}{[1 + (1.131)^2]} \left(\frac{48}{0.829}\right)^2} = 158.5$$

This value should be used in place of $K_y L/r_y$.

FIGURE 4.22 ≡

First, compute the flexural buckling strength. For $KL/r = 158.5$,

$$\lambda_c = \frac{KL}{r\pi} \sqrt{\frac{F_y}{E}} = \frac{158.5}{\pi} \sqrt{\frac{36}{29,000}} = 1.778 > 1.5$$

$$F_{cr} = \left[\frac{0.877}{\lambda_c^2}\right] F_y = \left[\frac{0.877}{(1.778)^2}\right](36) = 9.987 \text{ ksi}$$

$$\phi_c P_n = \phi_c A_g F_{cr} = 0.85(7.50)(9.987) = 63.7 \text{ kips}$$

Compute the flexural-torsional buckling strength.

$$F_{cry} = F_{cr} \text{ for } y\text{-axis} = 9.987 \text{ ksi}$$

$$F_{crz} = \frac{GJ}{A\bar{r}_0^2} = \frac{11,200(2 \times 0.3220)}{7.50(2.52)^2} = 151.4 \text{ ksi}$$

$$F_{cry} + F_{crz} = 9.987 + 151.4 = 161.4 \text{ ksi}$$

$$F_{crft} = \left(\frac{F_{cry} + F_{crz}}{2H}\right)\left[1 - \sqrt{1 - \frac{4F_{cry} F_{crz} H}{(F_{cry} + F_{crz})^2}}\right]$$

$$= \frac{161.4}{2(0.645)}\left[1 - \sqrt{1 - \frac{4(9.987)(151.4)(0.645)}{(161.4)^2}}\right] = 9.748 \text{ ksi}$$

$$\phi_c P_n = \phi_c A_g F_{crft} = 0.85(7.50)(9.748) = 62.1 \text{ kips}$$

This is less than the flexural buckling strength of 63.7 kips, so the flexural-torsional buckling strength controls. Note that this result compares favorably with the value of 62 kips given in the column load tables.

Answer   Design strength = 62 kips.

EXAMPLE 4.17

Design a 14-ft-long compression member to resist a factored load of 50 kips. Use a double-angle shape with the short legs back-to-back, separated by $\frac{3}{8}$ inch. The member will be braced at midlength against buckling about the x-axis (the axis parallel to the long legs). Specify the number of intermediate connectors needed (the midlength brace will provide one such connector). Use A36 steel.

SOLUTION   From the Column Load Tables, select 2L $3\frac{1}{2} \times 3 \times \frac{1}{4}$, weighing 10.8 lb/ft. The capacity of this shape is 51 kips, based on buckling about the y-axis with an effective length of 14 ft. (The strength corresponding to flexural buckling about the x-axis is 60 kips, based on an effective length of $\frac{14}{2} = 7$ ft).

Since bending about the y-axis subjects the fasteners to shear, a sufficient number of fasteners must be provided to account for this. From the table, we see that 3 intermediate connectors are required.

Answer   Use 2L $3\frac{1}{2} \times 3 \times \frac{1}{4}$ with three intermediate connectors within the 14-ft length.

Connection Requirements for Built-Up Members Composed of Plates or Both Plates and Shapes

When a built-up member consists of two or more rolled shapes separated by a substantial distance, plates must be used to connect the shapes. AISC E4 contains many details regarding the connection requirements and the proportioning of the plates. Additional connection requirements are given for other built-up compression members composed of plates or plates and shapes.

≡ **PROBLEMS** ≡

NOTES   Unless otherwise indicated, all given loads are *service* loads.

For Problems 4.1 through 4.4, use AISC Equation E2-2 or E2-3 to compute F_{cr}. Verify the design strength obtained from this value with (1) Table 3-36 or 3-50 and (2) with Table 4 in the Numerical Values section of the Specification.

4.1   Determine the compressive design strength for the column in Figure P4.1. Use A36 steel.

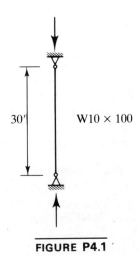

30'   W10 × 100

FIGURE P4.1

4.2   Use A36 steel and compute the compressive design strength for a W12 × 50 for effective lengths of (a) 12 ft, (b) 20 ft, and (c) 30 ft.

4.3   Find the compressive design strength for the column in Figure P4.3. Use A36 steel.

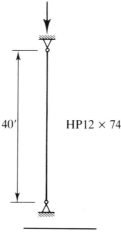

40′   HP12 × 74

FIGURE P4.3

4.4   What is the compressive design strength of a W12 × 58 with an effective length of 20 ft? Use $F_y = 50$ ksi.

For Problems 4.5 through 4.9, use A36 steel and compute the design strength for each column.

4.5   **4.6**

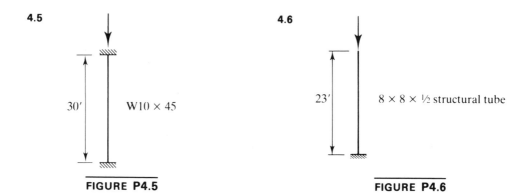

30′   W10 × 45

23′   8 × 8 × ½ structural tube

FIGURE P4.5   **FIGURE P4.6**

4.7

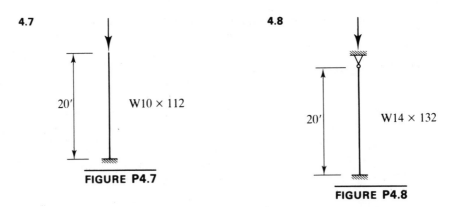

20′   W10 × 112

FIGURE P4.7

4.8

20′   W14 × 132

FIGURE P4.8

4.9

25′   W6 × 25

FIGURE P4.9

For Problems 4.10 through 4.14, use the basic trial-and-error design procedure.

4.10   Select a W21 shape for the conditions shown. Use A36 steel.

$D = 300^k$
$L = 90^k$

20′

FIGURE P4.10

4.11   Use A36 steel and select a W18 shape for the column.

$D = 110^k$
$L = 280^k$

30'

FIGURE P4.11

4.12   Select an S-shape. Use A36 steel.

$D = 10^k$
$L = 30^k$

15'

FIGURE P4.12

4.13   Select an A36 W-shape with a nominal depth of 21 inches for the column.

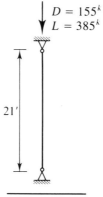

$D = 155^k$
$L = 385^k$

21'

FIGURE P4.13

4.14   Select a W16 of A36 steel.

$D = 58^k$
$L = 105^k$

15'

FIGURE P4.14

4.15   For the conditions in Figure P4.15, select

    a. W12, A36 steel
    b. a pipe
    c. a square structural tube
    d. a rectangular structural tube

$D = 90^k$
$L = 260^k$

15'-4"

FIGURE P4.15

4.16   Select a W10. $F_y = 50$ ksi.

FIGURE P4.16

4.17   Select a 12-inch W-shape to resist a factored axial compressive load of 780 kips. The member is 18-feet long, and the effective length factor K is 0.80. Use A36 steel.

4.18   Use A36 steel and select a W14.

FIGURE P4.18

4.19   A W18 × 76 is used as a 30-ft column, pinned at both ends. Lateral support is provided at mid-depth in the weak direction. Find the design strength. A36 steel.

4.20   The compression member shown has lateral support at the third points in the weak direction only. Find the design strength. A36 steel.

W14 × 132

FIGURE P4.20

4.21   What is the design strength? Use $F_y = 50$ ksi.

Braced in weak direction

W12 × 106

FIGURE P4.21

4.22   What is the design compressive strength of a 6-in. × 4-in. × $\frac{1}{2}$-in. structural tube if $K_x L = 15$ ft and $K_y L = 10$ ft? Use A36 steel.

4.23   A W10 × 112 is 30-feet long and pinned at both ends. Additional bracing is provided in the weak direction at a point 14 feet from the top. Determine the design axial compressive strength if A36 steel is used.

4.24   A W12 × 96 is used as a compression member. The effective length with respect to strong axis buckling is 25 feet. This member is also braced against buckling about the weak axis at a point 12 feet from the top. What factored axial compressive load can be supported? A36 steel.

4.25   Select a W10 of A36 steel to resist a factored load of 530 kips. $K_x L = 27$ ft and $K_y L = 15$ ft.

4.26   Select a W10 of A36 steel to resist a factored compressive load of 380 kips. $K_x L = 27$ ft and $K_y L = 15$ ft.

4.27   The rigid frame in Figure P4.27 is unbraced in the plane of the frame. In the direction perpendicular to the frame, the frame is braced at the joints. The connections at these points of bracing are simple (moment-free) connections. Roof girders are W14 × 30; floor girders are W16 × 36. Member *AB* is a W10 × 45 of A36 steel. Assume that the adjacent columns are also W10 × 45. Determine the axial compressive design strength of this member.

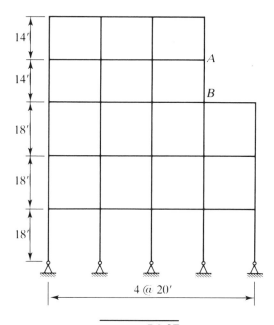

FIGURE P4.27

4.28   The rigid frame in Figure P4.28 is unbraced. The members are oriented so that bending is about the strong axis. Support conditions in the direction perpendicular to the plane of the frame are such that $K_y = 1.0$. The beams are W18 × 50, and the columns are W12 × 72. A36 steel is used. The factored axial compressive load on column *AB* is 300 kips.

   a. Determine K_x for column *AB*.
   b. Determine the axial compressive design strength of column *AB*.

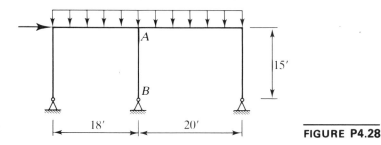

FIGURE P4.28

4.29   Determine the compressive design strength for a WT10.5 × 46.5 with an effective length of 16 ft with respect to each axis. Use A36 steel and the procedure of AISC Section E3 (*not* the appendix).

4.30   Use A36 steel and compute the design strength for the column.

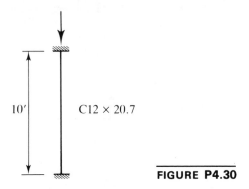

10'   C12 × 20.7

FIGURE P4.30

4.31   Select an American Standard Channel for the compression member. A36 steel.

$D = 30^k$
$L = 70^k$

10'

FIGURE P4.31

For each of the built-up compression members in Problems 4.32 through 4.36, compute the axial compressive design strength based on flexural buckling (no torsional or flexural-torsional buckling) and assume that enough connectors are provided so that the built-up section is fully effective.

4.32

36"

30"

40'

$F_y = 50$ ksi (all plates 4" thick)

FIGURE P4.32

4.33

$F_y = 50$ ksi   Section

FIGURE P4.33

4.34

2 ℄'s ⅜ × 10

2 C6 × 13

$F_y = 36$ ksi

FIGURE P4.34

4.35

2'-6"

2'-6"   40'

A36 (all plates 4" thick)

FIGURE P4.35

4.36

15'

5"

2" 1" 2"

1"

2"

Section

A36

FIGURE P4.36

4.37 Compute the design strength for the following double-angle shape: 2L8 × 4 × $\frac{3}{4}$, long legs $\frac{3}{8}$-in. back-to-back, $F_y = 50$ ksi; $KL = 20$ ft for all axes, and there are two intermediate connectors. Use the procedure of AISC Section E3 (*not* the appendix). Compare the flexural and the flexural-torsional buckling strengths.

4.38 For the conditions shown in Figure P4.38, select a double-angle section ($\frac{3}{8}$-in. gusset plate connection). Use A36 steel.

$D = 90^k$
$L = 260^k$

15'-04"

FIGURE P4.38

4.39 Select a double-angle section for the compression member shown in Figure P4.39. The load shown is the total service load, with a live-to-dead load ratio of 2.5 : 1. Use $\frac{3}{8}$-in. back-to-back of angles and $F_y = 50$ ksi.

560^k

21'

FIGURE P4.39

4.40 Select a double-angle shape for the top chord of the truss of Problem 3.32 (see AISC C2.1). Assume $\frac{3}{8}$-in. gusset plates and use A36 steel.

4.41 Write a computer program to determine the design strength of columns based on flexural buckling. Provide for any number of compression members.

Input: F_y (ksi), $K_x L$ (ft), $K_y L$ (ft), area (in.^{2}), I_x (in.^{4}), I_y (in.^{4})
Output: same as input, plus the design strength in kips

4.42 Write a computer program to produce tables of $\phi_c F_{cr}$ similar to Table 3-36 in the Numerical Values section of the AISC Specification.

Input: F_y (ksi)

Output: F_y, plus a table of $\phi_c F_{cr}$ for values of KL/r from 1.0 to 200.0 in increments of 0.1

Your table should have the following form:

FY = 36 KSI	
KL/R	*PHI*FCR (KSI)*
1.0	30.60
$\vdots$	$\vdots$
125.5	13.36
$\vdots$	$\vdots$
126.0	13.27
$\vdots$	$\vdots$

4.43 Write a computer program to generate a table of values of P_u/A and the stiffness reduction factor for any value of F_y — that is, a table similar to Table 3-1 in Part 3, "Column Design," of the *Manual of Steel Construction*. Do not print values when the stiffness reduction factor is less than 0.005.

Input data required: F_y (ksi)

5 BEAMS

5.1 ≡ INTRODUCTION

Beams are structural members that support transverse loads and are therefore subjected primarily to flexure, or bending. If a substantial amount of axial load is also present, the member is referred to as a *beam-column* (beam-columns will be considered in Chapter 6). Although some degree of axial load will be present in any structural member, there are many practical situations in which this effect is negligible and the member can be treated as a beam. Beams are usually thought of as being oriented horizontally and subjected to vertical loads, but this is not necessarily the case. A structural member is considered to be a beam if it is loaded in such a way as to cause bending.

Commonly used cross-sectional shapes include the W-, S-, and M-shapes. Channel shapes are sometimes used, as are beams built up from plates, in the form of I-, H-, or box shapes. For reasons to be discussed later, doubly symmetric shapes such as the standard rolled W-, M-, and S-shapes are the most efficient.

Shapes that are built up from plate elements are usually thought of as plate girders, but the AISC Specification distinguishes beams from plate girders on the basis of the width-thickness ratio of the web. Figure 5.1 shows both a hot-rolled shape and a built-up shape along with the dimensions to be used for the width-thickness ratios.
If

$$\frac{h}{t_w} \le \frac{970}{\sqrt{F_y}}$$

the member is to be treated as a beam, regardless of whether it is a rolled shape or built-up. This category is considered in Chapter F of the Specification,

FIGURE 5.1 ≡

"Beams and Other Flexural Members," and is the subject of the present chapter of this textbook. If

$$\frac{h}{t_w} > \frac{970}{\sqrt{F_y}}$$

the member is considered to be a plate girder and is covered in Chapter G of the Specification, "Plate Girders." In this textbook, plate girders will be covered in Chapter 10. Because of the slenderness of the web, plate girders require special consideration over and above that needed for beams.

All of the standard hot-rolled shapes found in the *Manual* are in the first category (beams). Most built-up shapes will be classified as plate girders, but some will be beams by the AISC definition.

For beams, the basic relationship between load effects and strength can be written as

$$M_u \le \phi_b M_n$$

where

M_u = controlling combination of factored load moments

ϕ_b = resistance factor for beams = 0.90

M_n = nominal moment strength

The design strength $\phi_b M_n$ is sometimes called the *design moment*.

5.2 ≣ BENDING STRESS AND THE PLASTIC MOMENT

To be able to determine the nominal moment strength M_n, we must first examine the behavior of beams throughout the full range of loading, from very small loads up to the point of collapse. Consider the beam shown in Figure 5.2a, which is oriented so that bending is about the major principal axis (for an I- or H-shape, this will be the x–x axis). For a linear elastic material and small deformations, the distribution of bending stress will be as shown in part b of the figure, with the stress assumed to be uniform across the width of the beam. (Shear will be considered separately in Section 5.7.) From elementary mechanics of materials, the stress at a given point can be found from the flexure formula:

$$f_b = \frac{My}{I_x} \tag{5.1}$$

where M is the bending moment at the cross section under consideration, y is the perpendicular distance from the neutral plane to the point of interest, and I_x is the moment of inertia of the area of the cross section with respect to the neutral axis. For a homogeneous material, the neutral axis coincides with the centroidal axis. Equation 5.1 is based on the assumption of a linear distribution

FIGURE 5.2 ≡

(a)

(b)

of strains from top to bottom, which in turn is based on the assumption that cross sections that are plane before bending remain plane after bending. In addition, the beam cross section must have a vertical axis of symmetry, and the loads must be in the plane containing this axis. Beams that do not satisfy these criteria will be considered in Section 5.13. The maximum stress will occur at the extreme fiber, where y is maximum. Thus, there are two maxima: maximum compressive stress in the top fiber and maximum tensile stress in the bottom fiber. If the neutral axis is an axis of symmetry, these two stresses will be equal in magnitude. For maximum stress, Equation 5.1 takes the following form:

$$f_{max} = \frac{Mc}{I_x} = \frac{M}{I_x/c} = \frac{M}{S_x} \tag{5.2}$$

where c is the perpendicular distance from the neutral axis to the extreme fiber, and S_x is the elastic section modulus of the cross section. For any given cross-sectional shape, the section modulus will be a constant. For an unsymmetrical cross section, there will be two values of S_x: one for the top extreme fiber and one for the bottom. Values of S_x for standard rolled shapes are tabulated in the Dimensions and Properties tables in the *Manual*.

Equations 5.1 and 5.2 are valid as long as the loads are small enough so that the material remains within its linear elastic range. For structural steel, this means that the stress f_{max} must not exceed F_y, and the bending moment must not exceed

$$M_y = F_y S_x$$

where M_y is the bending moment that brings the beam to the point of yielding.

In Figure 5.3, a simply supported beam with a concentrated load at midspan is shown at successive stages of loading. Once yielding begins, the distribution of stress on the cross section will no longer be linear, and yielding

FIGURE 5.3 ≡

FIGURE 5.4 ≡

will progress from the extreme fiber toward the neutral axis. At the same time, the yielded region will extend longitudinally from the center of the beam as the bending moment reaches M_y at more locations. These yielded regions are indicated by the dark areas in Figure 5.3c and d. In part b, yielding has just begun. In part c, the yielding has progressed into the web, and in part d the entire cross section has yielded. The additional moment required to bring the beam from stage b to stage d is, on the average, approximately 12% for W-shapes. When stage d has been reached, any further increase in the load will cause collapse, since all elements of the cross section have reached the yield plateau of the stress–strain curve and unrestricted plastic flow will occur. A *plastic hinge* is said to have formed at the center of the beam, and this hinge along with the actual hinges at the ends of the beam constitute an unstable mechanism. During the plastic collapse, the mechanism motion will be as shown in Figure 5.4. Structural analysis based on a consideration of collapse mechanisms is called *plastic analysis*. An introduction to plastic analysis and design is given in Appendix A of this book.

The plastic moment capacity, which is the moment required to form the plastic hinge, can easily be computed from a consideration of the corresponding stress distribution. In Figure 5.5, the compressive and tensile stress resultants are shown, where A_c is the cross-sectional area subjected to compression, and A_t is the area in tension. These are the areas above and below the plastic neutral axis, which is not necessarily the same as the elastic neutral axis. From equilibrium of forces,

$$C = T$$
$$A_c F_y = A_t F_y$$
$$A_c = A_t$$

Thus, the plastic neutral axis divides the cross section into two equal areas. For shapes that are symmetrical about the axis of bending, the elastic and plastic

FIGURE 5.5 ≡

neutral axes are the same. The plastic moment M_p is the resisting couple formed by the two equal and opposite forces:

$$M_p = F_y(A_c)a = F_y(A_t)a = F_y\left(\frac{A}{2}\right)a = F_y Z$$

where

A = total cross-sectional area

a = distance between the centroids of the two half-areas

$Z = \left(\dfrac{A}{2}\right)a$ = plastic section modulus

EXAMPLE 5.1

Compute the plastic moment M_p for a W10 × 60 of A36 steel.

SOLUTION From the Dimensions and Properties tables in Part 1 of the *Manual*,

$A = 17.6$ in.^{2}

$\dfrac{A}{2} = \dfrac{17.6}{2} = 8.8$ in.^{2}

The centroid of half of this area can be found in the tables for WT-shapes, which are cut from W-shapes. The relevant shape here is the WT5 × 30, and the distance from the outside face of the flange to the centroid is 0.884 inches (Figure 5.6).

$a = d - 2(0.884) = 10.22 - 2(0.884) = 8.452$ in.

$Z = \left(\dfrac{A}{2}\right)a = 8.8(8.452) = 74.38$ in.^{3}

This compares favorably with the value of 74.6 given in the Dimensions and Properties tables (the difference results from roundoff in the tabular values).

Answer $M_p = F_y Z = 36(74.38) = 2678$ in.-kips = 223 ft-kips.

FIGURE 5.6 ≡

0.884″

$d = 10.22''$

5.3 ≡ STABILITY

If a beam can be counted on to remain stable up to the fully plastic condition, the nominal moment strength can be taken as the plastic moment capacity; that is,

$$M_n = M_p$$

Otherwise, M_n will be less than M_p.

As with a compression member, instability can be in an overall sense or it can be local. Overall buckling is illustrated in Figure 5.7a. When a beam bends, the compression region (above the neutral axis) is analogous to a column and, in a manner similar to a column, will buckle if the member is slender enough. Unlike a column, however, the compression portion of the cross section is restrained by the tension portion, and the outward deflection (flexural buckling) is accompanied by twisting (torsion). This form of instability is called *lateral-torsional buckling* (*LTB*). Lateral-torsional buckling can be prevented by lateral bracing of the compression zone, preferably the compression flange, at sufficiently close intervals. This is shown schematically in Figure 5.7b. As we shall see, the moment strength depends in part on the unbraced length, which is the distance between points of lateral support.

Whether the beam can sustain a moment large enough to bring it to the fully plastic condition also depends on whether the cross-sectional integrity is maintained. This integrity will be lost if one of the compression elements of the cross section buckles. This can be either buckling of the compression flange, called *flange local buckling* (*FLB*), or buckling of the compression part of the web, called *web local buckling* (WLB). As discussed in Chapter 4, "Compression Members," this will depend on the width-thickness ratios of the compression elements of the cross section.

FIGURE 5.7 ≡

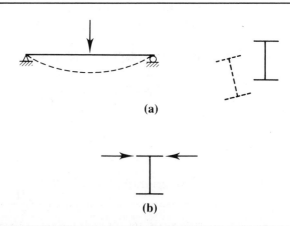

(a)

(b)

FIGURE 5.8 ≡

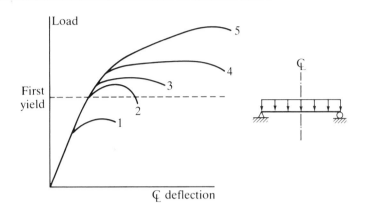

Figure 5.8 further illustrates the effects of local and lateral-torsional buckling. Five separate beams are represented on this graph of load versus central deflection. Curve 1 is the load-deflection curve of a beam that becomes unstable (in any way) and loses its load-carrying capacity before first yield (Figure 5.3b) is attained. Curves 2 and 3 correspond to beams that can be loaded past first yield but not far enough for the formation of a plastic hinge and the resulting plastic collapse. If plastic collapse can be reached, the load-deflection curve will have the appearance of either curve 4 or curve 5. Curve 4 is for the case of uniform moment over the full length of the beam, and curve 5 is for a beam with a variable bending moment (moment gradient). Safe designs can be achieved with beams corresponding to any of these curves, but curves 1 and 2 will represent inefficient use of material.

5.4 ≡ CLASSIFICATION OF SHAPES

AISC classifies cross-sectional shapes as compact, noncompact, or slender, depending on the values of the width-thickness ratios. For I- and H-shapes, the ratio for the projecting flange (an *unstiffened* element) is $b_f/2t_f$, and the ratio for the web (a *stiffened* element), is h/t_w. The classification of shapes is found in Section B5 of the Specification, "Local Buckling," in Table B5.1. It can be summarized in a general way as follows:

Let

λ = width : thickness ratio

λ_p = upper limit for compact category

λ_r = upper limit for noncompact category

TABLE 5.1 ≡ Width-Thickness Parameters*

Element	λ	λ_p	λ_r
Flange	$\dfrac{b_f}{2t_f}$	$\dfrac{65}{\sqrt{F_y}}$	$\dfrac{141}{\sqrt{F_y - 10}}$
Web	$\dfrac{h}{t_w}$	$\dfrac{640}{\sqrt{F_y}}$	$\dfrac{970}{\sqrt{F_y}}$

* For hot-rolled I- and H-shapes

Then

> if $\lambda \leq \lambda_p$ and the flange is continuously connected to the web, the shape is compact;
>
> if $\lambda_p < \lambda \leq \lambda_r$, the shape is noncompact; and
>
> if $\lambda > \lambda_r$, the shape is slender.

The category is based on the worst width-thickness ratio of the cross section. For example, if the web is compact and the flange is noncompact, the shape is classified as noncompact. Table 5.1 has been extracted from AISC Table B5.1 and contains width-thickness ratios for hot-rolled I- and H-shaped cross sections.

5.5 ≡ BENDING STRENGTH OF COMPACT SHAPES

A beam can fail by reaching M_p and becoming fully plastic, or it can fail by buckling in one of the following ways:

1. Lateral-torsional buckling (LTB), either elastically or inelastically
2. Flange local buckling (FLB), elastically or inelastically
3. Web local buckling (WLB), elastically or inelastically

If the maximum bending stress is less than the proportional limit when buckling occurs, the failure is said to be *elastic*. Otherwise, it is *inelastic*. (See the related discussion in Section 4.2, "Column Theory.")

It is convenient to first categorize beams as compact, noncompact, or slender, and then determine the moment resistance based on the degree of lateral support. The discussion in this section will apply to the following two types of beams: (1) hot-rolled I- and H-shapes bent about the strong axis and loaded in the plane of the weak axis and (2) channels bent about the strong axis and either loaded through the shear center or restrained against twisting. (The shear center is the point on the cross section through which a transverse load must pass if the beam is to bend without twisting.) Emphasis will be on I- and H-shapes. Hybrid beams (those with different grades of steel in the web and flanges) will not be considered, and some of the AISC equations will be slightly modified to reflect this specialization; F_{yf} and F_{yw}, the yield strengths of the flange and web, will be replaced by F_y.

We begin with *compact shapes,* defined as those whose webs are continuously connected to the flanges and that satisfy the following width-thickness ratio requirements for the flange and the web:

$$\frac{b_f}{2t_f} \le \frac{65}{\sqrt{F_y}} \quad \text{and} \quad \frac{h}{t_w} \le \frac{640}{\sqrt{F_y}}$$

The web criterion is met by all standard hot-rolled shapes listed in the *Manual* (see Problems 5.2 and 5.3), so only the flange ratio need be checked. Most shapes will also satisfy the flange requirement and will therefore be classified as compact. If the beam is compact and has continuous lateral support, or if the unbraced length is very short, the nominal moment strength M_n is the full plastic moment capacity of the shape M_p. For members with inadequate lateral support, the moment resistance is limited by the lateral-torsional buckling strength, either inelastic or elastic.

The first category, laterally supported compact beams, is quite common and is the simplest case. AISC F1.1 gives the nominal strength as

$$M_n = M_p \qquad \text{(AISC Equation F1-1)}$$

where

$$M_p = F_y Z \le 1.5 M_y$$

The limit of $1.5M_y$ for M_p is to prevent excessive working load deformations and is satisfied when

$$F_y Z \le 1.5 F_y S \quad \text{or} \quad \frac{Z}{S} \le 1.5$$

EXAMPLE 5.2

The beam shown in Figure 5.9 is a W16 × 31 of A36 steel. It supports a reinforced concrete floor slab that provides continuous lateral support of the compression flange. The service dead load, including the weight of the beam, is 500 lb/ft. The service live load is 550 lb/ft. Does this beam have adequate moment strength?

FIGURE 5.9 ≡

$W_D = 500 \text{ lb/ft}$
$W_L = 550 \text{ lb/ft}$

30'

SOLUTION   For a simply supported, uniformly loaded beam, the maximum bending moment occurs at midspan and is equal to

$$M_{max} = \frac{1}{8} wL^2$$

where w is the load in units of force per unit length, and L is the span length.

$$M_D = \frac{1}{8} w_D L^2 = \frac{0.500(30)^2}{8} = 56.25 \text{ ft-kips}$$

$$M_L = \frac{0.550(30)^2}{8} = 61.88 \text{ ft-kips}$$

Since the dead load is less than 8 times the live load, load combination A4-2 controls:

$$M_u = 1.2M_D + 1.6M_L = 1.2(56.25) + 1.6(61.88) = 166 \text{ ft-kips}$$

Alternatively, the loads can be factored at the outset:

$$w_u = 1.2w_D + 1.6w_L = 1.2(0.500) + 1.6(0.550) = 1.480 \text{ kips/ft}$$

$$M_u = \frac{1}{8} w_u L^2 = \frac{1.480(30)^2}{8} = 166 \text{ ft-kips}$$

Check for compactness:

$$\frac{b_f}{2t_f} = 6.3 \quad \text{(from Part 1 of the } \textit{Manual)}$$

$$\frac{65}{\sqrt{F_y}} = \frac{65}{\sqrt{36}} = 10.8 > 6.3 \quad \therefore \text{ the flange is compact.}$$

$$\frac{h}{t_w} < \frac{640}{\sqrt{F_y}} \quad \text{(for all shapes in the } \textit{Manual)}$$

$\therefore$ a W16 × 31 is compact for A36 steel.

Since the beam is compact and laterally supported,

$$M_n = M_p = F_y Z_x = 36(54.0) = 1944 \text{ in.-kips} = 162.0 \text{ ft-kips}$$

Check for $M_p \leq 1.5M_y$:

$$\frac{Z_x}{S_x} = \frac{54.0}{47.2} = 1.14 < 1.5 \quad \text{(OK)}$$

$$\phi_b M_n = 0.90(162.0) = 146 \text{ ft-kips} < 166 \text{ ft-kips} \quad \text{(N.G.)}$$

Answer   Since the design moment is less than the factored load moment, the W16 × 31 is unsatisfactory.

FIGURE 5.10 ≡

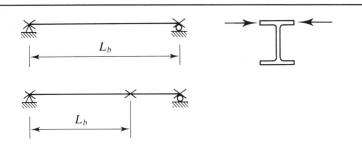

The moment strength of compact shapes is a function of the unbraced length L_b, defined as the distance between points of lateral support, or bracing. In this book, points of lateral support will be indicated by an "×," as shown in Figure 5.10. The relationship between nominal strength M_n and unbraced length is shown in Figure 5.11. If the unbraced length is no greater than L_p, to be defined presently, the beam is considered to have full lateral support, and $M_n = M_p$. If L_b is greater than L_p but less than or equal to the parameter L_r, the strength is based on inelastic LTB. If L_b is greater than L_r, the strength is based on elastic LTB.

The equation for the theoretical elastic lateral-torsional buckling strength can be found in the *Theory of Elastic Stability*[36] and, with some notational changes, is as follows:

$$M_n = \frac{\pi}{L_b} \sqrt{EI_y\, GJ + \left(\frac{\pi E}{L_b}\right)^2 I_y\, C_w} \qquad (5.3)$$

FIGURE 5.11 ≡

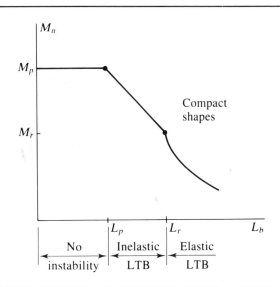

where

L_b = unbraced length (in.)

G = shear modulus = 11,200 ksi for structural steel

J = torsional constant (in.^{4})

C_w = warping constant (in.^{6})

If the moment is greater than that corresponding to first yield, the strength is based on inelastic behavior. The moment corresponding to first yield is taken as

$$M_r = F_L S_x \qquad \text{(AISC Equation F1-7)}$$

where F_L is the smaller of $(F_{yf} - F_r)$ or F_{yw}. In this expression, the flange yield stress is reduced by F_r, the residual stress. For a nonhybrid member, $F_{yf} = F_{yw} = F_y$, and F_L will always be equal to $F_y - F_r$. In the remainder of this chapter, F_L will be replaced by $F_y - F_r$. For example, AISC Equation F1-7 will be written as

$$M_r = (F_y - F_r)S_x \qquad \text{(AISC Equation F1-7)}$$

As shown in Figure 5.11, the boundary between elastic and inelastic behavior will be for an unbraced length of L_r, which is the value of L_b obtained from Equation 5.3 when M_n is set equal to M_r. The following equation is obtained:

$$L_r = \frac{r_y X_1}{(F_y - F_r)} \sqrt{1 + \sqrt{1 + X_2(F_y - F_r)^2}} \qquad \text{(AISC Equation F1-6)}$$

where

$$X_1 = \frac{\pi}{S_x}\sqrt{\frac{EGJA}{2}}$$

$$X_2 = \frac{4C_w}{I_y}\left(\frac{S_x}{GJ}\right)^2 \qquad \text{(AISC Equations F1-8 and F1-9)}$$

As with columns, inelastic behavior of beams is more complicated than elastic behavior, and empirical formulas are often used. With one minor modification, the following equation is used by AISC:

$$M_n = M_p - (M_p - M_r)\left(\frac{L_b - L_p}{L_r - L_p}\right) \qquad (5.4)$$

where

$$L_p = \frac{300 r_y}{\sqrt{F_y}} \qquad \text{(AISC Equation F1-4)}$$

The nominal bending strength of compact beams is completely described by Equations 5.3 and 5.4, subject to an upper limit of M_p for inelastic beams, provided the applied moment is uniform over the unbraced length L_b. Otherwise there is a *moment gradient*, and Equations 5.3 and 5.4 must be modified by a factor C_b. This factor is given in AISC F1.2 as

$$C_b = \frac{12.5M_{max}}{2.5M_{max} + 3M_A + 4M_B + 3M_C}$$ (AISC Equation F1-3)

where

M_{max} = absolute value of the maximum moment within the unbraced length (including the end points)

M_A = absolute value of the moment at the quarter point of the unbraced length

M_B = absolute value of the moment at the midpoint of the unbraced length

M_C = absolute value of the moment at the three-quarter point of the unbraced length

When the bending moment is uniform, the value of C_b is

$$C_b = \frac{12.5M}{2.5M + 3M + 4M + 3M} = 1.0$$

≡ **EXAMPLE 5.3**

Determine C_b for a uniformly loaded, simply supported beam with lateral support at its ends only.

SOLUTION

Because of symmetry, the maximum moment is at midspan:

$$M_{max} = M_B = \frac{1}{8}wL^2$$

Also because of symmetry, the moment at the quarter point is equal to the moment at the three-quarter point. From Figure 5.12,

$$M_A = M_C = \frac{wL}{2}\left(\frac{L}{4}\right) - \frac{wL}{4}\left(\frac{L}{8}\right) = \frac{wL^2}{8} - \frac{wL^2}{32} = \frac{3}{32}wL^2$$

$$C_b = \frac{12.5M_{max}}{2.5M_{max} + 3M_A + 4M_B + 3M_C}$$

$$= \frac{12.5\left(\frac{1}{8}\right)}{2.5\left(\frac{1}{8}\right) + 3\left(\frac{3}{32}\right) + 4\left(\frac{1}{8}\right) + 3\left(\frac{3}{32}\right)} = 1.14$$

Answer $C_b = 1.14$.

FIGURE 5.12 ≡

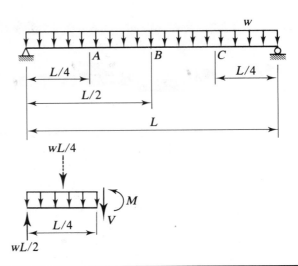

Figure 5.13 shows the value of C_b for several common cases of loading and lateral support (see Problem 5.10).

For unbraced cantilever beams, AISC specifies a value of C_b of 1.0. A value of 1.0 is always conservative, regardless of beam configuration or loading, but in some cases it may be excessively conservative. The complete specification of nominal moment strength for compact shapes can now be summarized.

For $L_b \leq L_p$,

$$M_n = M_p \leq 1.5M_y \qquad \text{(AISC Equation F1-1)}$$

For $L_p < L_b \leq L_r$,

$$M_n = C_b\left[M_p - (M_p - M_r)\left(\frac{L_b - L_p}{L_r - L_p}\right)\right] \leq M_p \qquad \text{(AISC Equation F1-2)}$$

For $L_b > L_r$,

$$M_n = M_{cr} \leq M_p \qquad \text{(AISC Equation F1-12)}$$

where

$$M_{cr} = C_b\frac{\pi}{L_b}\sqrt{EI_y\,GJ + \left(\frac{\pi E}{L_b}\right)^2 I_y C_w} \qquad \text{(AISC Equation F1-13)}$$

$$= \frac{C_b S_x X_1 \sqrt{2}}{L_b/r_y}\sqrt{1 + \frac{X_1^2 X_2}{2(L_b/r_y)^2}}$$

The constants X_1 and X_2 have been previously defined and are tabulated in the Dimensions and Properties Tables in the *Manual*.

The effect of C_b on the nominal strength is illustrated in Figure 5.14. Although the strength is directly proportional to C_b, this graph clearly shows the importance of observing the upper limit of M_p, regardless of which equation is used for M_n.

FIGURE 5.13 ≡

FIGURE 5.14 ≡

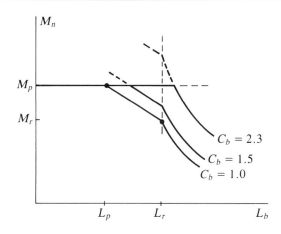

EXAMPLE 5.4

Determine the design strength $\phi_b M_n$ for a W14 × 68 of A242 steel subject to the following conditions:

a. continuous lateral support
b. unbraced length = 20 ft; $C_b = 1.0$
c. unbraced length = 20 ft; $C_b = 1.75$

SOLUTION

a. From Part 1 of the *Manual*, a W14 × 68 is seen to be in shape group 2 and is therefore available with a yield stress F_y of 50 ksi.

Determine whether shape is compact, noncompact, or slender:

$$\frac{b_f}{2t_f} = 7.0 < \frac{65}{\sqrt{50}}$$

∴   this shape is compact and

$$M_n = M_p = F_y Z_x = 50(115) = 5750 \text{ in.-kips} = 479.2 \text{ ft-kips}$$

Check for $M_p \le 1.5 M_y$:

$$\frac{Z_x}{S_x} = \frac{115}{103} = 1.12 < 1.5 \quad \text{(OK)}$$

Answer $\phi_b M_n = 0.90(479.2) = 431$ ft-kips

b. $L_b = 20$ ft and $C_b = 1.0$. Compute L_p and L_r:

$$L_p = \frac{300 r_y}{\sqrt{F_y}} = \frac{300(2.46)}{\sqrt{50}} = 104.4 \text{ in.} = 8.700 \text{ ft}$$

From the Torsion Properties tables in Part 1 of the *Manual*,

$$J = 3.02 \text{ in.}^4 \qquad C_w = 5380 \text{ in.}^6$$

Although X_1 and X_2 are tabulated in the Dimensions and Properties tables in Part 1 of the *Manual*, they will be computed here for illustration.

$$X_1 = \frac{\pi}{S_x} \sqrt{\frac{EGJA}{2}} = \frac{\pi}{103} \sqrt{\frac{29,000(11,200)(3.01)(20)}{2}} = 3021 \text{ ksi}$$

$$X_2 = 4 \frac{C_w}{I_y} \left(\frac{S_x}{GJ}\right)^2 = 4\left(\frac{5380}{121}\right)\left(\frac{103}{11,200 \times 3.02}\right)^2 = 0.001649 \text{ (ksi)}^{-2}$$

$$L_r = \frac{r_y X_1}{(F_y - F_r)} \sqrt{1 + \sqrt{1 + X_2(F_y - F_r)^2}}$$

$$= \frac{2.46(3021)}{(50 - 10)} \sqrt{1 + \sqrt{1 + 0.001649(50 - 10)^2}} = 316.8 \text{ in.} = 26.40 \text{ ft}$$

Since $L_p < L_b < L_r$, the strength is based on inelastic LTB.

$$M_r = (F_y - F_r)S_x = \frac{(50 - 10)(103)}{12} = 343.3 \text{ ft-kips}$$

$$M_n = C_b\left[M_p - (M_p - M_r)\left(\frac{L_b - L_p}{L_r - L_p}\right)\right]$$

$$= 1.0\left[479.2 - (479.2 - 343.3)\left(\frac{20 - 8.700}{26.40 - 8.700}\right)\right] = 392.4 \text{ ft-kips} < M_p$$

Answer $\phi_b M_n = 0.90(392.4) = 353 \text{ ft-kips}$

c. $L_b = 20$ ft and $C_b = 1.75$. The design strength for $C_b = 1.75$ is 1.75 times the design strength for $C_b = 1.0$.

$$\therefore \quad M_n = 1.75(392.4) = 686.7 \text{ ft-kips} > M_p = 479.2 \text{ ft-kips}$$

The nominal strength cannot exceed M_p; therefore, use a nominal strength of $M_n = 479.2$ ft-kips:

Answer $\phi_b M_n = 0.90(479.2) = 431 \text{ ft-kips}$

In most cases, the maximum factored load moment M_u will be obtained by an elastic structural analysis using factored loads. Under certain conditions, the required strength for a statically indeterminate structure may be found by plastic analysis. AISC authorizes the use of plastic analysis if the shape is compact and if

$$L_b \leq L_{pd}$$

where

$$L_{pd} = \frac{3600 + 2200(M_1/M_2)}{F_y} r_y \qquad \text{(AISC Equation F1-17)}$$

M_1 = smaller of the two end moments for the unbraced segment

M_2 = larger of the two end moments for the unbraced segment

The ratio M_1/M_2 is positive when the moments cause reverse curvature in the unbraced segment. In this context, L_b is the unbraced length adjacent to a plastic hinge that is part of the failure mechanism. If plastic analysis is used, however, the nominal moment strength M_n adjacent to the last hinge to form and in regions not adjacent to plastic hinges must be determined in the same way as for beams analyzed by elastic methods, and it may be less than M_p. For a more complete discussion of plastic analysis, refer to Appendix A.

Part 4 of the *Manual of Steel Construction*, "Beam and Girder Design," contains several useful tables and charts for the analysis and design of beams. For example, the Load Factor Design Selection Table lists shapes, in order of plastic section modulus Z_x (and consequently in order of plastic moment M_p), commonly used for beams. Of particular value is the tabulation of L_r, which is tedious to compute. The constant BF can be defined as follows: For $L_p < L_b \leq L_r$, inelastic lateral torsional buckling controls, and for $C_b = 1.0$, the design strength is given by

$$\phi_b M_n = \phi_b \left[M_p - (M_p - M_r)\left(\frac{L_b - L_p}{L_r - L_p}\right) \right]$$

$$= \phi_b M_p - \frac{\phi_b(M_p - M_r)}{(L_r - L_p)}(L_b - L_p)$$

or

$$\phi_b M_n = \phi_b M_p - BF(L_b - L_p) \tag{5.5}$$

where

$$BF = \frac{\phi_b(M_p - M_r)}{(L_r - L_p)}$$

Since $\phi_b M_p$ is also tabulated, the computation of design strength for compact rolled shapes is greatly simplified when inelastic lateral-torsional buckling controls. If C_b is not equal to 1.0, multiply the right side of Equation 5.5 by C_b.

≡ EXAMPLE 5.5

Compute the design strength for the beam of Example 5.4, part b.

SOLUTION

From Example 5.4, the beam is a W14 × 68, and $F_y = 50$ ksi. The unbraced length is 20 ft, and $C_b = 1.0$. From the Load Factor Design Selection Table,

$$\phi_b M_p = 431 \text{ ft-kips}, \qquad L_p = 8.7 \text{ ft}, \qquad L_r = 26.4 \text{ ft}, \qquad BF = 6.91 \text{ kips}$$

Since $L_p < L_b < L_r$,

$$\phi_b M_n = C_b[\phi_b M_p - BF(L_b - L_p)]$$
$$= 1.0[431 - 6.91(20 - 8.7)] = 353 \text{ ft-kips}$$

Answer   Design strength = 353 ft-kips

5.6 ≡ BENDING STRENGTH OF NONCOMPACT SHAPES

As previously noted, most standard W-, M-, and S-shapes are compact for $F_y = 36$ ksi and $F_y = 50$ ksi. A few are noncompact because of the flange width-thickness ratio, but none are slender. For these reasons, the AISC Specification deals with noncompact and slender flexural members in an appendix (Appendix F). In this textbook, we will defer consideration of slender flexural members until Chapter 10, "Plate Girders."

In general, a beam may fail by lateral-torsional buckling, flange local buckling, or web local buckling. Any of these failure modes can be in either the elastic range or the inelastic range. Since the webs of all rolled shapes in the *Manual* are compact, the noncompact shapes are subject only to the limit states of lateral-torsional buckling and flange local buckling. The strengths corresponding to both limit states must be computed and the smaller value will control. From AISC Appendix F, with

$$\lambda = \frac{b_f}{2t_f}$$

If $\lambda_p < \lambda \leq \lambda_r$, the flange is noncompact, buckling will be inelastic, and

$$M_n = M_p - (M_p - M_r)\left(\frac{\lambda - \lambda_p}{\lambda_r - \lambda_p}\right) \qquad \text{(AISC Equation A-F1-3)}$$

where

$$\lambda_p = \frac{65}{\sqrt{F_y}}$$

$$\lambda_r = \frac{141}{\sqrt{F_y - F_r}}$$

$$M_r = (F_y - F_r)S_x$$

$$F_r = \text{residual stress} = 10 \text{ ksi for rolled shapes}$$

(These terms have been specialized for nonhybrid beams.)

≡ EXAMPLE 5.6

A simply supported beam with a span length of 40 feet is laterally supported at its ends and is subjected to the following service loads:

dead load = 400 lb/ft (including the weight of the beam)

live load = 1000 lb/ft

If ASTM A572 Grade 50 steel is used, is a W14 × 90 adequate?

SOLUTION   Factored load moment:

$$w_u = 1.2w_D + 1.6w_L = 1.2(0.400) + 1.6(1.000) = 2.080 \text{ kips/ft}$$

$$M_u = \frac{1}{8} w_u L^2 = \frac{2.080(40)^2}{8} = 416.0 \text{ ft-kips}$$

Determine whether shape is compact, noncompact, or slender.

$$\lambda = \frac{b_f}{2t_f} = 10.2$$

$$\lambda_p = \frac{65}{\sqrt{F_y}} = \frac{65}{\sqrt{50}} = 9.19$$

$$\lambda_r = \frac{141}{\sqrt{F_y - F_r}} = \frac{141}{\sqrt{50 - 10}} = 22.3$$

Since $\lambda_p < \lambda < \lambda_r$, this is a noncompact shape.

Check the capacity based on the limit state of flange local buckling.

$$M_p = F_y Z_x = \frac{50(157)}{12} = 654.2 \text{ ft-kips}$$

$$M_r = (F_y - F_r)S_x = \frac{(50 - 10)(143)}{12} = 476.7 \text{ ft-kips}$$

$$M_n = M_p - (M_p - M_r)\left(\frac{\lambda - \lambda_p}{\lambda_r - \lambda_p}\right)$$

$$= 654.2 - (654.2 - 476.7)\left(\frac{10.2 - 9.19}{22.3 - 9.19}\right) = 640.5 \text{ ft-kips}$$

The design strength based on FLB is therefore

$$\phi_b M_n = 0.90(640.5) = 576 \text{ ft-kips}$$

Check the capacity based on the limit state of lateral-torsional buckling. From the Load Factor Design Selection Table,

$$L_p = 15, \quad L_r = 38.4 \text{ ft}$$

$$L_b = 40 \text{ ft} > L_r, \quad \therefore \text{ failure is by } \textit{elastic} \text{ LTB.}$$

From Part 1 of the *Manual*,

$$I_y = 362 \text{ in.}^4$$

$$J = 4.06 \text{ in.}^4$$

$$C_w = 16,000 \text{ in.}^6$$

For a uniformly loaded, simply supported beam with lateral support at the ends,

$$C_b = 1.14.$$

AISC Equation F1-13:

$$M_n = C_b \frac{\pi}{L_b} \sqrt{EI_y GJ + \left(\frac{\pi E}{L_b}\right)^2 I_y C_w} \leq M_p$$

$$= 1.14\left[\frac{\pi}{40(12)} \sqrt{29,000(362)(11,200)(4.06) + \left(\frac{\pi \times 29,000}{40 \times 12}\right)^2 (362)(16,000)}\right]$$

$$= 1.14(5421) = 6180 \text{ in.-kips} = 515.0 \text{ ft-kips}$$

$$M_p = 654.2 \text{ ft-kips} > 515.0 \text{ ft-kips} \quad \text{(OK)}$$

Since $515.0 < 640.5$, LTB controls:

$$\phi_b M_n = 0.90(515.0) = 464 \text{ ft-kips} > M_u = 416 \text{ ft-kips} \quad \text{(OK)}$$

Answer   Since $M_u < \phi_b M_n$, the beam has adequate moment strength.

The identification of noncompact shapes is facilitated by the Load Factor Design Selection Table. Noncompact shapes are marked by footnotes that specify whether the shape is noncompact for $F_y = 36$ ksi or $F_y = 50$ ksi. For noncompact shapes, the tabulated values of $\phi_b M_p$ *are actually values of the design strength based on flange local buckling.* In the preceding example, this value was computed to be 576 ft-kips and is tabulated as 577 ft-kips.

5.7 ≡ SHEAR STRENGTH

The shear strength of a beam must be sufficient to satisfy the relationship

$$V_u \leq \phi_v V_n$$

where

  V_u = maximum shear based on the controlling combination of factored loads

  ϕ_v = resistance factor for shear = 0.90

  V_n = nominal shear strength

Consider the simple beam of Figure 5.15. At a distance x from the left end and at the neutral axis of the cross section, the state of stress is as shown in part d. Since this element is located at the neutral axis, it is not subjected to flexural stress. From elementary mechanics of materials, the shearing stress is given by

$$f_v = \frac{VQ}{It} \tag{5.6}$$

FIGURE 5.15 ≡

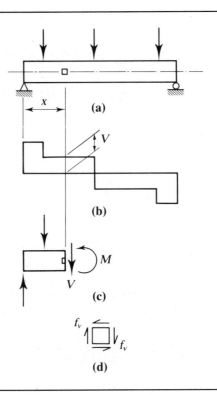

where

f_v = vertical and horizontal shearing stress at the point of interest

V = vertical shear force at the section under consideration

Q = first moment, about the neutral axis, of the area of the cross section between the point of interest and the top or bottom of the cross section

I = moment of inertia about the neutral axis

t = width of the cross section at the point of interest

Equation 5.6 is based on the assumption that the stress is constant across the thickness t, and it is therefore accurate only for small values of t. For a rectangular cross section of depth d and width t, the error for $d/t = 2$ is approximately 3%. For $d/t = 1$, the error is 12% and for $d/t = 1/4$, it is 100%.[37] For this reason, the equation cannot be applied to the flange of a W-shape in the same manner as for the web.

Figure 5.16 shows the shearing stress distribution for a W-shape. Superimposed on the actual distribution is the average stress in the web, V/A_w, which does not differ much from the maximum web stress. Clearly, the web will completely yield long before the flanges begin to yield. Because of this, yielding of

FIGURE 5.16 ≡

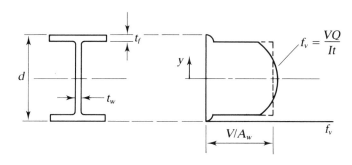

the web represents one of the shear limit states. Taking the shear yield stress as 60% of the tensile yield stress, the stress in the web at failure can be written as

$$f_v = \frac{V_n}{A_w} = 0.60F_y$$

where A_w = area of the web. The nominal strength corresponding to this limit state is

$$V_n = 0.6F_y A_w$$

This will be the nominal strength in shear provided there is no shear buckling of the web. Whether this occurs will depend on h/t_w, the width-thickness ratio of the web. If this ratio is too large — that is, if the web is too slender — the web can buckle in shear, either inelastically or elastically. The relationship between shear strength and width-thickness ratio is analogous to that between flexural strength and width-thickness ratio (for FLB or WLB) and between flexural strength and unbraced length (for LTB). This relationship is illustrated in Figure 5.17 and given in AISC F2.2 as follows:

For $h/t_w \leq 418/\sqrt{F_y}$, there is no web instability, and

$$V_n = 0.6F_y A_w \qquad \text{(AISC Equation F2-1)}$$

For $418/\sqrt{F_y} < h/t_w \leq 523/\sqrt{F_y}$, inelastic web buckling can occur, and

$$V_n = 0.6F_y A_w \frac{418/\sqrt{F_y}}{h/t_w} \qquad \text{(AISC Equation F2-2)}$$

For $523/\sqrt{F_y} < h/t_w \leq 260$, the limit state is elastic web buckling:

$$V_n = \frac{132,000A_w}{(h/t_w)^2} \qquad \text{(AISC Equation F2-3)}$$

where

A_w = area of the web = dt_w

d = overall depth of the beam

FIGURE 5.17 ≡

If h/t_w is greater than 260, web stiffeners are required, and the provisions of Appendix F2 (or Appendix G for plate girders) must be consulted.

AISC Equation F2-3 is based on elastic stability theory, and Equation F2-2 is an empirical equation for the inelastic region, providing a transition between the limit states of web yielding and elastic web buckling.

Shear is rarely a problem in rolled steel beams; the usual practice is to design a beam for flexure and then to check it for shear.

≡ **EXAMPLE 5.7**

Check the beam of Example 5.6 for shear.

SOLUTION

From Example 5.6, $w_u = 2.080$ kips/ft and $L = 40$ ft. A W14 × 90 with $F_y = 50$ ksi is used. For a simply supported, uniformly loaded beam, the maximum shear occurs at the support and is equal to the reaction:

$$V_u = \frac{w_u L}{2} = \frac{2.080(40)}{2} = 41.6 \text{ kips}$$

From the Dimensions and Properties Tables in Part 1 of the *Manual*, the web width-thickness ratio of a W14 × 90 is

$$\frac{h}{t_w} = 25.9$$

$$\frac{418}{\sqrt{F_y}} = \frac{418}{\sqrt{50}} = 59.11$$

Since h/t_w is less than $418/\sqrt{F_y}$, the strength is governed by shear yielding of the web:

$$V_n = 0.6F_y A_w = 0.6F_y(dt_w) = 0.6(50)(14.02)(0.440) = 185.1 \text{ kips}$$
$$\phi_v V_n = 0.90(185.1) = 167 \text{ kips} > 41.6 \text{ kips} \quad (OK)$$

Answer   Since the shear design strength is greater than the factored load shear, the beam is satisfactory.

Values of $\phi_v V_n$ are tabulated in the Factored Uniform Load tables in Part 4 of the *Manual*, so its computation is unnecessary for standard hot-rolled shapes.

Block Shear

Block shear, which was considered earlier in conjunction with tension member connections, can occur in certain types of beam connections. To facilitate the connection of beams to other beams so that the top flanges are at the same elevation, a short length of the top flange of one of the beams may be cut away, or *coped*. If a coped beam is connected with bolts as in Figure 5.18, there will be a tendency for segment ABC to tear out. The applied load in this case will be the vertical beam reaction, so shear will occur along line AB and there will be tension along BC. Thus, the block shear strength will be a limiting value of the reaction.

The computation of block shear strength has been covered in Chapter 3, "Tension Members," but it will be reviewed here. Failure can occur by a combination of shear yielding and tension fracture or by shear fracture and tension

FIGURE 5.18 ≡

yielding. AISC J4.3, "Block Shear Rupture Strength," gives two equations for the block shear design strength:

$$\phi R_n = \phi[0.6F_y A_{gv} + F_u A_{nt}] \qquad \text{(AISC Equation J4-3a)}$$
$$\phi R_n = \phi[0.6F_u A_{nv} + F_y A_{gt}] \qquad \text{(AISC Equation J4-3b)}$$

where

$\phi = 0.75$

A_{gv} = gross area in shear (in Figure 5.18, length AB times the web thickness)

A_{nv} = net area in shear

A_{gt} = gross area in tension (in Figure 5.18, length BC times the web thickness)

A_{nt} = net area in tension

The governing equation is the one that has the larger fracture term.

≣ EXAMPLE 5.8

Determine the maximum factored load reaction, based on block shear, that can be resisted by the beam shown in Figure 5.19.

SOLUTION   Effective hole diameter $= \frac{3}{4} + \frac{1}{8} = \frac{7}{8}$ in.

Gross and net shear areas:

$$A_{gv} = (2 + 3 + 3 + 3)t_w = 11(0.300) = 3.300 \text{ in.}^2$$
$$A_{nv} = (11 - 3.5 \times \tfrac{7}{8})(0.300) = 2.381 \text{ in.}^2$$

Gross and net tension areas:

$$A_{gt} = 1.25t_w = 1.25(0.300) = 0.3750 \text{ in.}^2$$
$$A_{nt} = (1.25 - 0.5 \times \tfrac{7}{8})(0.300) = 0.2438 \text{ in.}^2$$

AISC Equation J4-3a:

$$\phi R_n = \phi[0.6F_y A_{gv} + F_u A_{nt}] = 0.75[0.6(36)(3.300) + 58(0.2438)]$$
$$= 0.75[71.28 + 14.14] = 64.1 \text{ kips}$$

AISC Equation J4-3b:

$$\phi R_n = \phi[0.6F_u A_{nv} + F_y A_{gt}] = 0.75[0.6(58)(2.381) + 36(0.3750)]$$
$$= 0.75[82.86 + 13.50] = 72.3 \text{ kips}$$

The fracture term in AISC Equation J4-3b is the larger one (that is, $82.86 > 14.14$), so this equation governs.

Answer   Maximum factored load reaction = design strength for block shear = 72.3 kips.

FIGURE 5.19 ≡

1¼"

2"

3 @ 3"

$W18 \times 35$
A36 steel
¾ in. diameter bolts

5.8 ≡ DEFLECTION

In addition to being safe, a structure must be *serviceable*. A serviceable structure is one that performs in a satisfactory manner, not causing any discomfort or perceptions of unsafety for the occupants or users of the structure. For a beam, this usually means that the deformations, primarily the vertical sag, or deflection, must be limited. Excessive deflection is usually an indication of a very flexible beam, and this can lead to problems with vibrations. The deflection itself can cause problems if elements attached to the beam can be damaged by small distortions. In addition, users of the structure may view large deflections negatively and wrongly assume that the structure is unsafe.

For the common case of a simply supported, uniformly loaded beam such as that in Figure 5.20, the maximum vertical deflection is given by

$$\Delta = \frac{5}{384} \frac{wL^4}{EI}$$

Deflection formulas for a variety of beams and loading conditions can be found in Part 4, "Beam and Girder Design," of the *Manual of Steel Construction.* For more unusual situations, standard analytical methods such as the method of

FIGURE 5.20 ≡

w

$L/2$

Δ

L

$$\Delta = \frac{5}{384} \frac{wL^4}{EI}$$

virtual work may be used. Since deflection is a serviceability limit state, not one of strength, deflections should always be computed with *service* loads.

The appropriate limit for the maximum deflection depends on the function of the beam and the likelihood of damage resulting from the deflection. The AISC Specification furnishes little guidance other than a statement in Chapter L, "Serviceability Design Considerations," that deflections should be checked. Appropriate limits for deflection can usually be found from the governing building code. The following values are typical maximum allowable total (service dead load plus service live load) deflections:

Plastered construction: $\dfrac{L}{360}$

Unplastered floor construction: $\dfrac{L}{240}$

Unplastered roof construction: $\dfrac{L}{180}$

where L is the span length.

≡ **EXAMPLE 5.9**

Check the deflection of the beam shown in Figure 5.21. The maximum permissible total deflection is L/240.

SOLUTION

$$\text{maximum permissible deflection} = \frac{L}{240} = \frac{30(12)}{240} = 1.500 \text{ in.}$$

$$\text{total service load} = 500 + 550 = 1050 \text{ lb/ft} = 1.050 \text{ kips/ft}$$

Since it is more convenient to express the deflection in inches rather than feet, use units of inches in the following equation:

$$\text{maximum total deflection} = \frac{5}{384} \frac{wL^4}{EI} = \frac{5}{384} \frac{(1.050/12)(30 \times 12)^4}{29{,}000(510)}$$

$$= 1.294 \text{ in.} \ < 1.500 \text{ in.} \quad \text{(OK)}$$

Answer   The beam satisfies the deflection criterion.

FIGURE 5.21 ≡

$W_D = 500 \text{ lb/ft}$
$W_L = 550 \text{ lb/ft}$

W18 × 35

30′

Ponding is one deflection problem that does affect the safety of a structure. It is a potential hazard for flat roof systems that can trap rainwater. If drains become clogged during a storm, the weight of the water will cause the roof to deflect, thus providing a reservoir for still more water. If this process proceeds unabated, collapse can occur. The AISC specification requires that the roof system has sufficient stiffness to prevent ponding, and it prescribes limits on moments of inertia and other parameters in Section K2, "Ponding."

5.9 ≡ DESIGN

Beam design entails the selection of a cross-sectional shape that will have enough strength and will meet the serviceability requirements. As far as strength is concerned, flexure is almost always more critical than shear, so the usual practice is to design for flexure and then check shear. The design process can be outlined as follows:

1. Compute the factored load moment M_u. This will be the required design strength $\phi_b M_n$. The weight of the beam is part of the dead load but is unknown at this point. A value may be assumed, or the weight may be ignored initially and checked after a shape has been selected.

2. Select a shape that satisfies this strength requirement. This can be done in one of two ways:

   a. Assume a shape, compute the design strength, and compare it with the factored load moment. Revise if necessary. The trial shape can be easily selected in only a limited number of situations (Example 5.10).

   b. Use the beam design charts in Part 4 of the *Manual*. This is the preferred method and will be explained following Example 5.10.

3. Check the shear strength.

4. Check the deflection.

≡ EXAMPLE 5.10

Select a standard hot-rolled shape of A36 steel for the beam shown in Figure 5.22. The beam has continuous lateral support and must support a uniform service live load of 5 kips/ft. The maximum permissible *live* load deflection is $L/360$.

FIGURE 5.22 ≡

$5^{k/ft}$ live load

30'

SOLUTION   Assume a beam weight of 100 lb/ft.

$$w_u = 1.2w_D + 1.6w_L = 1.2(0.100) + 1.6(5.000) = 8.120 \text{ kips/ft}$$

$$M_u = \frac{1}{8}\,w_u L^2 = \frac{8.120(30)^2}{8}$$

$$= 913.5 \text{ ft-kips} = \text{required } \phi_b M_n$$

Assume a compact shape. For a compact shape and continuous lateral support,

$$M_n = M_p = Z_x F_y \le 1.5 M_y$$

From $\phi_b M_n \ge M_u$,

$$\phi_b F_y Z_x \ge M_u$$

$$Z_x \ge \frac{M_u}{\phi_b F_y} = \frac{913.5(12)}{0.90(36)} = 338.3 \text{ in.}^3$$

The Load Factor Design Selection Table lists rolled shapes normally used as beams in order of decreasing plastic section modulus. Furthermore, they are grouped in such a way that the shape at the top of the group (in bold type) is the lightest one that has enough section modulus to satisfy a required section modulus falling within the group. In the current case, the shape that comes closest to meeting the section modulus requirement is a W27 × 114, with $Z_x = 343$ in.^{3}, but the lightest one is a W30 × 108, with $Z_x = 346$ in.^{3} Since section modulus is not directly proportional to area, it is possible to have more section modulus with a smaller area and hence less weight.

Try a W30 × 108.

This shape is compact as assumed (noncompact shapes are marked as such in the table), and $Z_x/S_x = 346/299 = 1.16$. This is less than 1.5, therefore $M_n = M_p$ as assumed.

Since the weight is slightly larger than assumed, the required strength will be recomputed, although the W30 × 108 has more capacity than originally required, and it will almost certainly be adequate.

$$w_u = 1.2(0.108) + 1.6(5.000) = 8.130 \text{ kips/ft}$$

$$M_u = \frac{8.130(30)^2}{8} = 914.6 \text{ ft-kips}$$

From the Load Factor Design Selection Table,

$$\phi_b M_p = \phi_b M_n = 934 \text{ ft-kips} > 914.6 \text{ ft-kips} \quad \text{(OK)}$$

In lieu of basing the search on the required section modulus, the design strength $\phi_b M_p$ could be used, since it is directly proportional to Z_x and is also tabulated.

Check shear:

$$V_u = \frac{w_u L}{2} = \frac{8.130(30)}{2} = 122 \text{ kips}$$

From the Factored Uniform Load tables,

$$\phi_v V_n = 316 \text{ kips} > 122 \text{ kips} \quad \text{(OK)}$$

Check deflection:

The maximum permissible live load deflection is $L/360 = 30(12)/360 = 1$ in.

$$\Delta = \frac{5}{384} \frac{w_L L^4}{EI_x} = \frac{5}{384} \frac{(5.000/12)(30 \times 12)^4}{29,000(4470)} = 0.703 \text{ in.} < 1 \text{ in.} \quad \text{(OK)}$$

Answer   Use a W30 × 108.

Beam Design Charts

Many graphs, charts, and tables are available for the practicing engineer, and these aids can greatly simplify the design process. For the sake of efficiency, they are widely used in design offices, but the student should approach their use with caution and not allow basic principles to become obscured. It is not the purpose of this book to describe in detail all available design aids, but some are worthy of note, particularly the curves of design moment versus unbraced length given in Part 4 of the *Manual*.

These will be described with reference to Figure 5.23, which shows a graph of design moment $\phi_b M_n$ as a function of unbraced length L_b for a particular beam. Such a graph can be constructed for a given cross-sectional shape and specific values of F_y and C_b by using the appropriate equations for moment strength.

FIGURE 5.23 ≡

FIGURE 5.24 ≡

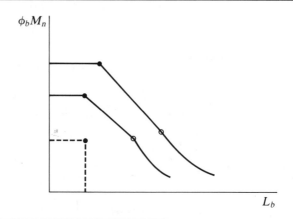

The *Manual* charts consist of a family of curves for various rolled shapes. All curves were generated with $C_b = 1.0$, but for other values of C_b, simply multiply the design moments from the charts by C_b (keeping in mind that $\phi_b M_n$ can never exceed $\phi_b M_p$). The use of the charts is illustrated in Figure 5.24, where two such curves are shown. Any point on this graph, such as the intersection of the two dashed lines, represents a design moment and an unbraced length. If the moment is a required moment capacity, then any curve above the point corresponds to a beam with a larger moment capacity, and any curve to the right is for a beam with exactly the required moment capacity, although for a larger unbraced length. In a design problem, therefore, if the charts are entered with a given unbraced length and a required design strength, curves above and to the right of the point correspond to acceptable beams. If a dashed portion of a curve is encountered, then a curve for a lighter shape lies above or to the right of the dashed curve. Points on the curves corresponding to L_p are indicated by a solid circle; L_r is represented by an open circle. Two sets of curves are available, one for $F_y = 36$ ksi and one for $F_y = 50$ ksi.

In the previous example, the required design strength (excluding the beam weight) was 913.5 ft-kips, and there was continuous lateral support. For continuous lateral support, L_b can be taken as zero. From the charts for $F_y = 36$ ksi, the first solid curve above the 913.5 ft-kip mark is for a W30 × 108, the same as was selected in Example 5.10. Although $L_b = 0$ is not on this particular chart, it can be seen that smallest value of L_b shown is less than L_p for all shapes on this page.

≡ **EXAMPLE 5.11**

The beam shown in Figure 5.25 must support two concentrated *live* loads of 20 kips each at the quarter points. The maximum deflection must not exceed $L/240$. Lateral support is provided at the ends of the beam. Use A572 Grade 50 steel and select a rolled shape.

FIGURE 5.25 ≡

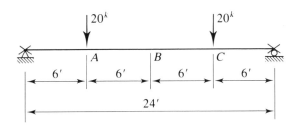

SOLUTION

If the weight of the beam is neglected, the central half of the beam is subjected to a uniform moment, and

$$M_A = M_B = M_C = M_{max}, \quad \therefore \ C_b = 1.0$$

Even if the weight is included, it will be negligible compared to the concentrated loads, and C_b can still be taken as 1.0, permitting the charts to be used without modification.

Temporarily ignoring the beam weight,

$$M_u = 6(1.6 \times 20) = 192 \text{ ft-kips}$$

From the charts, with $L_b = 24$ ft,

Try W12 × 53.

$$\phi_b M_n = 219 \text{ ft-kips} > 192 \text{ ft-kips} \quad \text{(OK)}$$

Account for the beam weight:

$$M_u = 192 + \frac{1}{8} (1.2 \times 0.053)(24)^2 = 197 \text{ ft-kips} < 219 \text{ ft-kips} \quad \text{(OK)}$$

Shear:

$$V_u = 1.6(20) + \frac{1.2(0.053)(24)}{2} = 32.8 \text{ ft-kips}$$

From the Factored Uniform Load tables,

$$\phi_v V_n = 112 \text{ kips} > 32.8 \text{ kips} \quad \text{(OK)}$$

Deflection:

$$\text{Maximum permissible deflection} = \frac{L}{240} = \frac{24(12)}{240} = 1.200 \text{ in.}$$

From the Beam Diagrams and Formulas section in Part 4 of the *Manual*, the maximum deflection (at midspan) for two equal and symmetrically placed loads is

$$\Delta = \frac{Pa}{24EI} (3L^2 - 4a^2)$$

where

P = magnitude of concentrated load

a = distance from support to load

L = span length

$$\Delta = \frac{20(6 \times 12)}{24EI}[3(24 \times 12)^2 - 4(6 \times 12)^2] = \frac{13.69 \times 10^6}{EI}$$

For the uniform beam weight, the maximum deflection is also at midspan:

$$\Delta = \frac{5}{384}\frac{wL^4}{EI} = \frac{5}{384}\frac{(0.053/12)(24 \times 12)^4}{EI} = \frac{0.04 \times 10^6}{EI}$$

Since both deflections occur at the same location, they can be added:

$$\text{Total } \Delta = \frac{(13.69 + 0.04) \times 10^6}{EI}$$

$$= \frac{13.73 \times 10^6}{29,000(425)} = 1.114 \text{ in.} < 1.200 \text{ in.} \quad (OK)$$

Answer   Use a W12 × 53.

Although the charts are based on $C_b = 1.0$, they can easily be used for design when C_b is not 1.0; simply divide the required design strength by C_b before entering the charts. This technique will be illustrated in the following example.

≡ **EXAMPLE 5.12**

Use A36 steel and select a rolled shape for the beam in Figure 5.26. The concentrated load is a service live load, and the uniform load is 30% dead load and 70% live load. Lateral bracing is provided at the ends and at midspan. There is no restriction on deflection.

FIGURE 5.26 ≡

SOLUTION Assume a beam weight of 100 lb/ft.

$$w_D = 0.30(3) + 0.100 = 1.000 \text{ kips/ft}$$
$$w_u = 1.2(1.000) + 1.6(0.70 \times 3) = 4.560 \text{ kips/ft}$$
$$P_u = 1.6(9) = 14.40 \text{ kips}$$

The factored loads and reactions are shown in Figure 5.27.

Moments required for the computation of C_b: The bending moment at a distance x from the left end is

$$M = 61.92x - 4.560x\left(\frac{x}{2}\right) = 61.92x - 2.280x^2 \quad \text{(for } x \leq 12 \text{ ft)}$$

For $x = 3$ ft, $M_A = 61.92(3) - 2.280(3)^2 = 165.2$ ft-kips
For $x = 6$ ft, $M_B = 61.92(6) - 2.280(6)^2 = 289.4$ ft-kips
For $x = 9$ ft, $M_C = 61.92(9) - 2.280(9)^2 = 372.6$ ft-kips
For $x = 12$ ft, $M_{max} = M_u = 61.92(12) - 2.280(12)^2 = 414.7$ ft-kips

$$C_b = \frac{12.5M_{max}}{2.5M_{max} + 3M_A + 4M_B + 3M_C}$$

$$= \frac{12.5(414.7)}{2.5(414.7) + 3(165.2) + 4(289.4) + 3(372.6)} = 1.36$$

Enter the charts with an unbraced length $L_b = 12$ ft and a bending moment of

$$\frac{M_u}{C_b} = \frac{414.7}{1.36} = 305 \text{ ft-kips}$$

Try W21 × 62.

$$\phi_b M_n = 343 \text{ ft-kips} > 305 \text{ ft-kips}$$
Beam weight = 62 lb/ft < 100 lb/ft assumed   (OK)

For $C_b = 1.36$,

$$\phi_b M_n = 1.36(343) = 466 \text{ ft-kips}$$

FIGURE 5.27 ≡

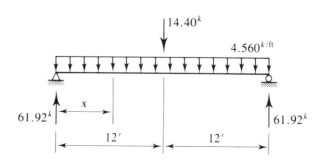

But the design strength cannot exceed $\phi_b M_p$, which is only 389 ft-kips (obtained from the chart).

$$\therefore \quad \phi_b M_n = 389 \text{ ft-kips} < 414.7 \text{ ft-kips} \quad \text{(N.G.)}$$

Select a shape with $\phi_b M_p \geq 414.7$ ft-kips:

Try W21 × 68.

$$\phi_b M_p = 432 \text{ ft-kips} > 414.7 \text{ ft-kips}$$

For $L_b = 12$ ft, the design strength from the chart is 385 ft-kips for $C_b = 1$. The strength for $C_b = 1.36$ is

$$\phi_b M_n = 1.36(385) = 524 \text{ ft-kips} > \phi_b M_p = 432 \text{ ft-kips}$$
$$\therefore \quad \phi_b M_n = \phi_b M_p = 432 \text{ ft-kips} > 414.7 \text{ ft-kips} \quad \text{(OK)}$$

Shear:

$$V_u = 61.92 \text{ kips}$$

(This is slightly conservative, since the beam weight is less than assumed.) From the Factored Uniform Load tables,

$$\phi_v V_n = 177 \text{ kips} > 61.92 \text{ kips} \quad \text{(OK)}$$

Answer   Use a W21 × 68.

If deflection requirements control the design of a beam, a minimum required moment of inertia is computed, and the lightest shape having this value is sought. This task is greatly simplified by the Moment of Inertia Selection Tables in Part 4 of the *Manual*. The use of these tables will be illustrated in the next example, which also explains the design procedure for a typical floor or roof system.

EXAMPLE 5.13

Part of a floor framing system is shown in Figure 5.28. A 4-in.-thick reinforced concrete floor slab is supported by *floor beams* spaced at 7 feet. The floor beams are supported by *girders*, which in turn are supported by the columns. In addition to the weight of the structure, loads consist of a uniform live load of 80 psf and movable partitions, to be accounted for by using a uniformly distributed load of 20 pounds per square foot of floor surface. The maximum total deflection must not exceed $\frac{1}{360}$ of the span length. Use A36 steel and design the floor beams. Assume that the slab provides continuous lateral support of the floor beams.

FIGURE 5.28 ≡

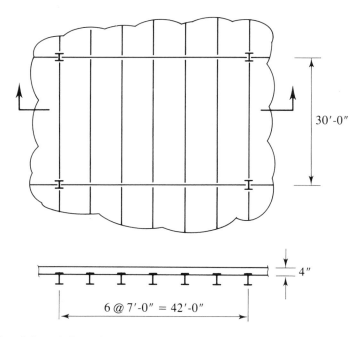

30'-0"

4"

6 @ 7'-0" = 42'-0"

SOLUTION   Normal-weight reinforced concrete is usually considered to weigh 150 pounds per cubic foot. The weight can be expressed as a load per square foot of floor surface by multiplying the unit weight by the volume of slab that has a surface area of one square foot. This requires multiplying 150 lb/ft^3 by the slab thickness in feet.

$$\text{Slab weight:} \quad 150\left(\frac{4}{12}\right) = 50 \text{ psf}$$

Assume that each beam supports a 7-ft width of floor:

50(7)	= 350 lb/ft
Beam weight	= 40 lb/ft (estimated)
Total	= 390 lb/ft = service dead load

Live load: In addition to the live load of 80 psf, the partitions will be considered to be live loads, since they are movable. (Some building codes treat movable partitions as dead load.)

$$(80 + 20)(7) = 700 \text{ lb/ft}$$
$$w_u = 1.2w_D + 1.6w_L = 1.2(0.390) + 1.6(0.700) = 1.588 \text{ kips/ft}$$

The typical floor-beam connection will provide virtually no moment restraint, and the beams can be treated as simply supported.

$$M_u = \frac{1}{8}w_u L^2 = \frac{1.588(30)^2}{8} = 178.6 \text{ ft-kips}$$

From the beam design charts, with $L_b = 0$,

Try a W18 × 35.

$$\phi_b M_n = 179.5 \text{ ft-kips} > 178.6 \text{ ft-kips} \quad (\text{OK})$$

Shear:

$$V_u \approx \frac{1.588(30)}{2} = 23.82 \text{ kips}$$

From the Factored Uniform Load tables,

$$\phi_v V_n = 103 \text{ kips} > 23.82 \text{ kips} \quad (\text{OK})$$

Deflection:

$$\frac{L}{360} = \frac{30(12)}{360} = 1 \text{ in.}$$

$$\Delta = \frac{5}{384} \frac{wL^4}{EI} = \frac{5}{384} \frac{(0.350 + 0.035 + 0.700)(30)^4(12)^3}{29,000(510)}$$
$$= 1.3 \text{ in.} > 1 \text{ in.} \quad (\text{N.G.})$$

The deflection equation can be solved for the required moment of inertia:

$$I_{\text{required}} = \frac{5wL^4}{384E\Delta_{\text{required}}} = \frac{5(1.085)(30)^4(12)^3}{384(29,000)(1)} = 682 \text{ in.}^4$$

The Moment of Inertia Selection Table is organized in the same way as the Load Factor Design Selection Table, so it is a simple matter to select the lightest shape with sufficient moment of inertia. From the I_x Table:

Try a W21 × 44.

$$I_x = 843 \text{ in.}^4 > 682 \text{ in.}^4 \qquad (\text{OK})$$
$$\phi_b M_n = 257.5 \text{ ft-kips} > 178.6 \text{ ft-kips} \quad (\text{OK})$$

(The weight of this shape is slightly more than the initial estimate, but the additional weight is insignificant compared to the excess moment capacity.)

$$\phi_v V_n = 141 \text{ kips} > 23.82 \text{ kips} \quad (\text{OK})$$

Answer   Use a W21 × 44.

5.10 ≡ HOLES IN BEAMS

If beam connections are made with bolts, holes will be punched or drilled in the beam web or flanges. In addition, relatively large holes are sometimes cut in beam webs to provide space for utilities such as electrical conduits and venti-lation ducts. Ideally, holes should be placed in the web only at sections of low shear, and holes should be made in the flanges at points of low bending moment. This will not always be possible, and we must account for the effect of the holes.

For relatively small holes such as those for bolts, the effect will be small, particularly for flexure. First, the reduction in the cross section is usually small. Second, adjacent cross sections are not reduced, and the change in cross section is actually more of a minor discontinuity than a "weak link."

For these reasons, AISC B10 permits bolt holes in flanges to be ignored when

$$0.75 F_u A_{fn} \geq 0.9 F_y A_{fg} \qquad \text{(AISC Equation B10-1)}$$

where

$$A_{fg} = \text{gross flange area}$$
$$A_{fn} = \text{net flange area}$$

If this condition is not met, the flexural properties must be based on an effective tension flange area of

$$A_{fe} = \frac{5}{6} \frac{F_u}{F_y} A_{fn} \qquad \text{(AISC Equation B10-3)}$$

≡ EXAMPLE 5.14

Compute the reduced elastic section modulus S_x for the shape shown in Figure 5.29. A36 steel is used, and the holes are for 1-in.-diameter bolts.

SOLUTION $A_{fg} = b_f t_f = 7.635(0.810) = 6.184 \text{ in.}^2$

The effective hole diameter is

$$d_h = 1 + \tfrac{1}{8} = 1\tfrac{1}{8} \text{ in.}$$

and the net flange area is

$$A_{fn} = A_{fg} - \sum d_h t_f = 6.184 - 2(1.125)(0.810) = 4.362 \text{ in.}^2$$

FIGURE 5.29 ≡

From AISC Equation B10-1,

$$0.75F_u A_{fn} = 0.75(58)(4.362) = 189.7 \text{ kips}$$

and

$$0.9F_y A_{fg} = 0.9(36)(6.184) = 200.4 \text{ kips}$$

Since $0.75F_u A_{fn} < 0.9F_y A_{fg}$, we must account for the holes. From AISC Equation B10-3, use an effective flange area of

$$A_{fe} = \frac{5}{6}\frac{F_u}{F_y} A_{fn} = \frac{5}{6}\left(\frac{58}{36}\right)(4.362) = 5.856 \text{ in.}^2$$

This corresponds to a reduction in flange area of $6.184 - 5.856 = 0.328$ in.^{2} The elastic neutral axis is located at a distance from the top of

$$\bar{y} = \frac{20.8(18.47/2) - 0.328(18.47 - 0.405)}{20.8 - 0.328} = 9.094 \text{ in.}$$

The reduced moment of inertia is

$$\bar{I}_x = 1170 + 20.8(9.094 - 9.235)^2 - 0.328(9.094 - 18.06)^2 = 1144 \text{ in.}^4$$

S_x referred to top fiber:

$$S_x = \frac{I_x}{\bar{y}} = \frac{1144}{9.094} = 125.8 \text{ in.}^3$$

S_x referred to bottom fiber:

$$S_x = \frac{I_x}{d - \bar{y}} = \frac{1144}{18.47 - 9.094} = 122.0 \text{ in.}^3$$

Answer   Reduced elastic section modulus referred to top is 125.8 in.^{3} and to bottom is 122.0 in.^{3}

Beams with large holes in their webs will require special treatment and are beyond the scope of this text. *Design of Steel and Composite Beams with Web Openings* is a useful guide to this topic.[38]

5.11 ≡ OPEN-WEB STEEL JOISTS

Open-web steel joists are relatively small, prefabricated trusses of the type shown in Figure 5.30. Many of the smaller ones use a continuous bent circular bar to form the web members and are commonly called *bar* joists. They are used in floor and roof systems in a wide variety of structures. For a given span, an open-web joist will be lighter in weight than a rolled shape, and the absence

FIGURE 5.30 ≡

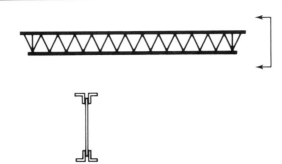

of a solid web allows for the easy passage of duct work and electrical conduits. Depending on the span length, they may be more economical than rolled shapes, although there are no general guidelines for making this determination.

Open-web joists are available in standard depths and lengths with service load capacities specified in pounds per linear foot. A 10K1 joist, for example, has a nominal depth of 10 inches and is available in span lengths of up to 20 feet. The total load capacity (dead plus live) for a 10K1 joist with a span length of 18 feet is 246 pounds per foot. The AISC Specification does not cover open-web steel joists; a separate organization, the Steel Joist Institute, exists for this purpose. All aspects of steel joist usage, including their design and manufacture, are addressed in the publication *Standard Specifications, Load Tables, and Weight Tables for Steel Joists and Joist Girders.*[39] At the time of this writing, the design specifications use an allowable stress approach rather than LRFD. These specifications are quite similar to the AISC Allowable Stresss Design Specification, and the requirements are essentially the same. Although this textbook is based on LRFD, this introduction to open-web steel joists is included because they are so widely used.

An open-web joist can be selected with the aid of the Standard Load tables,[39] one of which is reproduced in Figure 5.31. For each combination of span and joist, a pair of load values is given. The top number is the total service load capacity in pounds per foot, and the bottom number is the service live load per foot that will produce a deflection of $\frac{1}{360}$ of the span length. Steel fabricators who furnish open-web steel joists must certify that a particular joist of a given designation, such as the aforementioned 10K1 of span length 18 feet, will have a safe load capacity of at least the value given in the Standard Load Table. Different manufacturers' 10K1 joists may have different member cross sections, but they all must have a nominal depth of 10 inches and, for a span length of 18 feet, a load capacity of 246 pounds per foot.

The three Standard Load tables that correspond to the three categories of joists are Open-Web Steel Joists (K-series), Longspan Steel Joists (LH-series), and Deep Longspan Steel Joists (DLH-series).[39] As one moves up the series, the available span lengths and load-carrying capacities become greater. At the

FIGURE 5.31 ≡

Joist Designation	8K1	10K1	12K1	12K3	12K5	14K1	14K3	14K4	14K6	16K2	16K3	16K4	16K5	16K6	16K7	16K9
Depth (In.)	8	10	12	12	12	14	14	14	14	16	16	16	16	16	16	16
Approx. Wt. (lbs./ft.)	5.1	5.0	5.0	5.7	7.1	5.2	6.0	6.7	7.7	5.5	6.3	7.0	7.5	8.1	8.6	10.0
Span (ft.) ↓																
8	550															
	550															
9	550															
	550															
10	550	550														
	480	550														
11	532	550														
	377	542														
12	444	550	550	550	550											
	288	455	550	550	550											
13	377	479	550	550	550											
	225	363	510	510	510											
14	324	412	500	550	550	550	550	550	550							
	179	289	425	463	463	550	550	550	550							
15	281	358	434	543	550	511	550	550	550							
	145	234	344	428	434	475	507	507	507							
16	246	313	380	476	550	448	550	550	550	550	550	550	550	550	550	550
	119	192	282	351	396	390	467	467	467	550	550	550	550	550	550	550
17		277	336	420	550	395	495	550	550	512	550	550	550	550	550	550
		159	234	291	366	324	404	443	443	488	526	526	526	526	526	526
18		246	299	374	507	352	441	530	550	456	508	550	550	550	550	550
		134	197	245	317	272	339	397	408	409	456	490	490	490	490	490
19		221	268	335	454	315	395	475	550	408	455	547	550	550	550	550
		113	167	207	269	230	287	336	383	347	386	452	455	455	455	455
20		199	241	302	409	284	356	428	525	368	410	493	550	550	550	550
		97	142	177	230	197	246	287	347	297	330	386	426	426	426	426
21			218	273	370	257	322	388	475	333	371	447	503	548	550	550
			123	153	198	170	212	248	299	255	285	333	373	405	406	406
22			199	249	337	234	293	353	432	303	337	406	458	498	550	550
			106	132	172	147	184	215	259	222	247	289	323	351	385	385
23			181	227	308	214	268	322	395	277	308	371	418	455	507	550
			93	116	150	128	160	188	226	194	216	252	282	307	339	363
24			166	208	282	196	245	295	362	254	283	340	384	418	465	550
			81	101	132	113	141	165	199	170	189	221	248	269	298	346
25						180	226	272	334	234	260	313	353	384	428	514
						100	124	145	175	150	167	195	219	238	263	311
26						166	209	251	308	216	240	289	326	355	395	474
						88	110	129	156	133	148	173	194	211	233	276
27						154	193	233	285	200	223	268	302	329	366	439
						79	98	115	139	119	132	155	173	188	208	246
28						143	180	216	265	186	207	249	281	306	340	408
						70	88	103	124	106	118	138	155	168	186	220
29										173	193	232	261	285	317	380
										95	106	124	139	151	167	198
30										161	180	216	244	266	296	355
										86	96	112	126	137	151	178
31										151	168	203	228	249	277	332
										78	87	101	114	124	137	161
32										142	158	190	214	233	259	311
										71	79	92	103	112	124	147

Source: Standard Specifications, Load Tables, and Weight Tables for Steel Joists and Joist Girders. Myrtle Beach, SC: Steel Joist Institute, 1992. Reprinted with permission.

lower end, an 8K1 is available with a span length of 8 feet and a capacity of 550 pounds per foot, while a 72DLH19 can support a load of 497 pounds per foot on a span of 144 feet. In addition to these Standard Load tables, catalogs supplied by joist manufacturers contain tables that are more detailed and may be more convenient to use.

Although the structural engineer will not ordinarily be concerned with the actual design of open web steel joists, some of the Specification requirements are of interest. All joists are designed as simply supported trusses with uniformly distributed loads on the top chord. If concentrated loads must be supported, a special analysis should be requested from the manufacturer. To ensure stability of the top chord, the floor or roof deck must be attached in such a way that continuous lateral support is provided. Both top and bottom chord members of K-series joists must be made of steel with a yield stress of 50 ksi, and the web members may have a yield stress of either 36 ksi or 50 ksi. All members of LH- and DLH-series joists can be made with steel of any yield stress between 36 ksi and 50 ksi inclusive. The load capacity of K-series joists must be verified by the manufacturer by testing, and a minimum factor of safety of 1.65 must be demonstrated. No such testing program is required for LH- or DLH-series joists.

A fourth category, Joist Girders, comprises large joists that are designed to support open-web steel joists. The engineer designates a joist girder by specifying its depth, the number of joist spaces, and the load in kips at each loaded top-chord panel point of the joist girder. A 54G9N10K, for example, is 54 inches deep, provides for 9 equal joist spaces on the top chord, and will support 10 kips at each joist location. The weight tables in *Standard Specifications*[39] give the weight in pounds per linear foot for the specified joist girder for a specific span length.

≡ **EXAMPLE 5.15**

Use the load table given in Figure 5.31 to select an open-web steel joist for the following floor system and loads:

> Joist spacing = 3 ft 0 in.
> Span length = 20 ft 0 in.

Loads:

> 3 in. floor slab
> Other dead load: 20 psf
> Live load: 50 psf

SOLUTION   Loads:

> Slab: $150(3/12)$ = 37.5 psf
> Other dead load = 20 psf
> Live load = 50 psf
> Joist weight = 3 psf (estimated)
> Total: 110.5 psf

Load per joist = $110.5(3) = 331.5$ lb/ft

From Figure 5.31, the following joists are seen to satisfy the load requirement: a 12K5, weighing approximately 7.1 lb/ft; a 14K3, weighing approximately 6.0

lb/ft; and a 16K2, weighing approximately 5.5 lb/ft. Since no restriction was placed on the depth, choose the lighter joist.

Answer   Use a 16K2.

Standard Specifications[39] also contains a K-series economy table, which facilitates the selection of the lightest joist for a given load.

5.12 ≡ BEAM BEARING PLATES AND COLUMN BASE PLATES

The design procedure for column base plates is similar to that for beam bearing plates, and for that reason they will be considered together. In addition, the determination of the thickness of a column base plate requires a consideration of flexure, so it logically belongs in this chapter rather than the previous one. In both cases, the function of the plate is to distribute a concentrated load to the supporting medium.

Two types of beam bearing plates will be considered: one that transmits the beam reaction to a support such as a concrete wall and one that transmits a load to the top flange of a beam. Consider first the beam support shown in Figure 5.32. Although many beams are connected to columns or other beams, the type of support shown here is occasionally used, particularly at bridge abutments. The design of the bearing plate consists of three steps:

1. Determine dimension N so that web yielding and web crippling are prevented.
2. Determine dimension B so that the area $B \times N$ is sufficient to prevent the supporting material (usually concrete) from being crushed in bearing.
3. Determine the thickness t so that the plate has sufficient bending strength.

FIGURE 5.32 ≡

Web yielding and web crippling are addressed by AISC in Chapter K of the Specification, "Strength Design Considerations." Bearing strength of concrete is covered in Chapter J, "Connections, Joints, and Fasteners."

Web Yielding

Web yielding is the compressive crushing of a beam web caused by the application of a compressive force to the flange directly above or below the web. This force could be an end reaction from a support of the type shown in Figure 5.32, or it could be a load delivered to the top flange by a column or another beam. Yielding occurs when the compressive stress on a horizontal section through the web reaches the yield point. When the load is transmitted through a plate, web yielding is assumed to take place on the nearest section of width t_w. In a rolled shape, this will be the toe of the fillet, a distance k from the outside face of the flange (this dimension is tabulated in the Dimensions and Properties tables in the *Manual*). If the load is assumed to distribute itself at a slope of $1:2.5$ as shown in Figure 5.33, the area at the support subject to yielding is $(2.5k + N)t_w$. If this is multiplied by the yield stress, we obtain the nominal strength for web yielding at the support:

$$R_n = (2.5k + N)F_y t_w \qquad \text{(AISC Equation K1-3)}$$

The bearing length N at the support should not be less than k.

At the interior load, the length of the section subject to yielding is $2(2.5k) + N = 5k + N$, and the nominal strength is

$$R_n = (5k + N)F_y t_w \qquad \text{(AISC Equation K1-2)}$$

The design strength is ϕR_n, where $\phi = 1.0$.

Web Crippling

Web crippling is buckling of the web caused by the compressive force delivered through the flange. For an interior load, the nominal strength for web crippling is

$$R_n = 135t_w^2 \left[1 + 3\left(\frac{N}{d}\right)\left(\frac{t_w}{t_f}\right)^{1.5} \right] \sqrt{\frac{F_y t_f}{t_w}} \qquad \text{(AISC Equation K1-4)}$$

For a load at or near the support (no greater than half the beam depth from the end), the nominal strength is

$$R_n = 68t_w^2 \left[1 + 3\left(\frac{N}{d}\right)\left(\frac{t_w}{t_f}\right)^{1.5} \right] \sqrt{\frac{F_y t_f}{t_w}} \quad \text{for } \frac{N}{d} \leq 0.2$$

$$\text{(AISC Equation K1-5a)}$$

FIGURE 5.33 ≡

or

$$R_n = 68t_w^2\left[1 + \left(4\,\frac{N}{d} - 0.2\right)\left(\frac{t_w}{t_f}\right)^{1.5}\right]\sqrt{\frac{F_y t_f}{t_w}} \quad \text{for } \frac{N}{d} > 0.2$$

(AISC Equation K1-5b)

The resistance factor for this limit state is $\phi = 0.75$.

Concrete Bearing Strength

The material used for a beam support can be concrete, brick, or some other material, but it usually will be concrete. This material must resist the bearing load applied by the steel plate. The nominal bearing strength specified in AISC J9 is the same as that given in the American Concrete Institute's Building Code[22] and may be used if no other building code requirements are in effect. If the plate covers the full area of the support, the nominal strength is

$$P_p = 0.85 f_c' A_1 \qquad\qquad \text{(AISC Equation J9-1)}$$

If the plate does not cover the full area of the support,

$$P_p = 0.85 f_c' A_1 \sqrt{A_2/A_1} \qquad\qquad \text{(AISC Equation J9-2)}$$

where

$f_c' = $ 28-day compressive strength of the concrete

$A_1 = $ bearing area

$A_2 = $ full area of the support

If area A_2 is not concentric with A_1, then A_2 should be taken as the largest concentric area that is geometrically similar to A_1, as illustrated in Figure 5.34. AISC also requires that

$$\sqrt{A_2/A_1} \leq 2$$

FIGURE 5.34 ≡

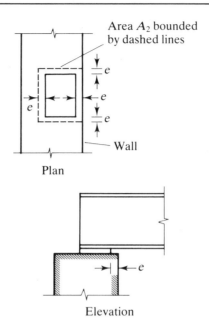

Plan

Elevation

so AISC Equation J9-2 can be written as

$$P_p = 0.85f_c' A_1 \sqrt{A_2/A_1} \leq 0.85f_c' A_1(2) = 0.85f_c' A_1 \sqrt{A_2/A_1} \leq 1.7f_c' A_1$$

The design bearing strength is $\phi_c P_p$, where $\phi_c = 0.60$.

Plate Thickness

Once the length and width of the plate have been determined, the average bearing pressure is treated as a uniform load on the bottom of the plate, which is assumed to be supported at the top over a central width of $2k$ and length N (Figure 5.35). The plate is then considered to bend about an axis parallel to the beam span. Thus, the plate is treated as a cantilever of span length $n = (B - 2k)/2$ and a width of N. For convenience, a 1-in. width is considered, with a uniform load in pounds per linear inch numerically equal to the bearing pressure in pounds per square inch.

From Figure 5.35, the maximum bending moment in the plate is

$$M_u = \frac{R_u}{BN} \times n \times \frac{n}{2} = \frac{R_u n^2}{2BN}$$

where R_u/BN is the average bearing pressure between the plate and the concrete. For a rectangular cross section bent about the minor axis, the nominal

FIGURE 5.35 ≡

moment strength M_n is equal to the plastic moment capacity M_p. For a rectangular cross section of unit width and depth t, the plastic moment is (refer to Figure 5.36)

$$M_p = F_y\left(1 \times \frac{t}{2}\right)\left(\frac{t}{2}\right) = F_y\,\frac{t^2}{4}$$

Since $\phi_b M_n$ must be at least equal to M_u,

$$\phi_b M_p \geq M_u$$

$$0.90 F_y\,\frac{t^2}{4} \geq \frac{R_u n^2}{2NB}$$

$$t \geq \sqrt{\frac{2.222 R_u n^2}{BNF_y}} \qquad\qquad (5.7)$$

FIGURE 5.36 ≡

The foregoing procedure, which will be demonstrated in the following example, is essentially the same as that given in Part 11 of the *Manual* (Volume II), "Connections for Tension and Compression."

EXAMPLE 5.16

Design a bearing plate to distribute the reaction of a W21 × 68 with a span length of 15 ft 10 in. center-to-center of supports. The total service load, including the beam weight, is 9 kips per foot, with equal parts dead and live load. The beam is to be supported on reinforced concrete walls with $f'_c = 3500$ psi. Both the plate and the beam are A36 steel.

SOLUTION

Reaction:

$$w_u = 1.2w_D + 1.6w_L = 1.2(4.5) + 1.6(4.5) = 12.60 \text{ kips/ft}$$

$$R_u = \frac{w_u L}{2} = \frac{12.60(15.83)}{2} = 99.73 \text{ kips}$$

Determine length of bearing N required to prevent web yielding. From AISC Equation K1-3, the design strength for this limit state is

$$R_n = (2.5k + N)F_y t_w$$

For $\phi R_n \geq R_u$,

$$1.0[2.5(1.438) + N](36)(0.430) \geq 99.73$$
$$N \geq 2.85 \text{ in.}$$

Use AISC Equation K1-5 to determine the value of N required to prevent web crippling. Assume $N/d > 0.2$ and try the second form of the equation:

For $\phi R_n \geq R_u$,

$$\phi 68 t_w^2 \left[1 + \left(4\frac{N}{d} - 0.2 \right)\left(\frac{t_w}{t_f} \right)^{1.5} \right] \sqrt{\frac{F_y t_f}{t_w}} \geq R_u$$

$$0.75(68)(0.430)^2 \left[1 + \left(\frac{4N}{21.13} - 0.2 \right)\left(\frac{0.430}{0.685} \right)^{1.5} \right] \sqrt{\frac{36(0.685)}{0.430}} \geq 99.73$$

$$N \geq 5.27 \text{ in.} \quad \text{(controls)}$$

Check assumption:

$$\frac{N}{d} = \frac{5.268}{21.13} = 0.25 > 0.2 \quad \text{(OK)}$$

Try $N = 6$ in.

Determine dimension B from a consideration of bearing strength. If we conservatively assume that the full area of the support is used, the required plate area A_1 can be found as follows:

$$\phi_c(0.85)f'_c A_1 \geq R_u$$
$$0.6(0.85)(3.5)A_1 \geq 99.73$$
$$A_1 \geq 55.87 \text{ in.}^2$$

The minimum value of dimension B is

$$B = \frac{A_1}{N} = \frac{55.87}{6} = 9.31 \text{ in.}$$

The flange width of a W21 × 68 is 8.270 in., making the plate slightly wider than the flange, and this is desirable. Rounding up to the nearest inch:

Try $B = 10$ in.

Compute the required plate thickness:

$$n = \frac{B - 2k}{2} = \frac{10 - 2(1.438)}{2} = 3.562 \text{ in.}$$

From Equation (5.7),

$$t = \sqrt{\frac{2.222 R_u n^2}{BNF_y}} = \sqrt{\frac{2.222(99.73)(3.562)^2}{10(6)(36)}} = 1.14 \text{ in.}$$

Answer   Use a PL $1\frac{1}{4}$ × 6 × 10.

If the beam is not laterally braced at the load point (in such a way as to prevent relative lateral displacement between the loaded compression flange and the tension flange), the Specification requires that *sidesway web buckling* be investigated (AISC K1.5). When loads are applied to both flanges, *compression buckling* must be checked (AISC K1.6).

Column Base Plates

As with beam bearing plates, the design of column base plates requires a consideration of bearing pressure on the supporting material and bending of the plate, but web crippling and web yielding are not factors. In contrast to beam bearing plates, in which bending is in one direction, column base plates are subjected to two-way bending.

There are two categories of column base plates: large and small. If the plate is large, the thickness is determined by bending of the portions of the plate extending beyond the column outline. Bending is assumed to take place about axes at mid-depth of the plate near the edges of the column flanges. Two of the axes are parallel to the web and $0.80b_f$ apart, and two axes are in the perpendicular direction $0.95d$ apart. Of the two 1-in. cantilever strips, labeled m and n

FIGURE 5.37 ≡

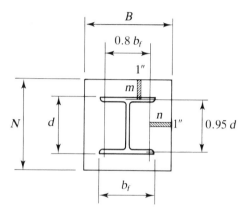

in Figure 5.37, the larger is used in place of n in Equation 5.7 to compute the plate thickness.

If the plate is small — approximately the same size as the column outline — a different approach is used. The portion of the column load that falls within the confines of the column cross section — that is, over an area $b_f d$ — is assumed to be uniformly distributed over the H-shaped area shown in Figure 5.38. The plate thickness is determined on the basis of a flexural analysis of a

FIGURE 5.38 ≡

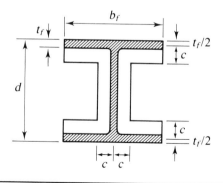

cantilever strip of length c, bending about either the midthickness of the flange or midthickness of the web. If the total factored load on area BN is P_u, the load on area $b_f d$, and hence the H-shaped area, is

$$P_0 = P_u \times \frac{b_f d}{BN}$$

To determine the area of the H-shaped portion, consider the bearing strength of the concrete under this area. From the nominal bearing strength given by

$$P_p = 0.85 f'_c A_1 \sqrt{A_2/A_1} \leq 1.7 f'_c A_1$$

The strength for an area $A_1 = b_f d$ is

$$P_p = 0.85 f'_c b_f d \sqrt{A_2/b_f d} \leq 1.7 f'_c b_f d$$

The strength corresponding to the H-shaped area can be found by proportion. Denoting this area by A_H, multiply through by the ratio $A_H/b_f d$ to obtain

$$P_p = 0.85 f'_c \sqrt{A_2/b_f d}\, A_H \leq 1.7 f'_c A_H$$

From the basic relation between loads and strength,

$$P_0 \leq \phi_c P_p$$
$$P_0 \leq \phi_c(0.85) f'_c \sqrt{A_2/b_f d}\, A_H \leq \phi_c(1.7) f'_c A_H$$
$$\therefore \quad A_H = \frac{P_0}{\phi_c(0.85) f'_c \sqrt{A_2/b_f d}} \leq \frac{P_0}{\phi_c(1.7) f'_c} \tag{5.8}$$

To determine the distance c, use the dimensions given in Figure 5.38 to compute the area A_H:

$$A_H = b_f\left(c + \frac{t_f}{2}\right)(2) + 2c(d - 2c - t_f)$$

Expanding and solving for the smallest root of the resulting quadratic equation,

$$c = \frac{1}{4}\left[b_f + d - t_f - \sqrt{(b_f + d + t_f)^2 - 4(A_H - b_f t_f)}\right] \tag{5.9}$$

To determine the required plate thickness, use Equation (5.7) but substitute c and P_0/A_H for n and P_u/BN to obtain

$$t \geq \sqrt{\frac{2.222 P_0 c^2}{A_H F_y}} \tag{5.10}$$

It is not necessary to ascertain in advance whether the column base plate should be classified as large or small. Simply compute the thickness required for each case and use the larger value.

EXAMPLE 5.17

A W10 × 49 is used as a column and is supported by a concrete pier as shown in Figure 5.39. The top surface of the pier is 18 in. by 18 in. Design an A36 base plate for a column dead load of 98 kips and a live load of 145 kips. The concrete strength is $f'_c = 3000$ psi.

SOLUTION   Factored load:

$$P_u = 1.2D + 1.6L = 1.2(98) + 1.6(145) = 349.6 \text{ kips}$$

Required bearing area:

$$\phi_c P_p \geq P_u$$
$$\phi_c(0.85)f'_c A_1 \sqrt{A_2/A_1} \geq P_u$$
$$0.6(0.85)(3)A_1 \sqrt{18(18)/A_1} \geq 349.6$$
$$A_1 \geq 161.1 \text{ in.}^2$$

Check:

$$\sqrt{A_2/A_1} = \sqrt{18(18)/161.1} = 1.41 < 2 \quad \text{(OK)}$$

Also, the plate must be at least as large as the column:

$$b_f d = 10.00(9.98) = 99.8 \text{ in.}^2 < 161.1 \text{ in.}^2 \quad \text{(OK)}$$

Let $B = N = 13$ in., ∴ A_1 provided $= 13(13) = 169$ in.^{2}

Determine required plate thickness based on "large" plate behavior. From Figure 5.39, it can be seen that the longer cantilever strip will be

$$n = \frac{13 - 8}{2} = 2.5 \text{ in.}$$

From Equation 5.7

$$t \geq \sqrt{\frac{2.222 P_u n^2}{BN F_y}} = \sqrt{\frac{2.222(349.6)(2.5)^2}{13(13)(36)}} = 0.893 \text{ in.}$$

FIGURE 5.39 ≡

Check "small" plate behavior. The load on the H-shaped area is

$$P_0 = P_u \times \frac{b_f d}{BN} = 349.6 \times \frac{10.0(9.98)}{13(13)} = 206.5 \text{ kips}$$

From Equation (5.8),

$$A_H = \frac{P_0}{\phi_c(0.85)f_c'\sqrt{A_2/b_f d}}$$

$$= \frac{206.5}{0.6(0.85)(3)\sqrt{(13)^2/(9.98 \times 10.0)}} = 103.7 \text{ kips}$$

Check upper limit:

$$A_H \leq \frac{P_0}{\phi_c(1.7)f_c'} = \frac{206.5}{0.6(1.7)(3)} = 67.48 \text{ kips} \quad \text{(controls)}$$

Determine c from Equation 5.9:

$$c = \frac{1}{4}\left[b_f + d - t_f - \sqrt{(b_f + d - t_f)^2 - 4(A_H - b_f t_f)}\right]$$

$$= \frac{1}{4}\Big[10.0 + 9.98 - 0.560$$

$$\qquad - \sqrt{(10.0 + 9.98 - 0.560)^2 - 4(67.48 - 10.0 \times 0.560)}\Big]$$

$$= 2.009 \text{ in.}$$

From Equation 5.10,

$$t \geq \sqrt{\frac{2.222 P_0 c^2}{A_H F_y}} = \sqrt{\frac{2.222(206.5)(2.009)^2}{67.48(36)}} = 0.616 \text{ in.}$$

Since $0.893 > 0.616$, the thickness from the large plate analysis controls.

Answer Use a PL $1 \times 13 \times 13$.

An alternate method, which combines the large and small plate procedures, is given in Part 11 of the *Manual* (Volume II), "Connections for Tension and Compression."

5.13 ≡ BIAXIAL BENDING

Biaxial bending occurs when a beam is subjected to a loading condition that produces bending about both the major (strong) axis and the minor (weak) axis. Such a case is illustrated in Figure 5.40, where a single concentrated load acts normal to the longitudinal axis of the beam but is inclined with respect to each of the principal axes of the cross section. Although this is a more general loading than previously considered, it is still a special case: The load passes through the shear center of the cross section. The shear center is that point

FIGURE 5.40 ≡

through which the loads must act if there is to be no twisting, or torsion, of the beam. The location of the shear center can be determined from elementary mechanics of materials by equating the internal resisting torsional moment, derived from the shear flow on the cross section, to the external torque.

The location of the shear center for several common cross sections is given in Figure 5.41a, where the shear center is indicated by an "o." The value of e_0, which locates the shear center for channel shapes, is tabulated in the *Manual of Steel Construction*. The shear center is always located on an axis of symmetry; thus, the shear center will be at the centroid of a cross section with two axes of symmetry. Figure 5.41b shows the deflected position of two different beams when loads are applied through the shear center and when they are not.

Case I: Loads Applied through the Shear Center

If loads act through the shear center, the problem is one of simple bending in two perpendicular directions. As illustrated in Figure 5.42, the load can be resolved into rectangular components in the x- and y-directions, each producing bending about a different axis.

To deal with this combined loading, we temporarily look ahead to Chapter H of the Specification, "Members Under Combined Forces and Torsion." This is the subject of the next chapter of this textbook, "Beam-Columns." The Specification deals with combined loading primarily through the use of *interaction formulas*, which account for the relative importance of each load effect in relation to the strength corresponding to that effect. For example, if there is

FIGURE 5.41 ≡

(a)

(b)

bending about the x-axis only, it is required that

$$M_{ux} \le \phi_b M_{nx}$$

or

$$\frac{M_{ux}}{\phi_b M_{nx}} \le 1.0$$

FIGURE 5.42 ≡

where

M_{ux} = factored load bending moment about the x-axis

M_{nx} = nominal moment strength about the x-axis

Similarly, if there were bending about the y-axis only, the requirement would be

$$\frac{M_{uy}}{\phi_b M_{ny}} \le 1.0$$

where

M_{uy} = factored load bending moment about the y-axis

M_{ny} = nominal moment strength about the y-axis

When both types of bending are present, the interaction formula approach requires that the sum of the two ratios be less than 1; that is,

$$\frac{M_{ux}}{\phi_b M_{nx}} + \frac{M_{uy}}{\phi_b M_{ny}} \le 1.0 \tag{5.11}$$

This in effect allows the designer to allocate to one direction what has not been "used up" in the other direction. AISC Section H1 incorporates a comparable ratio for axial loads and gives two interaction formulas, one for small axial loads and one for large axial loads. (The reason for this will be discussed in the next chapter of this textbook.) With biaxial bending and no axial load, the formula for small axial load is the appropriate one:

$$\frac{P_u}{2\phi P_n} + \left(\frac{M_{ux}}{\phi_b M_{nx}} + \frac{M_{uy}}{\phi_b M_{ny}} \right) \le 1.0 \qquad \text{(AISC Equation H1-1b)}$$

If the axial load P_u is zero, this reduces to Equation 5.11 above.

Up to this point, the strength of I- and H-shaped cross sections bent about their weak axis has not been considered. This is a relatively simple matter. Since any shape bent about its weak axis cannot buckle in the other direction, lateral-torsional buckling is not a limit state. Therefore, only local buckling, either FLB or WLB, need be considered. Since all shapes tabulated in the *Manual* have compact webs, and most have compact flanges, the nominal moment strength about the weak axis will usually be equal to the plastic moment capacity; that is,

$$M_{ny} = M_{py} = F_y Z_y$$

As with bending about the strong axis, M_n cannot exceed $1.5M_y$, or, equivalently, Z_y/S_y cannot exceed 1.5.

≣ **EXAMPLE 5.18**

A W21 × 68 is used as a simply supported beam with a span length of 12 ft. Lateral support of the compression flange is provided only at the ends. Loads act through the shear center with factored load moments of $M_{ux} = 200$ ft-kips and $M_{uy} = 25$ ft-kips. If A36 steel is used, does this beam satisfy the provisions of the AISC Specification? Assume that both moments are uniform over the length of the beam.

SOLUTION The following data for a W21 × 68 are obtained from the Load Factor Design Selection Table:

The shape is compact (no footnote to indicate otherwise):

$L_p = 7.5$ ft, $L_r = 22.8$ ft
$\phi_b M_p = 432$ ft-kips
$BF = 10.4$ kips

The unbraced length $L_b = 12$ ft, ∴ $L_p < L_b < L_r$, and the controlling limit state is inelastic lateral torsional buckling:

$$\phi_b M_{nx} = C_b[\phi_b M_p - BF(L_b - L_p)]$$

Because the bending moment is uniform, $C_b = 1.0$.

$$\phi_b M_{nx} = 1.0[432 - 10.4(12.0 - 7.5)] = 385.2 \text{ ft-kips}$$

Since the shape is compact, there is no flange local buckling, and

$$\phi_b M_{ny} = \phi_b M_{py} = \phi_b Z_y F_y = 0.90(24.4)(36) = 790.6 \text{ in.-kips} = 65.88 \text{ ft-kips}$$

Check:

$$\frac{Z_y}{S_y} = \frac{24.4}{15.7} = 1.55 > 1.5$$

∴ use $M_{ny} = 1.5M_y = 1.5F_y S_y = 1.5(36)(15.7) = 847.8$ in.-kips
$$= 70.75 \text{ ft-kips}$$

$$\phi_b M_{ny} = 0.90(70.65) = 63.59 \text{ ft-kips}$$

From Equation 5.11,

$$\frac{M_{ux}}{\phi_b M_{nx}} + \frac{M_{uy}}{\phi_b M_{ny}} = \frac{200}{385.2} + \frac{25}{63.59} = 0.912 < 1.0 \quad \text{(OK)}$$

Answer The W21 × 68 is satisfactory.

Case II: Loads Not Applied through the Shear Center

When loads are not applied through the shear center of a cross section, the result is flexure plus torsion. If possible, the structure or connection geometry should be modified to remove the eccentricity. The problem of torsion in rolled shapes is a complex one, and we will resort to approximate, although conservative, methods for dealing with it. A more detailed treatment of this topic, complete with design aids, can be found in *Torsional Analysis of Steel Members*.[40] A typical loading condition that gives rise to torsion is shown in Figure 5.43a. The resultant load is applied to the center of the top flange, but its line of action does not pass through the shear center of the section. As far as equilibrium is concerned, the force can be moved to the shear center provided that a couple is added. The equivalent system thus obtained will consist of the given force acting through the shear center plus a twisting moment as shown. In part b of the figure, there is only one component of load to contend with, but the concept is the same.

Figure 5.44 outlines a simplified scheme for treating these two cases. In Figure 5.44a, the top flange is assumed to provide the total resistance to the horizontal component of the load. In Figure 5.44b, the twisting moment Pe is resisted by a couple consisting of equal forces acting at each flange. As an approximation, each flange can be considered to resist each of these forces independently. Consequently, the problem is reduced to a case of bending of two shapes, each one loaded through its shear center. In each of the two situations depicted in Figure 5.44, only about half of the cross section is considered effective with respect to its y-axis; therefore, when considering the strength of a single flange, use half of the tabulated value of Z_y for the shape.

FIGURE 5.43 ≡

(a)

(b)

FIGURE 5.44 ≡

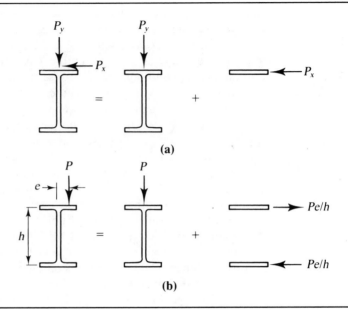

(a)

(b)

Design of Roof Purlins

Roof purlins that are part of a sloping roof system can be subjected to biaxial bending of the type just described. For the roof purlin shown in Figure 5.45, the load is vertical, but the axes of bending are inclined. This corresponds to the

FIGURE 5.45 ≡

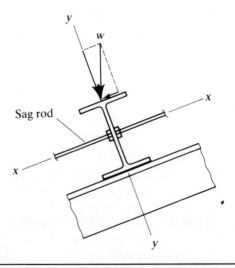

loading of Figure 5.44a. The component of load normal to the roof will cause bending about the x-axis, and the parallel component bends the beam about its y-axis. If the purlins are simply supported at the trusses (or rigid frame rafters), the maximum bending moment about each axis is $wL^2/8$, where w is the appropriate component of load. If sag rods are used, they will provide lateral support with respect to x-axis bending and will act as transverse supports for y-axis bending, requiring that the purlin be treated as a continuous beam. For uniform sag-rod spacings, the formulas for continuous beams in Part 4 of the *Manual* can be used.

EXAMPLE 5.19

A roof system consists of trusses of the type shown in Figure 5.46 spaced 15 feet apart. Purlins are to be placed at the joints and at the midpoint of each top chord member. Sag rods will be located at the center of each purlin. The total gravity load, including an estimated purlin weight, is 30 pounds per square foot of roof surface, with a live load-to-dead load ratio of 1.0. Assuming that this is the critical loading condition, use A36 steel and select a 6-in.-deep W-shape for the purlins.

SOLUTION

For this loading condition, dead load plus a roof live load with no wind or snow, load combination (A4-3) will control:

$$w_u = 1.2w_D + 1.6L_r = 1.2(15) + 1.6(15) = 42 \text{ psf}$$

Width of roof surface tributary to each purlin:

$$\frac{15}{2}\frac{\sqrt{10}}{3} = 7.906 \text{ ft}$$

$$\text{purlin load} = 42(7.906) = 332.1 \text{ lb/ft}$$

$$\text{normal component} = \frac{3}{\sqrt{10}}(332.1) = 315.1 \text{ lb/ft}$$

$$\text{parallel component} = \frac{1}{\sqrt{10}}(332.1) = 105.0 \text{ lb/ft}$$

$$M_{ux} = \frac{1}{8}(0.3151)(15)^2 = 8.862 \text{ ft-kips}$$

With sag rods placed at the midpoint of each purlin, the purlins are two-span continuous beams with respect to weak axis bending. From the Beam Diagrams and Formulas section in Part 4 of the *Manual*, the bending moment at the interior support with only one span loaded is

$$M = \frac{1}{16}wL^2$$

FIGURE 5.46 ≡

where

w = uniform load intensity

L = span length (two equal spans)

With both spans loaded, the moment can be obtained by superposition:

$$M = M_{max} = \frac{1}{16} wL^2 (2) = \frac{1}{8} wL^2$$

$$\therefore \quad M_{uy} = \frac{1}{8} (0.1050)(15/2)^2 = 0.7382 \text{ ft-kips}$$

To select a trial shape, use the Beam Design charts and choose a shape with a relatively large margin of strength with respect to major axis bending. For an unbraced length of $\frac{15}{2} = 7.5$ ft:

Try a W6 × 9.

For $C_b = 1.0$, $\phi_b M_{nx} = 14.0$ ft-kips. From Figure 5.13b, C_b is 1.30 for the load and lateral support conditions of this beam.

$$\therefore \quad \phi_b M_{nx} = 1.30(14.0) = 18.20 \text{ ft-kips}$$

But $\phi_b M_{px} = 16.8$ ft-kips < 18.20 $\therefore$ use $\phi_b M_{nx} = 16.8$ ft-kips.

Since this shape is compact (no footnote in the Factored Uniform Load tables),

$$\phi_b M_{ny} = \phi_b M_{py} = \phi_b Z_y F_y = 0.90(1.72)(36) = 55.73 \text{ in.-kips} = 4.644 \text{ ft-kips}$$

$$\text{but} \quad \frac{Z_y}{S_y} = \frac{1.72}{1.11} = 1.55 > 1.5$$

$$\therefore \quad \phi_b M_{ny} = \phi_b(1.5 M_y) = \phi_b(1.5)(F_y S_y)$$

$$= 0.9(1.5)(36)(1.11) = 53.95 \text{ in.-kips} = 4.496 \text{ ft-kips}$$

Because the load is applied to the top flange, use only half of this capacity to account for the torsional effects. From Equation 5.11,

$$\frac{M_{ux}}{\phi_b M_{nx}} + \frac{M_{uy}}{\phi_b M_{ny}} = \frac{8.862}{16.8} + \frac{0.7382}{4.496/2} = 0.856 < 1.0 \quad \text{(OK)}$$

Shear:

$$V_u = \frac{0.3151(15)}{2} = 2.363 \text{ kips}$$

From the Factored Uniform Load tables,

$$\phi_v V_n = 19.5 \text{ kips} > 2.363 \text{ kips} \quad (OK)$$

Answer   Use a W6 × 9.

5.14 ≡ BENDING STRENGTH OF VARIOUS SHAPES

W-, S-, and M-shapes are the most commonly used hot-rolled shapes for beams, and their bending strength has been covered in the preceding sections. Other shapes are sometimes used as flexural members, however, and this section provides a summary of the relevant AISC provisions. All equations are from Chapter F or Appendix F of the Specification. The nominal strength will be given for compact and noncompact hot-rolled shapes, but not for slender shapes or shapes built up from plate elements. No numerical examples are given in this section, but Example 6.11 in the next chapter includes the computation of the flexural strength of a structural tee-shape.

As in previous coverage, equations are specialized for nonhybrid sections ($F_{yw} = F_{yf} = F_y$) and for the case of bending only (no axial load).

I. Channels

   A. Width-thickness parameters for flexure:

      1. Flange:

$$\lambda = \frac{b_f}{t_f}, \quad \lambda_p = \frac{65}{\sqrt{F_y}}, \quad \lambda_r = \frac{141}{\sqrt{F_y - 10}}$$

      2. Web:

$$\lambda = \frac{h}{t_w}, \quad \lambda_p = \frac{640}{\sqrt{F_y}}, \quad \lambda_r = \frac{970}{\sqrt{F_y}}$$

   B. Bending about the major axis [with either (1) the load applied through the shear center and in a plane parallel to the web or (2) restraint against twisting at the point of load application and at the supports]: M_n is the same as for I-shapes (see Sections 5.5 and 5.6).

   C. Bending about the minor axis: M_n is the same as for I-shapes (see Section 5.13).

II. Rectangular Structural Tubes

   A. Width-thickness parameters (see Figure 5.47):

      1. Flange:

$$\lambda = \frac{b}{t}, \quad \lambda_p = \frac{190}{\sqrt{F_y}}, \quad \lambda_r = \frac{238}{\sqrt{F_y}}$$

2. Web:

$$\lambda = \frac{h}{t_w}, \quad \lambda_p = \frac{640}{\sqrt{F_y}}, \quad \lambda_r = \frac{970}{\sqrt{F_y}}$$

If the actual dimensions b and h, shown in Figure 5.47, are not known, they may be estimated as the total width or depth minus three times the thickness.

B. Bending about the major axis (loaded in the plane of symmetry):

1. Compact shapes:

For compact shapes, the strength will be based on the limit state of lateral-torsional buckling (LTB).

For $L_b \leq L_p$,

$$M_n = M_p \leq 1.5M_y \qquad \text{(AISC Equation F1-1)}$$

For $L_p < L_b \leq L_r$,

$$M_n = C_b\left[M_p - (M_p - M_r)\left(\frac{L_b - L_p}{L_r - L_p}\right)\right] \leq M_p$$

$$\text{(AISC Equation F1-2)}$$

For $L_b > L_r$,

$$M_n = M_{cr} \leq M_p \qquad \text{(AISC Equation F1-12)}$$

where

$$M_{cr} = \frac{57,000C_b\sqrt{JA}}{L_b/r_y} \qquad \text{(AISC Equation F1-14)}$$

$$L_p = \frac{3750r_y}{M_p}\sqrt{JA} \qquad \text{(AISC Equation F1-5)}$$

$$L_r = \frac{57,000r_y\sqrt{JA}}{M_r} \qquad \text{(AISC Equation F1-10)}$$

$$M_r = F_y S_x \qquad \text{(AISC Equation F1-11)}$$

FIGURE 5.47 ≡

2. Noncompact shapes: The nominal strength M_n will be the smallest of the values computed for the following limit states: Lateral torsional buckling (LTB), flange local buckling (FLB), or web local buckling (WLB). For each of the two local buckling states, the strength is determined from the following equation:

$$M_n = M_p - (M_p - M_r)\left(\frac{\lambda - \lambda_p}{\lambda_r - \lambda_p}\right) \qquad \text{(AISC Equation A-F1-3)}$$

C. Bending about the minor axis: There is no LTB limit state for any shape bent about its minor axis.

1. Compact shapes:

$$M_n = M_p \le 1.5M_y \qquad \text{(AISC Equation F1-1)}$$

2. Noncompact shapes: Check FLB and WLB with AISC Equation A-F1-3.

III. Square Structural Tubes

A. Width-thickness parameters:

$$\lambda = \frac{b}{t}, \quad \lambda_p = \frac{190}{\sqrt{F_y}}, \quad \lambda_r = \frac{238}{\sqrt{F_y}}$$

B. Nominal bending strength:

There is no LTB limit state for square (or circular) shapes.

1. Compact shapes:

$$M_n = M_p \le 1.5M_y \qquad \text{(AISC Equation F1-1)}$$

2. Noncompact shapes: The strength is limited by local buckling, either FLB or WLB, whichever yields the smaller value of M_n:

$$M_n = M_p - (M_p - M_r)\left(\frac{\lambda - \lambda_p}{\lambda_r - \lambda_p}\right) \qquad \text{(AISC Equation A-F1-3)}$$

IV. Solid Rectangular Bars

For rectangular bars, the applicable limit state is LTB for major axis bending; local buckling is not a limit state for either major or minor axis bending.

A. Bending about the major axis:

For $L_b \le L_p$,

$$M_n = M_p \le 1.5M_y \qquad \text{(AISC Equation F1-1)}$$

For $L_p < L_b \le L_r$,

$$M_n = C_b\left[M_p - (M_p - M_r)\left(\frac{L_b - L_p}{L_r - L_p}\right)\right] \le M_p$$

$$\text{(AISC Equation F1-2)}$$

For $L_b > L_r$,

$$M_n = M_{cr} \leq M_p \qquad \text{(AISC Equation F1-12)}$$

$$M_{cr} = \frac{57,000C_b\sqrt{JA}}{L_b/r_y} \qquad \text{(AISC Equation F1-14)}$$

where

$$L_p = \frac{3750r_y}{M_p}\sqrt{JA} \qquad \text{(AISC Equation F1-5)}$$

$$L_r = \frac{57,000r_y\sqrt{JA}}{M_r} \qquad \text{(AISC Equation F1-10)}$$

$$M_r = F_y S_x \qquad \text{(AISC Equation F1-11)}$$

B. Bending about the minor axis:

$$M_n = M_p \leq 1.5M_y \qquad \text{(AISC Equation F1-1)}$$

V. Tees and Double-Angle Shapes
   A. Width-thickness parameters:
      1. Tees:
         a. Flange:

$$\lambda = \frac{b_f}{2t_f}, \quad \lambda_r = \frac{95}{\sqrt{F_y}} \quad (\lambda_p \text{ does not apply})$$

         b. Web:

$$\lambda = \frac{d}{t_w}, \quad \lambda_r = \frac{127}{\sqrt{F_y}} \quad (\lambda_p \text{ does not apply})$$

      2. Double angles with separators, either leg:

$$\lambda = \frac{b}{t}, \quad \lambda_r = \frac{76}{\sqrt{F_y}} \quad (\lambda_p \text{ does not apply})$$

      3. Double angles in continuous contact, outstanding leg:

$$\lambda = \frac{b}{t}, \quad \lambda_r = \frac{95}{\sqrt{F_y}} \quad (\lambda_p \text{ does not apply})$$

   B. With loading in the plane of symmetry:
      For nonslender shapes ($\lambda \leq \lambda_r$ for both the flange and the web), the
      strength is based on the limit state of LTB.

$$M_n = M_{cr} = \frac{\pi\sqrt{EI_y GJ}}{L_b}\left[B + \sqrt{1 + B^2}\right] \quad \text{(AISC Equation F1-15)}$$

where

$$M_n \leq 1.5M_y \text{ for stems in tension}$$
$$M_n \leq 1.0M_y \text{ for stems in compression}$$

$$B = \pm 2.3 \left(\frac{d}{L_b} \right) \sqrt{\frac{I_y}{J}} \qquad \text{(AISC Equation F1-16)}$$

$$M_y = F_y S_x$$

The positive sign is used for B when the stem is in tension, and the negative sign is used when the stem is in compression *anywhere* along the unbraced length.

C. Bending about the minor axis:
For nonslender shapes ($\lambda \leq \lambda_r$),

$$M_n = M_p \leq 1.5M_y$$

VI. Solid circular and square shapes:

$$M_n = M_p \leq 1.5M_y$$

VII. Hollow circular shapes:
A. Width-thickness parameters:

$$\lambda = \frac{D}{t}, \quad \lambda_p = \frac{2070}{F_y}, \quad \lambda_r = \frac{8970}{F_y}$$

where D is the outer diameter.

B. Nominal bending strength:
There is no LTB limit state for circular (or square) shapes. The strength is limited by local buckling:
For $\lambda \leq \lambda_p$,

$$M_n = M_p \leq 1.5M_y$$

For $\lambda_p < \lambda \leq \lambda_r$,

$$M_n = \left(\frac{600}{D/t} + F_y \right) S \qquad \text{(AISC Appendix F, Table A-F1.1)}$$

≡ PROBLEMS ≡

5.1   A built-up beam consists of two parallel flange plates, 16 in. × 1 in., and a 48 in. × $\frac{3}{8}$ in. web. A36 steel is used.

a. Compute Z and M_p for bending about the strong axis.

b. Would this member be classified as a beam or as a plate girder by the AISC Specification?

5.2   Classify the following shapes as compact, noncompact, or slender,

    a. the shape used in Problem 5.1

    b. M5 × 18.9, F_y = 50 ksi

    c. S24 × 80, A36 steel

    d. W6 × 15, A36 steel

5.3   **a.** For each of the following shape groups tabulated in Part 1 of the *Manual*, determine the shape with the largest flange width-thickness ratio: W-, M-, and S-shapes. For what value of F_y do each of these shapes become *slender*?

    b. Repeat for the web width-thickness ratio.

    c. What conclusions can be drawn from the results?

5.4   **a.** For what values of F_y do the shapes selected in Problem 5.3a become *noncompact*?

    b. For A36 steel, what is the limiting value of $b_f/2t_f$ for compact shapes? List those shapes that exceed it (i.e., list the noncompact shapes).

    c. Repeat part b for F_y = 50 ksi.

5.5   Compute the design moment for a W36 × 150 of A36 steel with continuous lateral support.

5.6   The beam in Figure P5.6 has continuous lateral support. How much service live load and dead load can be supported if the ratio of live load to dead load is 2.5? All loads will be uniformly distributed.

FIGURE P5.6

5.7   What is the maximum safe span length for the beam shown in Figure P5.7? The 40 kip load is a service live load. Use A572 Grade 50 steel and assume continuous lateral support of the top flange.

FIGURE P5.7

5.8   The beam shown in Figure P5.8 has continuous lateral support of both flanges. The uniform load is a service load that includes the weight of the beam and consists of 50% dead load and 50% live load. If A36 steel is used, is a W14 × 26 adequate?

FIGURE P5.8

5.9   A simply supported beam with a span length of 20 ft supports a uniformly distributed service live load of 5 kips/ft. The compression flange will have full lateral support. If A36 steel is used, is a W12 × 96 adequate?

5.10   Verify the values of C_b given in Figure 5.13, parts b–f.

5.11   Compute C_b for the beam in Figure P5.11. Lateral support is provided only at the ends. The uniform load is a service dead load, and the concentrated load is a service live load.

FIGURE P5.11

5.12   Compute the design moment for a W12 × 58 with $F_y = 50$ ksi and $C_b = 1.0$. The unbraced length is 25 feet.

   a. Use the Specification equations.

   b. Use the constants from the Load Factor Design Selection Table in the *Manual*.

5.13   Repeat Problem 5.12 for an unbraced length of 30 feet.

5.14   The beam shown in Figure P5.14 has continuous lateral support. If A572 Grade 50 steel is used, what is the maximum service live load *P* that can be supported?

FIGURE P5.14

5.15   A W10 × 12 of A242 steel is used as a simply supported, uniformly loaded beam. The span length is 20 ft, and there is lateral support at the ends and at midspan. If the ratio of live load to dead load is 3.0, determine the service dead load and live load capacities.

5.16   Check the beam in Figure P5.16 for flexure and shear. The uniform load is the total dead load including the beam weight. The concentrated loads are live loads. Lateral support of the compression flange is provided at the ends and at the load points. A572 Grade 50 steel is used.

FIGURE P5.16

5.17   Compute the design moment $\phi_b M_n$ and the web shear strength $\phi_v V_n$ for a W6 × 15 of A36 steel with an unbraced length $L_b = 10$ ft and $C_b = 1.75$.

5.18   The beam shown in Figure P5.18 is A36 steel and has continuous lateral support. The two concentrated loads are service live loads. Is the beam adequate?

FIGURE P5.18

5.19   The beam in Figure P5.19 has continuous lateral support. The uniform load of 1.8 kips/ft is the superimposed service dead load; there is no live load. Use A36 steel and select a W-shape. The maximum permissible total deflection is $L/180$.

FIGURE P5.19

5.20   A simply supported beam with a span length of 20 ft must support a superimposed dead load of 1.0 kips/ft and a live load of 3.0 kips/ft. Lateral support is provided only at the ends. Use $F_y = 50$ ksi and select a W-shape. The maximum *live load* deflection must not exceed $L/360$.

For Problems 5.21 through 5.25, use A36 steel and select the most economical rolled shape without regard to deflection. Assume that each beam has continuous lateral support of both flanges. In each case, the ratio of live load to dead load is 2.0. Service loads are shown.

5.21

FIGURE P5.21

5.22

FIGURE P5.22

5.23

FIGURE P5.23

5.24

FIGURE P5.24

5.25

FIGURE P5.25

5.26 The given beam has continuous lateral support of both flanges. The service load shown is made up of equal parts dead and live load. Use A36 steel and select one shape for *AB* and *CD*; select a different shape for *BC*. Assume there is no deflection limitation.

FIGURE P5.26

In Problems 5.27 and 5.28, the beam has continuous lateral support of both flanges. The concentrated loads are service live loads. Use A36 steel and select the lightest available W-shape without regard to deflection.

5.27

FIGURE P5.27

5.28

FIGURE P5.28

For Problems 5.29 and 5.30, use A36 steel and select the most economical shape. Each beam is laterally supported at the ends and at points where concentrated loads are applied. All given loads are live loads. In each case, use a deflection limit of $L/240$.

5.29

FIGURE P5.29

5.30

FIGURE P5.30

5.31   Select a shape for a simply supported beam with a span length of 35 feet to support a service dead load of 500 lb/ft and a service live load of 1500 lb/ft. The maximum total deflection must not exceed $L/360$. The beam will have continuous lateral support. Use A36 steel.

5.32   The beam in Figure P5.32 has continuous lateral support, and the concentrated load is a service live load. Use A36 steel and select a shape The maximum permissible total deflection is $L/360$.

FIGURE P5.32

5.33   Select an A36 rolled shape for beam *AB* of the floor system shown in Figure P5.33. In addition to the weight of the beam, the dead load consists of a 5-in.-thick reinforced concrete slab that provides continuous lateral support of the

compression flange. The live load is 50 psf. The total deflection must not exceed
$L/240$.

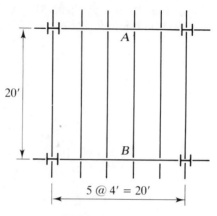

20'

5 @ 4' = 20'

FIGURE P5.33

5.34   Simply supported floor beams are spaced at 9 ft on centers as shown in Figure
P5.34. A 6-in.-thick reinforced concrete floor slab provides continuous lateral
support of the top flange. The live load is 200 psf. Use A36 steel and select the
most economical W-shape without exceeding a beam depth of 21 in. or a total
deflection of $L/240$.

30'

9'   9'   9'

FIGURE P5.34

5.35   **a.** Design a typical floor beam *AB* for the floor system shown in Figure P5.35.
Continuous lateral support is furnished by a floor slab. Use A36 steel and the
following loads:

Live load: 80 psf

Floor tiles: 1 psf

Suspended ceiling with light fixtures: 5 psf

3-in.-thick reinforced concrete floor slab

The total deflection must not exceed L/360.

b. Design girder *CD*. Use the girder weight and the reactions from the beams in part a as loads.

c. Design girder *CD* by treating the entire load as a uniformly distributed load. Be sure to include the weight of the floor beams designed in part a.

FIGURE P5.35

5.36 Floor beams spanning 20 ft are spaced at 8 feet on centers. The beams support a 3-in.-thick reinforced concrete slab and a live load of 125 psf. The slab furnishes partial lateral support for the compression flanges. The amount of support is estimated to be equivalent to lateral bracing at midspan. Use A36 steel and select the lightest rolled shape. The maximum deflection is not to exceed L/240.

5.37 The built-up beam in Figure P5.37 consists of a W12 × 16 with a $\frac{5}{16}$ × 5 cover plate at each flange connected with $\frac{5}{8}$-in.-diameter bolts.

a. Compute the elastic section modulus.

b. Compute the plastic section modulus.

FIGURE P5.37

5.38   Select the most economical K-series open-web steel joist for spacings of (a) 2 ft 6 in., (b) 3 ft 0 in., and (c) 3 ft 9 in. In each case, compute the total weight of the joists within the panel shown in Figure P5.38.

Live load = 80 psf.

Reinforced concrete floor slab thickness = 4 inches.

Weight of suspended ceiling and light fixtures = 5 psf.

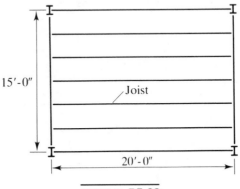

15'-0"

Joist

20'-0"

FIGURE P5.38

5.39   The beam in Figure P5.39 has continuous lateral support. The uniform load is a dead load, and the concentrated load is a live load.

a. Use A36 steel and design the beam. The total deflection is limited to $\frac{1}{240}$ of the span length.

b. Design bearing plates for the supports and the concentrated load. Assume that the bearing plate at the support rests on concrete with a surface area larger than the bearing area by an amount equal to 1 inch of concrete on all sides of the plate. Use A36 steel and a concrete strength of $f'_c = 3000$ psi.

40^k

$2^{k/ft}$

20'   20'

FIGURE P5.39

5.40   Design a bearing plate to support the concentrated load on the beam of Problem 5.7. The beam is A572 Grade 50 steel, but use A36 for the bearing plate.

5.41   Design a base plate for a W12 × 87 column supporting a factored axial load of 600 kips. The support will be a 16 in. × 16 in. concrete pier. Use A36 steel and $f'_c = 3500$ psi.

5.42 Design a column base plate for a W10 × 33 with a factored axial load of 100 kips. The column is supported by a 12 in. × 12 in. concrete pier. Use A36 steel and $f'_c = 3000$ psi.

5.43 The 20 kip concentrated load in Figure P5.43 is a live load. Neglect the weight of the beam and determine whether the beam satisfies the AISC Specification if A36 steel is used. Lateral support is provided at the ends only.

FIGURE P5.43

5.44 The beam shown in Figure P5.44 is a W14 × 90 of A36 steel and has lateral support only at the ends. Check it for compliance with the AISC Specification.

FIGURE P5.44

5.45 Horizontal live loads of 10 kips each are applied at *B* and *C* as shown in Figure P5.45. Lateral support is provided only at the ends. Use A36 steel and select a W-shape.

FIGURE P5.45

5.46   The beam shown in Figure P5.46 is simply supported and has lateral support only at its ends. Neglect the beam weight and determine whether it is satisfactory for each of the loading conditions shown. A36 steel is used, and the 1.5 kips/ft is a live load.

FIGURE P5.46

5.47   Check the beam shown in Figure P5.47 for compliance with the AISC Specification. Lateral support is provided only at the ends, and A36 steel is used. The given service loads are 30% dead load and 70% live load.

FIGURE P5.47

5.48   The truss shown in Figure P5.48 is part of a roof system supporting a total gravity load of 40 pounds per square foot of roof surface, half dead load and half snow. Trusses are spaced at 10 feet on centers. Assume that wind load is not a factor and investigate the adequacy of a W6 × 12 of A36 steel for use as a purlin. No sag rods are used. Assume lateral support at the ends only.

FIGURE P5.48

5.49   The truss shown in Figure P5.49 is one of several roof trusses spaced 18 feet apart. Purlins are located at the joints and halfway between the joints. The weight of the roofing materials is 8 psf, and the snow load is 20 pounds per square foot of horizontal projection of the roof surface. Use A36 steel and select W-shapes for the purlins in the following cases:

a. Sag rods midway between the trusses

b. Sag rods at the third points

FIGURE P5.49

5.50   Write a computer program to determine required design moments for floor beams.

Input: Span length, beam spacing, slab thickness, live load in psf, other dead load in psf.

Output: Factored load moment M_u.

5.51   Write a computer program to determine the maximum permissible uniform service dead and live loads for simply supported beams that are loaded in the plane of the weak axis and have full lateral support of the compression flange. Follow all provisions of the AISC Specification.

Input: Span length in feet, F_y, name of the shape (e.g., W14 × 34), S_x, Z_x, t_f, b_f, and the ratio of live load to dead load.

Output: Same as input, plus the maximum permissible dead and live loads in kips per foot.

5.52   Combine the programs of Problems 5.50 and 5.51 to obtain a program that will evaluate given shapes as floor beams.

Input: 1. Span length, beam spacing, slab thickness, live load in psf, other dead load in psf.

2. F_y, name of shape, S_x, Z_x, t_f, and b_f. Repeat for each shape to be checked.

Output: Same as input, plus M_u and $\phi_b M_n$ for each shape to be checked.

BEAM-COLUMNS

6.1 ☰ DEFINITION

While many structural members can be treated as either axially loaded columns or beams with only flexural loading, most beams and columns are subjected to some degree of both bending and axial load. This is especially true of statically indeterminate structures. Even the roller support of a simple beam can experience friction that restrains the beam longitudinally, inducing axial tension when transverse loads are applied. In this particular case, however, the secondary effects are usually small and can be neglected. Many columns can be treated as pure compression members with negligible error. If the column is a one-story member and can be treated as pinned at both ends, the only bending will result from minor accidental eccentricity of the load.

For many structural members, however, there will be a significant amount of both effects, and such members are called *beam-columns*. Consider the rigid frame in Figure 6.1. For the given loading condition, the horizontal member *AB* must not only support the vertical uniform load but must also assist the vertical members in resisting the concentrated lateral load P_1. Member *CD* is a more critical case, because it must resist the load $P_1 + P_2$ without any assistance from the vertical members. This is because of the x-bracing, indicated by dashed lines, whose purpose is to prevent sidesway in the lower story. For the direction of P_2 shown, member *ED* will be in tension and member *CF* will be slack, provided the bracing elements have been designed to resist only tension. For this to take place, however, member *CD* must transmit the load P_2 from *C* to *D*.

FIGURE 6.1 ☰

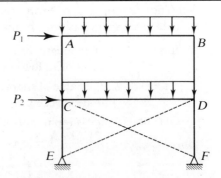

The vertical members of this frame must also be treated as beam-columns. In the upper story, members AC and BD will bend under the influence of P_1. In addition, at A and B, bending moments are transmitted from the horizontal member through the rigid joints. This also takes place at C and D and is true in any rigid frame, although these moments are usually smaller than those resulting from lateral loads. Most columns in rigid frames are actually beam-columns, and the effects of bending should not be ignored. On the other hand, many isolated one-story columns can be realistically treated as axially loaded compression members.

Another example of beam-columns can sometimes be found in roof trusses. Although the top chord is normally treated as an axially loaded compression member, if purlins are placed between the joints, their reactions will cause bending, which must be accounted for. Methods for handling this problem will be discussed later in this chapter.

6.2 ≡ INTERACTION FORMULAS

The inequality of Equation 2.3 may be written in the following form:

$$\frac{\sum \gamma_i Q_i}{\phi R_n} \le 1.0 \tag{6.1}$$

or

$$\frac{\sum \text{load effects}}{\text{resistance}} \le 1.0$$

If more than one type of resistance is involved, Equation 6.1 can be used to form the basis of an interaction formula. As discussed in Chapter 5 in conjunction with biaxial bending, the sum of the load-to-resistance ratios must be limited to unity. For example, if both bending and axial compression are acting, the interaction formula would be

$$\frac{P_u}{\phi_c P_n} + \frac{M_u}{\phi_b M_n} \le 1.0 \tag{6.2}$$

where

$$P_u = \text{factored axial compressive load}$$
$$\phi_c P_n = \text{compressive design strength}$$
$$M_u = \text{factored bending moment}$$
$$\phi_b M_n = \text{design moment}$$

For biaxial bending, there will be two bending ratios:

$$\frac{P_u}{\phi_c P_n} + \left(\frac{M_{ux}}{\phi_b M_{nx}} + \frac{M_{uy}}{\phi_b M_{ny}} \right) \le 1.0 \tag{6.3}$$

where the x and y subscripts refer to bending about the x- and y-axes. The following example illustrates the application of interaction formulas for bending and axial load.

≡ EXAMPLE 6.1

The beam-column in Figure 6.2 is pinned at both ends and subjected to the factored loads shown. Bending is about the strong axis. Determine whether this member satisfies Equation 6.2.

SOLUTION

As we will see shortly, the AISC requirements for beam-columns are based on interaction equations similar to Equations 6.2 and 6.3, but there is a difference. It is important to realize that a member satisfying Equations 6.2 or 6.3 may or may not satisfy the AISC requirements. This example is merely an illustration of how interaction formulas work.

From the Column Load Tables, the axial compressive design strength of a W8 × 58 of A36 steel with an effective length of $K_y L = 1.0 \times 17 = 17$ feet is

$$\phi_c P_n = 318 \text{ kips}$$

Since bending is about the strong axis, the design moment $\phi_b M_n$ for $C_b = 1.0$ can be obtained from the beam design charts in Part 4 of the *Manual*.

For an unbraced length $L_b = 17$ ft,

$$\phi_b M_n = 151.5 \text{ ft-kips}$$

For the end conditions and loading of this problem, $C_b = 1.32$ (Figure 5.13c).
For $C_b = 1.32$,

$$\phi_b M_n = 1.32(151.5) = 200.0 \text{ ft-kips}$$

FIGURE 6.2 ≡

But this is larger than $\phi_b M_p = 161.5$ ft-kips (also obtained from the beam design charts), so the design moment must be limited to $\phi_b M_p$.

$$\therefore \quad \phi_b M_n = 161.5 \text{ ft-kips}$$

The maximum bending moment occurs at midheight:

$$M_u = \frac{14(17)}{4} = 59.5 \text{ ft-kips}$$

Equation 6.2:

$$\frac{P_u}{\phi_c P_n} + \frac{M_u}{\phi_b M_n} = \frac{175}{318} + \frac{59.5}{161.5} = 0.919 < 1.0 \quad \text{(OK)}$$

Answer   This member satisfies Equation 6.2.

Equation 6.3 is the basis for the AISC formulas for members subject to bending plus axial compressive load. Two formulas are given in the Specification: one for small axial load and one for large axial load. If the axial load is small, the axial load term is reduced. For large axial load, the bending term is slightly reduced. The AISC requirements are given in Chapter H, "Members Under Combined Forces and Torsion," and can be summarized as follows:

For $\dfrac{P_u}{\phi_c P_n} \geq 0.2$,

$$\frac{P_u}{\phi_c P_n} + \frac{8}{9}\left(\frac{M_{ux}}{\phi_b M_{nx}} + \frac{M_{uy}}{\phi_b M_{ny}}\right) \leq 1.0 \qquad \text{(AISC Equation H1-1a)}$$

For $\dfrac{P_u}{\phi_c P_n} < 0.2$,

$$\frac{P_u}{2\phi_c P_n} + \left(\frac{M_{ux}}{\phi_b M_{nx}} + \frac{M_{uy}}{\phi_b M_{ny}}\right) \leq 1.0 \qquad \text{(AISC Equation H1-1b)}$$

6.3 ≡ MOMENT AMPLIFICATION

The foregoing approach to the analysis of members subjected to both bending and axial load is satisfactory so long as the axial load is not too large. The presence of the axial load produces secondary moments, and unless the axial load is relatively small, we must account for these additional moments. This can be explained with reference to Figure 6.3, which shows a beam-column with an axial load and a transverse uniform load. At an arbitrary point O, there is a bending moment because of the uniform load and an additional moment Py,

FIGURE 6.3 ≡

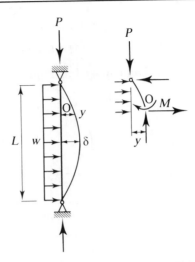

caused by the axial load acting at an eccentricity from the longitudinal axis of the beam. This secondary moment is largest where the deflection is largest — in this case, at the centerline, where the total moment is $wL^2/8 + P\delta$. Of course, the additional moment causes an additional deflection over and above that resulting from the transverse load. Since the total deflection cannot be found directly, this is a nonlinear problem, and without knowing the deflection, the moment cannot be computed.

Ordinary structural analysis methods that do not take the displaced geometry into account are referred to as *first-order* methods. Iterative numerical techniques, called *second-order* methods, can be used to find the deflections and secondary moments, but these methods are impractical for manual calculations and are usually implemented with a computer program. Most current design codes and specifications, including the AISC Specification, permit the use of either a second-order analysis or the *moment amplification method*. This method entails computing the maximum bending moment resulting from flexural loading (transverse loads or member end moments) by a first-order analysis, then multiplying by a *moment amplification factor* to account for the secondary moment. An expression for this factor will now be developed.

Figure 6.4 shows a simply supported member with an axial load and an initial out-of-straightness. This initial crookedness will be approximated by

$$y_0 = e \sin \frac{\pi x}{L}$$

where e is the maximum initial displacement, occurring at midspan. For the coordinate system shown, the moment–curvature relationship can be written as

$$\frac{d^2y}{dx^2} = -\frac{M}{EI}$$

FIGURE 6.4 ≣

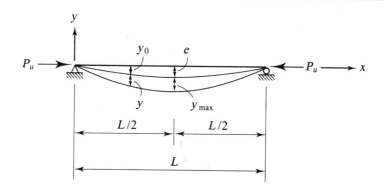

The bending moment M is caused by the eccentricity of the axial load P_u with respect to the axis of the member. This eccentricity consists of the initial crookedness y_0 plus additional deflection y resulting from bending. At any location, the moment is

$$M = P_u(y_0 + y)$$

Substituting this into the differential equation, we obtain

$$\frac{d^2y}{dx^2} = -\frac{P_u}{EI}\left(e \sin\frac{\pi x}{L} + y\right)$$

Rearranging,

$$\frac{d^2y}{dx^2} + \frac{P_u}{EI}\,y = -\frac{P_u e}{EI}\sin\frac{\pi x}{L}$$

This is an ordinary, nonhomogeneous differential equation. Since it is a second-order equation, there are two boundary conditions. For the support conditions shown, the boundary conditions are as follows:

At $x = 0$, $y = 0$

and

At $x = L$, $y = 0$

That is, the displacement is zero at each end. A function that satisfies both the differential equation and the boundary conditions is

$$y = B \sin\frac{\pi x}{L}$$

where B is a constant. Substituting into the differential equation,

$$-\frac{\pi^2}{L^2}B\sin\frac{\pi x}{L} + \frac{P_u}{EI}B\sin\frac{\pi x}{L} = -\frac{P_u e}{EI}\sin\frac{\pi x}{L}$$

Solving for the constant,

$$B = \frac{-\dfrac{P_u e}{EI}}{\dfrac{P_u}{EI} - \dfrac{\pi^2}{L^2}} = \frac{-e}{1 - \dfrac{\pi^2 EI}{P_u L^2}} = \frac{e}{\dfrac{P_e}{P_u} - 1}$$

where

$$P_e = \frac{\pi^2 EI}{L^2} = \text{the Euler buckling load}$$

$$\therefore \quad y = B \sin \frac{\pi x}{L} = \left(\frac{e}{\dfrac{P_e}{P_u} - 1} \right) \sin \frac{\pi x}{L}$$

$$M = P_u(y_0 + y)$$

$$= P_u \left[e \sin \frac{\pi x}{L} + \left(\frac{e}{\dfrac{P_e}{P_u} - 1} \right) \sin \frac{\pi x}{L} \right]$$

The maximum moment occurs at $x = L/2$:

$$M_{\max} = P_u \left(e + \frac{e}{\dfrac{P_e}{P_u} - 1} \right)$$

$$= P_u e \left(\frac{\dfrac{P_e}{P_u} - 1 + 1}{\dfrac{P_e}{P_u} - 1} \right)$$

$$= M_0 \left(\frac{1}{1 - \dfrac{P_u}{P_e}} \right)$$

where M_0 is the unamplified maximum moment. In this case, it results from initial crookedness, but in general it can be the result of transverse loads or end moments. The moment amplification factor is therefore

$$\frac{1}{1 - \dfrac{P_u}{P_e}} \tag{6.4}$$

≡ **EXAMPLE 6.2**

Use Expression 6.4 to compute the amplification factor for the beam-column of Example 6.1.

SOLUTION

Since the Euler load P_e is part of an amplification factor for a *moment*, it must be computed for the axis of bending, which in this case is the x-axis. In terms of effective length and slenderness ratio, the Euler load can be written as

$$P_e = \frac{\pi^2 E A_g}{\left(\dfrac{KL}{r}\right)^2}$$

For the axis of bending,

$$\frac{KL}{r} = \frac{K_x L}{r_x} = \frac{1.0(17)(12)}{3.65} = 55.89$$

$$P_e = \frac{\pi^2 E A_g}{(KL/r)^2} = \frac{\pi^2(29,000)(17.1)}{(55.89)^2} = 1567 \text{ kips}$$

From Expression 6.4,

$$\frac{1}{1 - \dfrac{P_u}{P_e}} = \frac{1}{1 - \dfrac{175}{1567}} = 1.13$$

This represents a 13% increase in bending moment. The amplified primary moment is

$$1.13 \times M_u = 1.13(59.5) = 67.2 \text{ ft-kips}$$

Answer   Amplification factor = 1.13.

6.4 ≡ WEB LOCAL BUCKLING IN BEAM-COLUMNS

The determination of the design moment requires that the cross section be checked for compactness. The web is compact for all tabulated shapes as long as there is no axial load. In the presence of axial load, this may not be the case. Using the notation $\lambda = h/t_w$,

If $\lambda \leq \lambda_p$, the shape is compact.

If $\lambda_p < \lambda \leq \lambda_r$, the shape is noncompact.

If $\lambda > \lambda_r$, the shape is slender.

AISC B5, in Table B5.1, prescribes the following limits:

$$\text{For } \frac{P_u}{\phi_b P_y} \le 0.125, \quad \lambda_p = \frac{640}{\sqrt{F_y}}\left(1 - \frac{2.75P_u}{\phi_b P_y}\right)$$

$$\text{For } \frac{P_u}{\phi_b P_y} > 0.125, \quad \lambda_p = \frac{191}{\sqrt{F_y}}\left(2.33 - \frac{P_u}{\phi_b P_y}\right) \ge \frac{253}{\sqrt{F_y}}$$

$$\text{For any value of } \frac{P_u}{\phi_b P_y}, \quad \lambda_r = \frac{970}{\sqrt{F_y}}\left(1 - 0.74 \frac{P_u}{\phi_b P_y}\right)$$

where $P_y = A_g F_y$, the axial load required to reach the limit state of yielding.

Since P_u is variable, compactness of the web cannot be checked and tabulated in advance. Some rolled shapes satisfy the worst case limit of $253/\sqrt{F_y}$, however, which means that these shapes have compact webs no matter what the axial load. Shapes listed in the Column Load Tables in Part 3 of the *Manual* that do not satisfy this criterion are marked, and only these need be checked for compactness of the web. Shapes whose flanges are not compact are also marked, so if there is no indication to the contrary, shapes found in the Column Load Tables are compact.

≡ **EXAMPLE 6.3**

A W12 × 65 of A36 steel is subjected to a bending moment and a factored axial load of 300 kips. Check the web for compactness.

SOLUTION

This shape is compact for any value of axial load since there is no footnote in the Column Load Table to indicate otherwise. For illustration, however, the web width-thickness ratio will be checked.

$$\frac{P_u}{\phi_b P_y} = \frac{P_u}{\phi_b(A_g F_y)} = \frac{300}{0.90(19.1)(36)} = 0.4848 > 0.125$$

$$\therefore \quad \lambda_p = \frac{191}{\sqrt{F_y}}\left(2.33 - \frac{P_u}{\phi_b P_y}\right) = \frac{191}{\sqrt{36}}(2.33 - 0.4848) = 58.74$$

$$\frac{253}{\sqrt{F_y}} = \frac{253}{\sqrt{36}} = 42.17 < 58.74 \quad \therefore \quad \lambda_p = 58.74$$

From the Dimensions and Properties tables,

$$\lambda = \frac{h}{t_w} = 24.9 < \lambda_p$$

The web is therefore compact. Note that for any available F_y, h/t_w will be less than $253/\sqrt{F_y}$, the smallest possible value of λ_p, so the web of a W12 × 65 will always be compact.

6.5 ≡ BRACED VERSUS UNBRACED FRAMES

The AISC Specification covers moment amplification in Chapter C, "Frames and Other Structures." Two amplification factors are used in LRFD: one to account for amplification resulting from the member deflection and one to account for the effect of sway when the member is part of an unbraced frame. This is the same approach used in the ACI Building Code for reinforced concrete.[22] Figure 6.5 illustrates these two components. In part a of the figure, the member is restrained against sidesway, and the maximum secondary moment is $P\delta$, which is added to the maximum moment within the member. If the frame is actually unbraced, there is an additional component of the secondary moment, shown in Figure 6.5b, that is caused by sidesway. This secondary moment has a maximum value of $P\Delta$, which represents an amplification of the *end* moment.

To approximate these two effects, two amplification factors, B_1 and B_2, are used for the two types of moments. The amplified moment to be used in design is computed from the factored loads and moments as follows (x and y subscripts are not used here; amplified moments must be computed in the following manner for each axis about which there are moments):

$$M_u = B_1 M_{nt} + B_2 M_{\ell t}$$

(AISC Equation C1-1)

where

M_{nt} = maximum moment assuming that no sidesway occurs, whether the frame is actually braced or not (the subscript "*nt*" is for "no translation").

$M_{\ell t}$ = maximum moment caused by sidesway (the subscript "*ℓt*" is for "lateral translation"). This moment can be caused by lateral loads or by unbalanced gravity loads. Gravity loads can produce sidesway if

FIGURE 6.5 ≡

(a)   (b)

the frame is unsymmetrical or if the gravity loads are unsymmetrically placed. $M_{\ell t}$ will be zero if the frame is actually braced.

B_1 = amplification factor for the moments occurring in the member when it is braced against sidesway.

B_2 = amplification factor for the moments resulting from sidesway.

The evaluation of B_1 and B_2 will be covered in the following sections.

6.6 ☰ MEMBERS IN BRACED FRAMES

The amplification factor given by Expression 6.4 was derived for a member braced against sidesway — that is, one whose ends cannot translate with respect to each other. Figure 6.6 shows a member of this type subjected to equal end moments producing *single-curvature bending* (bending that produces tension or compression on one side throughout the length of the member). Maximum moment amplification occurs at the center, where the deflection is largest. For equal end moments, the moment is constant throughout the length of the member, so the maximum primary moment also occurs at the center. Thus, the maximum secondary moment and maximum primary moment are additive. Even if the end moments are not equal, as long as one is clockwise and the other is counterclockwise there will be single-curvature bending, and the maximum primary and secondary moments will occur near each other.

This is not the case if applied end moments produce reverse-curvature bending as shown in Figure 6.7. Here the maximum primary moment is at one of the ends, and maximum moment amplification occurs between the ends.

FIGURE 6.6 ☰

FIGURE 6.7 ≡

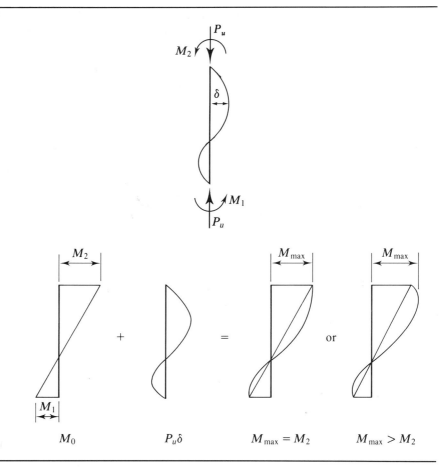

Depending on the value of the axial load P_u, the amplified moment can be either larger or smaller than the end moment.

The maximum moment in a beam-column therefore depends on the distribution of bending moment within the member. This is accounted for by a factor C_m applied to the amplification factor B_1. Since the amplification factor given by Expression 6.4 was derived for the worst case, C_m will never be greater than 1.0. The final form of the amplification factor is

$$B_1 = \frac{C_m}{(1 - P_u/P_{e1})} \geq 1 \qquad \text{(AISC Equation C1-2)}$$

where

$$P_{e1} = \frac{A_g F_y}{\lambda_c^2} = \frac{\pi^2 E A_g}{(KL/r)^2}$$

When computing P_{e1}, use KL/r for the axis of bending and an effective length factor K less than or equal to 1.0 (corresponding to the braced condition).

Evaluation of C_m

The factor C_m applies only to the braced condition. There are two categories of members: those with transverse loads applied between the ends and those with no transverse loads. Figure 6.8b and c illustrate these two cases (member AB is the beam-column under consideration).

1. If there are no transverse loads acting on the member,

$$C_m = 0.6 - 0.4\left(\frac{M_1}{M_2}\right) \qquad \text{(AISC Equation C1-3)}$$

M_1/M_2 is a ratio of the bending moments at the ends of the member. M_1 is the end moment that is smaller in absolute value, M_2 is the larger, and the ratio is positive for members bent in reverse curvature and negative for single-curvature bending (Figure 6.9). Reverse curvature (a positive ratio) occurs when M_1 and M_2 are both clockwise or both counterclockwise.

FIGURE 6.8 ≡

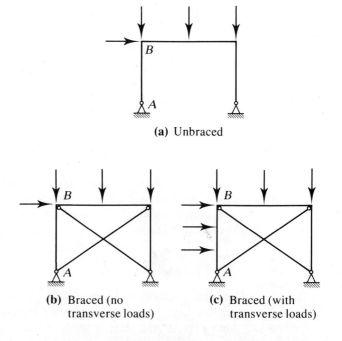

(a) Unbraced

(b) Braced (no transverse loads)   (c) Braced (with transverse loads)

FIGURE 6.9 ≡

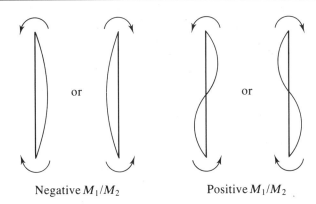

Negative M_1/M_2 Positive M_1/M_2

2. For transversely loaded members, C_m can be taken as 0.85 if the ends are restrained against rotation and 1.0 if the ends are unrestrained against rotation (pinned). End restraint will usually result from the stiffness of members connected to the beam-column. The pinned end condition is the one used in the derivation of the amplification factor; hence, there is no reduction for this case, and this corresponds to $C_m = 1.0$. Although the actual end condition may lie between full fixity and a frictionless pin, use of one of the two values given here will give satisfactory results.

A more refined procedure for transversely loaded members (the second category above) is provided in Section C1 of the Commentary to the Specification. The reduction factor is given as

$$C_m = 1 + \Psi \frac{P_u}{P_{e1}}$$

For simply supported members,

$$\Psi = \frac{\pi^2 \delta_0 EI}{M_0 L^2} - 1$$

where δ_0 is the maximum deflection resulting from transverse loading, and M_0 is the maximum moment between supports resulting from the transverse loads. The factor Ψ has been evaluated for several common situations and is given in Commentary Table C-C1.1.

≡ **EXAMPLE 6.4**

The member shown in Figure 6.10 is part of a braced frame. The given load and moments have been computed with factored loads, and bending is about the strong axis. If A36 steel is used, is this member adequate? $K_x L = K_y L = 14$ feet.

SOLUTION   Determine which interaction formula to apply:

$$\text{maximum } \frac{KL}{r} = \frac{K_y L}{r_y} = \frac{14(12)}{3.02} = 55.63$$

From AISC Table 3-36, $\phi_c F_{cr} = 26.00$ ksi.

$$\phi_c P_n = A_g(\phi_c F_{cr}) = 19.1(26.00) = 496.6 \text{ kips}$$

$$\frac{P_u}{\phi_c P_n} = \frac{300}{496.6} = 0.6041 > 0.2 \quad \therefore \quad \text{use AISC Equation H1-1a.}$$

In the plane of bending,

$$\frac{KL}{r} = \frac{K_x L}{r_x} = \frac{14(12)}{5.28} = 31.82$$

$$P_{e1} = \frac{A_g F_y}{\lambda_c^2} = \frac{\pi^2 E A_g}{(K_x L/r_x)^2} = \frac{\pi^2(29,000)(19.1)}{(31.82)^2} = 5399 \text{ kips}$$

$$C_m = 0.6 - 0.4\left(\frac{M_1}{M_2}\right) = 0.6 - 0.4\left(-\frac{50}{60}\right) = 0.9333$$

$$B_1 = \frac{C_m}{(1 - P_u/P_{e1})} = \frac{0.9333}{(1 - 300/5399)} = 0.988 < 1.0 \quad \therefore \quad \text{use 1.0}$$

Moment strength:
    From the Beam Design charts, with $C_b = 1.0$ and $L_b = 14$ feet,

$$\phi_b M_n = 257 \text{ ft-kips}$$

For the actual value of C_b, refer to the moment diagram shown in Figure 6.10:

FIGURE 6.10 ≡

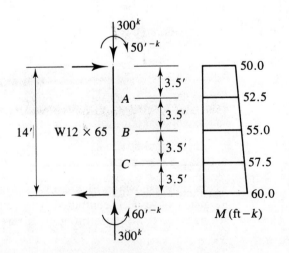

$$C_b = \frac{12.5 M_{max}}{2.5 M_{max} + 3M_A + 4M_B + 3M_C} = \frac{12.5(60)}{2.5(60) + 3(52.5) + 4(55.0) + 3(57.5)}$$

$$= 1.07$$

$$\therefore \quad \phi_b M_n = C_b(257) = 1.07(257) = 275 \text{ ft-kips}$$

But $\phi_b M_p = 261$ ft-kips (from charts) < 275 ft-kips

$$\therefore \quad \text{use } \phi_b M_n = 261 \text{ ft-kips}$$

Factored load moment:

$$M_{nt} = 60 \text{ ft-kips} \qquad M_{\ell t} = 0$$

From AISC Equation C1-1

$$M_u = B_1 M_{nt} + B_2 M_{\ell t} = 1.0(60) + 0 = 60 \text{ ft-kips} = M_{ux}$$

AISC Equation H1-1a:

$$\frac{P_u}{\phi_c P_n} + \frac{8}{9} \left(\frac{M_{ux}}{\phi_b M_{nx}} + \frac{M_{uy}}{\phi_b M_{ny}} \right) = 0.6041 + \frac{8}{9} \left(\frac{60}{261} \right)$$

$$= 0.808 < 1.0 \quad (\text{OK})$$

Answer   This member is satisfactory.

EXAMPLE 6.5

The horizontal beam-column in Figure 6.11 is subjected to the service live loads shown. This member is laterally braced at its ends, and bending is about the x-axis. Check for compliance with the AISC Specification.

SOLUTION   **Factored loads and moments:**

$$P_u = 1.6(20) = 32.0 \text{ kips}$$

The maximum moment is

$$M_{nt} = \frac{(1.6 \times 20)(10)}{4} + \frac{(1.2 \times 0.035)(10)^2}{8} = 80.52 \text{ ft-kips}$$

Since this member is braced against end translation, $M_{\ell t} = 0$.

Moment amplification factor:
    For a member braced against sidesway, transversely loaded, and with unrestrained ends, C_m can be taken as 1.0. A more accurate value from AISC Commentary Table C-C1.1 is

$$C_m = 1 - 0.2 \frac{P_u}{P_{e1}}$$

FIGURE 6.11 ≡

For the axis of bending,

$$\frac{KL}{r} = \frac{K_x L}{r_x} = \frac{1.0(10)(12)}{3.51} = 34.19$$

$$P_{e1} = \frac{\pi^2 E A_g}{(KL/r)^2} = \frac{\pi^2 (29,000)(10.3)}{(34.19)^2} = 2522 \text{ kips}$$

$$C_m = 1 - 0.2\left(\frac{32.0}{2522}\right) = 0.9975$$

The amplification factor is

$$B_1 = \frac{C_m}{1 - \dfrac{P_u}{P_{e1}}} = \frac{0.9975}{1 - \dfrac{32.0}{2522}} = 1.010 > 1.0$$

For the axis of bending,

$$M_u = B_1 M_{nt} + B_2 M_{\ell t} = 1.010(80.52) + 0 = 81.33 \text{ ft-kips}$$

Design strengths:

From the Column Load Tables in Part 3 of the *Manual*,

$$\phi_c P_n = 262 \text{ kips}$$

From the Beam Design charts in Part 4 of the *Manual*, for $L_b = 10$ ft and $C_b = 1.0$,

$$\phi_b M_n = 91.8 \text{ ft-kips}$$

Since the beam weight is very small in relation to the concentrated live load, C_b may be taken from Figure 5.13c as 1.32. This results in a design moment of

$$\phi_b M_n = 1.32(91.8) = 121 \text{ ft-kips}$$

This is greater than $\phi_b M_p = 93.6$ ft-kips, also taken from the Beam Design charts, so the design strength must be limited to this value;

$$\therefore \quad \phi_b M_n = 93.6 \text{ ft-kips}$$

Interaction formula:

$$\frac{P_u}{\phi_c P_n} = \frac{32.0}{262} = 0.1221 < 0.2 \quad \therefore \quad \text{use AISC Equation H1-1b:}$$

$$\frac{P_u}{2\phi_c P_n} + \left(\frac{M_{ux}}{\phi_b M_{nx}} + \frac{M_{uy}}{\phi_b M_{ny}}\right) = \frac{0.1221}{2} + \left(\frac{81.33}{93.6} + 0\right)$$

$$= 0.930 < 1.0 \quad \text{(OK)}$$

Answer   A W8 × 35 is adequate.

EXAMPLE 6.6

The member shown in Figure 6.12 is a W12 × 65 of A242 steel and must support a factored axial compressive load of 300 kips. One end is pinned, and the other is subjected to factored load moments of 135 ft-kips about the strong axis and 30 ft-kips about the weak axis. Use $K_x = K_y = 1.0$, and investigate for compliance with the AISC Specification.

SOLUTION   Determine the yield stress F_y:

From Table 1-2, Part 1 of the *Manual*, a W12 × 65 is seen to be a Group 2 shape. From Table 1-1, Group 2 shapes are available in only one strength in A242 steel, with $F_y = 50$ ksi.

Compressive strength:

For $KL = 1.0(15) = 15$ ft, the axial compressive design strength from the Column Load Tables is

$$\phi_c P_n = 626 \text{ kips}$$

Note that the table indicates that the flange of a W12 × 65 is noncompact for $F_y = 50$ ksi.

FIGURE 6.12 ≡

$P_u = 300^k$
$M_{ntx} = 135'^{-k}$
$M_{nty} = 30'^{-k}$

15'

W12 × 65
A242 steel

300^k

Strong axis bending:

$$C_{mx} = 0.6 - 0.4 \frac{M_1}{M_2} = 0.6 - 0.4(0) = 0.6$$

$$\frac{K_x L}{r_x} = \frac{15(12)}{5.28} = 34.09$$

$$P_{e1x} = \frac{\pi^2 E A_g}{(K_x L/r_x)^2} = \frac{\pi^2 (29,000)(19.1)}{(34.09)^2} = 4704 \text{ kips}$$

$$B_{1x} = \frac{C_{mx}}{1 - \dfrac{P_u}{P_{e1x}}} = \frac{0.6}{1 - \dfrac{300}{4704}} = 0.641 < 1.0 \quad \therefore \quad \text{use } B_{1x} = 1.0.$$

$$M_{ux} = B_{1x} M_{ntx} + B_{2x} M_{\ell tx} = 1.0(135) + 0 = 135 \text{ ft-kips}$$

From the Beam Design charts with $L_b = 15$ ft, $\phi_b M_{nx} = 342$ ft-kips for $C_b = 1.0$, and $\phi_b M_{px} = 357.8$ ft-kips.

For a linear distribution of bending moment with $M = 0$ at one end (Figure 6.13),

$$C_b = \frac{12.5 M_{max}}{2.5 M_{max} + 3 M_A + 4 M_B + 3 M_C} = \frac{12.5(1.0)}{2.5(1.0) + 3(0.25) + 4(0.50) + 3(0.75)}$$

$$= 1.67$$

(This also corresponds to the moment distribution of Figure 5.13d.)

$$C_b \times (\phi_b M_{nx} \text{ for } C_b = 1.0) = 1.67(342) = 571 \text{ ft-kips}$$

This is larger than $\phi_b M_{px}$, $\quad \therefore \quad$ use $\phi_b M_{nx} = \phi_b M_{px} = 357.8$ ft-kips.

Weak-axis bending:

$$C_{my} = 0.6 - 0.4 \frac{M_1}{M_2} = 0.6 - 0.4(0) = 0.6$$

$$\frac{K_y L}{r_y} = \frac{15(12)}{3.02} = 59.60$$

$$P_{e1y} = \frac{\pi^2 E A_g}{(K_y L/r_y)^2} = \frac{\pi^2 (29,000)(19.1)}{(59.60)^2} = 1539 \text{ kips}$$

$$B_{1y} = \frac{C_{my}}{1 - \dfrac{P_u}{P_{ey}}} = \frac{0.6}{1 - \dfrac{300}{1539}} = 0.745 < 1.0 \quad \therefore \quad \text{use } B_{1y} = 1.0.$$

$$M_{uy} = B_{1y} M_{nty} + B_{2y} M_{\ell ty} = 1.0(30) + 0 = 30 \text{ ft-kips}$$

FIGURE 6.13 ≡

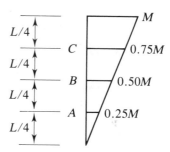

Because the flange of this shape is noncompact, the weak-axis bending strength is limited by FLB.

$$\lambda = \frac{b_f}{2t_f} = 9.9$$

$$\lambda_p = \frac{65}{\sqrt{F_y}} = \frac{65}{\sqrt{50}} = 9.192$$

$$\lambda_r = \frac{141}{\sqrt{F_y - 10}} = \frac{141}{\sqrt{50 - 10}} = 22.29$$

Since $\lambda_p < \lambda < \lambda_r$,

$$M_n = M_p - (M_p - M_r)\left(\frac{\lambda - \lambda_p}{\lambda_r - \lambda_p}\right) \qquad \text{(AISC Equation A-F1-3)}$$

$$M_p = M_{py} = F_y Z_y = \frac{50(44.1)}{12} = 183.8 \text{ ft-kips}$$

$$M_r = M_{ry} = (F_y - F_r)S_y = (50 - 10)(29.1) = 1164 \text{ in.-kips} = 97.00 \text{ ft-kips}$$

Substituting into AISC Equation A-F1-3,

$$M_n = M_{ny} = 183.8 - (183.8 - 97.0)\left(\frac{9.9 - 9.192}{22.29 - 9.192}\right) = 179.1 \text{ ft-kips}$$

$$\phi_b M_{ny} = 0.90(179.1) = 161.2 \text{ ft-kips}$$

Interaction:

$$\frac{P_u}{\phi_c P_n} = \frac{300}{626} = 0.4792 > 0.2 \quad \therefore \quad \text{use AISC Equation H1-1a:}$$

$$\frac{P_u}{\phi P_n} + \frac{8}{9}\left(\frac{M_{ux}}{\phi_b M_{nx}} + \frac{M_{uy}}{\phi_b M_{ny}}\right) = 0.4792 + \frac{8}{9}\left(\frac{135}{357.8} + \frac{30}{161.2}\right)$$

$$= 0.980 < 1.0 \quad \text{(OK)}$$

Answer   The W12 × 65 is satisfactory.

6.7 ≡ MEMBERS IN UNBRACED FRAMES

In a beam-column whose ends are free to translate, the maximum primary moment resulting from the sidesway is almost always at one end. As was illustrated in Figure 6.5, the maximum secondary moment from the sidesway is *always* at the end. This means that the maximum primary and secondary moments are usually additive and there is no need for the factor C_m; in effect, $C_m = 1.0$. Even when there is a reduction, it will be slight and can be neglected. Consider the beam-column shown in Figure 6.14. Here the equal end moments are caused by the sidesway (from the horizontal load), and the axial load, which may partly result from loads not causing the sidesway, is carried along and amplifies the end moment.

The amplification factor for the sidesway moments, B_2, is given by two expressions. Either may be used; the choice is usually one of convenience:

$$B_2 = \frac{1}{1 - \sum P_u \left(\dfrac{\Delta_{oh}}{\sum HL} \right)} \qquad \text{(AISC Equation C1-4)}$$

FIGURE 6.14 ≡

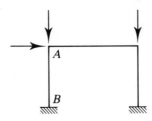

or

$$B_2 = \cfrac{1}{1 - \cfrac{\sum P_u}{\sum P_{e2}}}$$ (AISC Equation C1-5)

where

$\sum P_u$ = sum of factored loads on all columns in the story under consideration

Δ_{oh} = drift (sidesway displacement) of the story under consideration

$\sum H$ = sum of all horizontal forces causing Δ_{oh}

L = story height

$\sum P_{e2}$ = sum of the Euler loads for all columns in the story (when computing P_{e2}, use KL/r for the axis of bending and a value of K corresponding to the unbraced condition)

The summations for P_u and P_{e2} apply to all columns that are in the same story as the column under consideration. The rationale for using the summations is that B_2 applies to unbraced frames, and if sidesway is going to occur, all columns in the story must sway simultaneously. In most cases, the structure will be made up of plane frames, so $\sum P_u$ and $\sum P_{e2}$ are for the columns within a story of the frame, and the lateral loads H are the lateral loads acting on the frame at and above the story. Since Δ_{oh} is caused by $\sum H$, the ratio $\Delta_{oh}/\sum H$ can be based on either factored or unfactored loads. The alternate form of B_2 given by AISC Equation C1-5 is almost the same as the equation for B_1 except for the summations.

AISC Equations C1-4 and C1-5 were derived using two different methods, but in most cases they will give almost identical results.[41] In those cases where the two values of B_2 would be significantly different, the axial load term of the interaction formula will be dominant, and the end results will not differ by much. As mentioned earlier, the choice is based on convenience; it depends on which terms in the equations are readily available.

In situations where M_{nt} and $M_{\ell t}$ act at two different points on the member as in Figure 6.5, AISC Equation C1-1 will produce conservative results.

Figure 6.15 further illustrates the superposition concept. Part a shows an unbraced frame subject to both gravity and lateral loads. The moment M_{nt} in member AB is computed by using only the gravity loads. Because of symmetry, no bracing is needed to prevent sidesway from these loads. This moment is amplified with the factor B_1 to account for the $P\delta$ effect. $M_{\ell t}$, the moment corresponding to the sway (caused by the horizontal load H), will be amplified by B_2 to account for the $P\Delta$ effect.

In part b of Figure 6.15, the unbraced frame supports only a vertical load. Because of the unsymmetrical placement of this load, there will be a small amount of sidesway. The moment M_{nt} is computed by considering the frame to

FIGURE 6.15 ≡

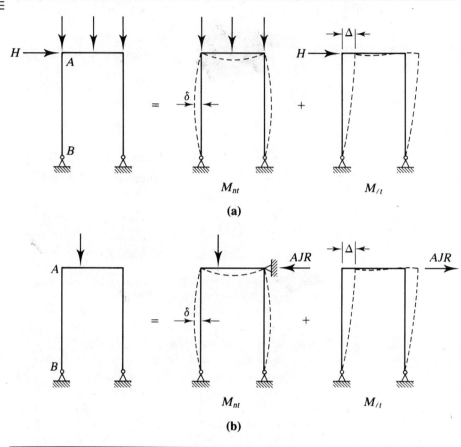

M_{nt} M_{lt}

(a)

M_{nt} M_{lt}

(b)

be braced — in this case, by a fictitious horizontal support and corresponding reaction called an artificial joint restraint (AJR). To compute the sidesway moment, the fictitious support is removed, and a force equal to the artificial joint restraint, but opposite in direction, is applied to the frame. In cases such as this one, the secondary moment $P\Delta$ will be very small, and M_{lt} can usually be neglected.

If both lateral loads and unsymmetrical gravity loads are present, the AJR force should be added to the actual lateral loads when computing M_{lt}.

≡ **EXAMPLE 6.7**

A W12 × 65 of A36 steel, 15 feet long, is to be investigated for use as a column in an unbraced frame. The axial load and end moments obtained from a first-order analysis of the gravity loads (dead load and live load) are shown in Figure 6.16a. The frame is symmetric, and the gravity loads are symmetrically placed.

FIGURE 6.16 ≡

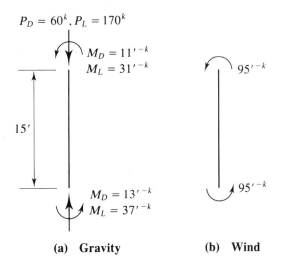

$P_D = 60^k, P_L = 170^k$

$M_D = 11'^{-k}$
$M_L = 31'^{-k}$

$95'^{-k}$

15'

$M_D = 13'^{-k}$
$M_L = 37'^{-k}$

$95'^{-k}$

(a) Gravity   (b) Wind

Figure 6.16b shows the wind load moments obtained from a first-order analysis. All bending moments are about the strong axis. Effective length factors are $K_x = 1.2$ for the sway case, $K_x = 1.0$ for the nonsway case, and $K_y = 1.0$. Determine whether this member is in compliance with the AISC Specification.

SOLUTION   All of the load combinations given in AISC A4.1 involve dead load, and except for the first one, all combinations also involve live load or wind load, or both. If load types not present in this example (E, L_r, S, and R) are omitted, the load conditions can be summarized as follows:

$1.4D$	(A4-1)
$1.2D + 1.6L$	(A4-2)
$1.2D + (0.5L \text{ or } 0.8W)$	(A4-3)
$1.2D + 1.3W + 0.5L$	(A4-4)
$1.2D + 0.5L$	(A4-5)
$0.9D \pm 1.3W$	(A4-6)

Since the dead load is less than eight times the live load, combination (A4-1) can be ruled out. Load combination (A4-4) will be more critical than (A4-3), so (A4-3) can be eliminated. Combination (A4-5) can be eliminated since it will be less critical than (A4-2). Finally, combination (A4-6) will not be as critical as (A4-4) and can be removed from consideration, leaving only two load combinations to be investigated:

$$1.2D + 1.6L \qquad \qquad (A4-2)$$

and

$$1.2D + 1.3W + 0.5L \qquad \qquad (A4-4)$$

Figure 6.17 shows the axial loads and bending moments as calculated for these two combinations.

Load condition (A4-2):

For this load condition, $P_u = 344.0$ kips, $M_{nt} = 74.80$ ft-kips, and $M_{\ell t} = 0$ (because of symmetry, there are no sidesway moments).

Bending:

$$C_m = 0.6 - 0.4\left(\frac{M_1}{M_2}\right) = 0.6 - 0.4\left(\frac{62.80}{74.80}\right) = 0.2642$$

For the axis of bending,

$$\frac{KL}{r} = \frac{K_x L}{r_x} = \frac{1.0(15)(12)}{5.28} = 34.09$$

(Since this is the no-sidesway case, K_x for the braced condition is used.)

FIGURE 6.17 ≡

344^k

$62.8'^{-k}$

$74.8'^{-k}$

(a) Load combination A4-2 (1.2D + 1.6L)

157^k

$28.7'^{-k}$

$123.5'^{-k}$

$34.1'^{-k}$

$123.5'^{-k}$

Gravity (1.2D + 0.5L)   Wind (1.3W)

(b) Load combination A4-4

$$P_{e1} = \frac{\pi^2 E A_g}{(KL/r)^2} = \frac{\pi^2 (29{,}000)(19.1)}{(34.09)^2} = 4704 \text{ kips}$$

Amplification factor for nonsway moments:

$$B_1 = \frac{C_m}{1 - \dfrac{P_u}{P_{e1}}} = \frac{0.2642}{1 - \dfrac{344.0}{4704}} = 0.285 < 1.0 \quad \therefore \quad \text{use } B_1 = 1.0.$$

$$M_u = B_1 M_{nt} + B_2 M_{\ell t} = 1.0(74.80) + 0 = 74.80 \text{ ft-kips}$$

From the Beam Design charts, with $L_b = 15$ ft,

$\phi_b M_n = 254.5$ ft-kips (for $C_b = 1.0$)
$\phi_b M_p = 261.4$ ft-kips

Figure 6.18 shows the bending moment diagram for the gravity load moments. (Since the computation of C_b is based on absolute values, a sign convention for the diagram is not necessary.)

$$C_b = \frac{12.5 M_{\max}}{2.5 M_{\max} + 3 M_A + 4 M_B + 3 M_C}$$

$$= \frac{12.5(74.80)}{2.5(74.80) + 3(28.40) + 4(6.002) + 3(40.40)} = 2.24$$

For $C_b = 2.24$,

$\phi_b M_n = 2.24(254.5) > \phi_b M_p = 261.4$ ft-kips
$\therefore$ use $\phi_b M_n = 261.4$ ft-kips.

Determine the critical axis for axial compressive strength:

$K_y L = 15$ ft
$\dfrac{K_x L}{r_x / r_y} = \dfrac{1.2(15)}{1.75} = 10.29$ ft < 15 ft $\therefore$ use $KL = 15$ ft.

FIGURE 6.18 ≣

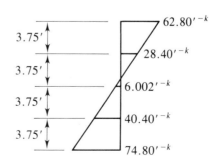

3.75'

3.75'

3.75'

3.75'

62.80$'^{-k}$

28.40$'^{-k}$

6.002$'^{-k}$

40.40$'^{-k}$

74.80$'^{-k}$

From the Column Load Tables, with $KL = 15$ ft, $\phi_c P_n = 485$ kips.

$$\frac{P_u}{\phi_c P_n} = \frac{344.0}{485} = 0.7093 > 0.2 \quad \therefore \quad \text{use AISC Equation H1-1a:}$$

$$\frac{P_u}{\phi_c P_n} + \frac{8}{9}\left(\frac{M_{ux}}{\phi_b M_{nx}} + \frac{M_{uy}}{\phi_b M_{ny}}\right) = 0.7093 + \frac{8}{9}\left(\frac{74.80}{261.4} + 0\right)$$

$$= 0.964 < 1.0 \quad \text{(OK)}$$

Load condition (A4-4):

For this load combination, $P_u = 157.0$ kips, $M_{nt} = 34.10$ ft-kips, and $M_{\ell t} = 123.5$ ft-kips.

For the braced condition,

$$C_m = 0.6 - 0.4\left(\frac{M_1}{M_2}\right) = 0.6 - 0.4\left(\frac{28.70}{34.10}\right) = 0.2633$$

$P_{e1} = 4704$ kips   (P_{e1} is independent of the loading condition)

$$B_1 = \frac{C_m}{1 - \dfrac{P_u}{P_{e1}}} = \frac{0.2633}{1 - \dfrac{157.0}{4704}} = 0.272 < 1.0 \quad \therefore \quad \text{use } B_1 = 1.0.$$

Amplification factor for sidesway moments:

Not enough data are available to accurately compute B_2 from either AISC Equation C1-4 or C1-5. If we assume that the ratio of applied axial load to Euler load capacity is the same for all of the columns in the story as for the column under consideration, we can use Equation C1-5:

$$B_2 = \frac{1}{1 - \dfrac{\sum P_u}{\sum P_{e2}}} \approx \frac{1}{1 - \dfrac{P_u}{P_{e2}}}$$

For P_{e2}, use K_x corresponding to the unbraced condition:

$$\frac{KL}{r} = \frac{K_x L}{r_x} = \frac{1.2(15)(12)}{5.28} = 40.91$$

$$P_{e2} = \frac{\pi^2 E A_g}{(KL/r)^2} = \frac{\pi^2(29,000)(19.1)}{(40.91)^2} = 3266 \text{ kips}$$

From AISC Equation C1-5,

$$B_2 \approx \frac{1}{1 - \dfrac{157.0}{3266}} = 1.050$$

The total amplified moment is

$$M_u = B_1 M_{nt} + B_2 M_{\ell t} = 1.0(34.10) + 1.050(123.5) = 163.8 \text{ ft-kips}$$

Although the moments M_{nt} and $M_{\ell t}$ are different, they are distributed similarly, and C_b will be roughly the same; at any rate, they are large enough so that $\phi_b M_p = 261.4$ ft-kips will be the design strength regardless of which moment is considered.

$$\frac{P_u}{\phi_c P_n} = \frac{157.0}{485.0} = 0.3237 > 0.2 \quad \therefore \quad \text{use AISC Equation H1-1a:}$$

$$\frac{P_u}{\phi_c P_n} + \frac{8}{9}\left(\frac{M_{ux}}{\phi_b M_{nx}} + \frac{M_{uy}}{\phi_b M_{ny}}\right) = 0.3237 + \frac{8}{9}\left(\frac{163.8}{261.4} + 0\right)$$

$$= 0.881 < 1.0 \quad \text{(OK)}$$

Answer   This member satisfies the AISC Specification requirements.

6.8 ≡ DESIGN OF BEAM-COLUMNS

Because of the many variables in the interaction formulas, the design of beam-columns is essentially a trial-and-error process. A trial shape is selected and then reviewed for satisfaction of the governing interaction formula. Obviously, the closer the trial shape is to the final selection, the better. A very efficient procedure for choosing a trial shape, originally developed for allowable stress design,[42] has been adapted for LRFD and is given in Part 3 of the *Manual*, "Column Design." The essence of this method is to "convert" the bending moments to equivalent axial loads. These fictitious loads are added to the actual loads, and a shape that will support the total load is selected from the Column Load Tables. This selection must then be investigated with AISC Equation H1-1a or H1-1b. The total effective axial load is given by

$$P_{u\,eq} = P_u + M_{ux}m + M_{uy}\,mu$$

where

P_u = actual factored axial load (kips)

M_{ux} = factored moment about the x-axis (ft-kips)

M_{uy} = factored moment about the y-axis (ft-kips)

m = a tabulated constant

u = a tabulated constant

The basis of this procedure can be examined by rewriting Equation 6.3 as follows. First, multiply both sides by $\phi_c P_n$:

$$P_u + \frac{\phi_c P_n M_{ux}}{\phi_b M_{nx}} + \frac{\phi_c P_n M_{uy}}{\phi_b M_{ny}} \leq \phi_c P_n$$

or

$$P_u + (M_{ux} \times \text{a constant}) + (M_{uy} \times \text{a constant}) \le \phi_c P_n$$

The right side of this inequality is the design strength of a member under consideration, and the left side can be thought of as the applied factored load to be resisted. Each of the three terms on the left must have units of force, so the constants "convert" the bending moments M_{ux} and M_{uy} to axial load components.

Average values of the constant m have been computed for different groups of W-shapes and are given in Table 3-2 in Part 3 of the *Manual*. Values of u are given in the Column Load Tables for each shape listed therein. To select a trial shape for a member with axial load and bending about both axes, proceed as follows:

1. Select a trial value of m based on the effective length KL. Let $u = 2$.
2. Compute an effective axial compressive load:

$$P_{u\,eq} = P_u + M_{ux}m + M_{uy}\,mu$$

   Use this load to select a shape from the Column Load Tables.

3. Use the value of u given in the Column Load Tables and an improved value of m from Table 3-2 to compute an improved value of $P_{u\,eq}$. Select another shape.

4. Repeat until there is no change in $P_{u\,eq}$.

EXAMPLE 6.8

A certain structural member in a braced frame must support a factored axial compressive load of 150 kips and factored end moments of 75 ft-kips about the strong axis and 30 ft-kips about the weak axis. Both of these moments occur at one end; the other end is pinned. The effective length with respect to each axis is 15 feet. There are no transverse loads on the member. Use A36 steel and select the lightest W-shape.

SOLUTION

The amplification factor B_1 can be estimated as 1.0 for purposes of making a trial selection. For each of the two axes,

$$M_{ux} = B_1 M_{ntx} \approx 1.0(75) = 75 \text{ ft-kips}$$
$$M_{uy} = B_1 M_{nty} \approx 1.0(30) = 30 \text{ ft-kips}$$

From Table 3-2, Part 3 of the *Manual*, $m = 1.75$ by interpolation.
Use an initial value of $u = 2.0$.

$$P_{u\,eq} = P_u + M_{ux}m + M_{uy}\,mu = 150 + 75(1.75) + 30(1.75)(2.0) = 386 \text{ kips}$$

Beginning with the smaller shapes in the Column Load Tables,

Try W8 × 67 ($\phi_c P_n = 412$ kips, $u = 2.03$).

$$m = 2.1$$
$$P_{u\,eq} = 150 + 75(2.1) + 30(2.1)(2.03) = 435 \text{ kips}$$

This value is larger than the design strength of 412 kips, so another shape must be tried.

Try W10 × 60 ($\phi_c P_n = 416$ kips, $u = 2.00$).

$$m = 1.85$$
$$P_{u\,eq} = 150 + 75(1.85) + 30(1.85)(2.00) = 400 \text{ kips} < 416 \text{ kips} \quad \text{(OK)}$$

The W10 × 60 is therefore a potential trial shape. Check the W12s and W14s for other possibilities.

Try W12 × 58 ($\phi_c P_n = 397$ kips, $u = 2.41$).

$$m = 1.55$$
$$P_{u\,eq} = 150 + 75(1.55) + 30(1.55)(2.41) = 378 \text{ kips} < 397 \text{ kips} \quad \text{(OK)}$$

A W12 × 58 is therefore a potential trial shape. The lightest W14 with a chance of working is a W14 × 61, and it is heavier than the W12 × 58.

∴ **Use a W12 × 58 as a trial shape.**

$$\frac{P_u}{\phi_c P_n} = \frac{150}{397} = 0.3778 > 0.2 \quad \therefore \quad \text{use AISC Equation H1-1a.}$$

Bending about the x-axis:

$$\frac{K_x L}{r_x} = \frac{15(12)}{5.28} = 34.09$$

$$P_{e1} = \frac{\pi^2 E A_g}{(KL/r)^2} = \frac{\pi^2 (29{,}000)(17.0)}{(34.09)^2} = 4187 \text{ kips}$$

$$C_m = 0.6 - 0.4(M_1/M_2) = 0.6 - 0.4(0/M_2) = 0.6 \text{ (for both axes)}$$

$$B_1 = \frac{C_m}{1 - P_u/P_{e1}} = \frac{0.6}{1 - (150/4187)} = 0.622 < 1 \quad \therefore \quad \text{use } B_1 = 1.0.$$

$$M_{ux} = B_1 M_{ntx} = 1.0(75) = 75 \text{ ft-kips}$$

Design strength: From the Beam Design curves, for $C_b = 1$ and $L_b = 15$ ft, $\phi_b M_n = 220$ ft-kips. The moment gradient for this beam-column is the same as for the one in Example 6.6 and shown in Figure 6.13; therefore, $C_b = 1.67$.

For $C_b = 1.67$, the design strength is

$$C_b \times 220 = 1.67(220) = 367 \text{ ft-kips}$$

This is greater than $\phi_b M_p = 233$ ft-kips, ∴ use $\phi_b M_n = 233$ ft-kips.

Bending about the y-axis:

$$\frac{K_y L}{r_y} = \frac{15(12)}{2.51} = 71.71$$

$$P_{e1} = \frac{\pi^2 E A_g}{(KL/r)^2} = \frac{\pi^2 (29{,}000)(17.0)}{(71.71)^2} = 946.2 \text{ kips}$$

$$B_1 = \frac{C_m}{1 - \dfrac{P_u}{P_{e1}}} = \frac{0.6}{1 - 150/946.2} = 0.713 < 1.0 \quad \text{Use } B_1 = 1.0.$$

$$M_{uy} = B_1 M_{nty} = 1.0(30) = 30 \text{ ft-kips}$$

A W12 × 58 is compact for any value of P_u (there is no footnote in the Column Load Tables), so the design strength is

$$\phi_b M_{ny} = \phi_b M_{py} = \phi_b Z_y F_y = 0.90(32.5)(36) = 1053 \text{ in.-kips}$$
$$= 87.75 \text{ ft-kips}$$

But $Z_y/S_y = 32.5/21.4 = 1.52 > 1.5$, which means that $\phi_b M_{ny}$ should be taken as

$$\phi_b(1.5 M_y) = \phi_b(1.5 F_y S_y) = 0.90(1.5)(36)(21.4) = 1040 \text{ in.-kips} = 86.67 \text{ ft-kips}$$

AISC Equation H1-1a:

$$\frac{P_u}{\phi_c P_n} + \frac{8}{9}\left(\frac{M_{ux}}{\phi_b M_{nx}} + \frac{M_{uy}}{\phi_b M_{ny}}\right) = 0.3778 + \frac{8}{9}\left(\frac{75}{233} + \frac{30}{86.67}\right)$$
$$= 0.972 < 1.0 \quad \text{(OK)}$$

Answer   Use a W12 × 58.

Although the method just presented for selecting a trial shape converges quickly, a somewhat simpler approach is suggested by Yura.[41] The equivalent axial load to be used is given by

$$P_{equiv} = P + \frac{2M_x}{d} + \frac{7.5 M_y}{b} \tag{6.5}$$

where

$\quad P$ = factored axial load

$\quad M_x$ = factored moment about x-axis

$\quad M_y$ = factored moment about y-axis

$\quad d$ = beam depth

$\quad b$ = beam width

All terms in Equation 6.5 must have consistent units.

≡ **EXAMPLE 6.9**

Use Yura's method to select a W12 shape for the beam-column of Example 6.8.

SOLUTION   From Equation 6.5, the equivalent axial load is

$$P_{equiv} = P + \frac{2M_x}{d} + \frac{7.5M_y}{b} = 150 + \frac{2(75 \times 12)}{12} + \frac{7.5(30 \times 12)}{12} = 525 \text{ kips}$$

where the width b is assumed to be 12 inches. From the Column Load Tables, **try a W12 × 72 ($\phi_c P_n = 537$ kips).**

Since the flange width is 12 inches as assumed, no iteration is required. For this particular problem, Yura's method produces a more conservative trial shape than the *Manual* method, although this may not always be true.

When the bending terms dominate (i.e., the member is more a beam than a column), Yura recommends that the axial load be converted to an equivalent bending moment about the x-axis. A trial shape can then be selected from the Beam Design charts in Part 3 of the *Manual*. This equivalent moment is given by

$$M_{equiv} = M_x + P\frac{d}{2}$$

Design of Unbraced Beam-Columns

The preliminary design of beam columns in braced frames has been illustrated. The amplification factor B_1 was assumed to be equal to 1.0 for purposes of selecting a trial shape; B_1 could then be evaluated for this trial shape. For beam-columns subject to sidesway, the amplification factor B_2 is based on several quantities that may not be known until all columns in the frame have been selected. If AISC Equation C1-4 is used for B_2, the sidesway deflection Δ_{oh} may not be available for a preliminary design. If AISC Equation C1-5 is used, $\sum P_{e2}$ may not be known. The following methods are suggested for evaluating B_2.

Method 1. Assume $B_2 = 1.0$. After a trial shape has been selected, compute B_2 from AISC Equation C1-5 by assuming that $\sum P_u / \sum P_{e2}$ is the same as P_u/P_{e2} for the member under consideration (as was done in Example 6.7).

Method 2. Use a predetermined limit for the *drift index* Δ_{oh}/L, the ratio of story drift to story height. The use of a maximum permissible drift index is a serviceability requirement that is similar to a limit on beam deflection. Although no building code or other standard used in the United States contains a limit on drift, values of $\frac{1}{500}$ to $\frac{1}{200}$ are commonly used.[43] Remember that Δ_{oh} is the drift caused by $\sum H$, so if the drift index is based on service loads, then the lateral loads H must also be service loads. Use of a prescribed drift index enables the designer to determine the final value of B_2 at the outset.

≣ EXAMPLE 6.10

Figure 6.19 shows a single-story unbraced frame subjected to dead load, roof live load, and wind. The service gravity loads are shown in part a, and the service wind load (including an uplift, or suction, on the roof) is shown in part b. Use A36 steel and select a W12 shape for the columns (vertical members). Design for a drift index of $\frac{1}{400}$ based on service wind load. Bending is about the strong axis, and each column is laterally braced at the top and bottom.

SOLUTION

This frame is statically indeterminate to the first degree. The analysis of statically indeterminate structures is not a prerequisite for the use of this book, so the frame will not be analyzed here. The results of an approximate analysis, which is adequate for the early stages of a structural design, are summarized in Figure 6.20. The axial load and end moment are given separately for dead load, live load, wind acting on the roof, and lateral wind load. All vertical loads are symmetrically placed and contribute only to the M_{nt} moments. The lateral load produces an $M_{\ell t}$ moment.

Load combinations involving dead load (D), roof live load (L_r), and wind (W) are as follows:

A4-2: $1.2D + 0.5L_r$

$$P_u = 1.2(14) + 0.5(26) = 29.8 \text{ kips}$$

$$M_{nt} = 1.2(50) + 0.5(94) = 107 \text{ ft-kips}$$

$$M_{\ell t} = 0$$

A4-3: $1.2D + 1.6L_r + 0.8W$

$$P_u = 1.2(14) + 1.6(26) + 0.8(-9 + 1) = 52.0 \text{ kips}$$

$$M_{nt} = 1.2(50) + 1.6(94) + 0.8(-32) = 184.8 \text{ ft-kips}$$

$$M_{\ell t} = 0.8(20) = 16.0 \text{ ft-kips}$$

A4-4: $1.2D + 1.3W + 0.5L_r$

$$P_u = 1.2(14) + 1.3(-9 + 1) + 0.5(26) = 19.4 \text{ kips}$$

$$M_{nt} = 1.2(50) + 1.3(-32) + 0.5(94) = 65.4 \text{ ft-kips}$$

$$M_{\ell t} = 1.3(20) = 26 \text{ ft-kips}$$

Load Combination A4-3 will obviously govern. It produces the largest axial load and the largest total moment. (Combination A4-4 could not control unless B_2, the amplification factor for $M_{\ell t}$, was unrealistically large.)

FIGURE 6.19 ≡

(a) Gravity loads   (b) Wind load

For purposes of selecting a trial shape, assume $B_1 = 1.0$. The value of B_2 can be computed with AISC Equation C1-4 and the design drift index:

$$B_2 = \cfrac{1}{1 - \sum P_u \left(\cfrac{\Delta_{oh}}{\sum HL} \right)} = \cfrac{1}{1 - \cfrac{\sum P_u}{\sum H} \left(\cfrac{\Delta_{oh}}{L} \right)}$$

$$= \cfrac{1}{1 - \cfrac{2(52.0)}{2.7} \left(\cfrac{1}{400} \right)} = 1.107$$

The unfactored horizontal load $\sum H$ is used because the drift index is based on the maximum drift caused by *service* loads.

$$M_u = B_1 M_{nt} + B_2 M_{\ell t} = 1.0(184.8) + 1.107(16) = 202.5 \text{ ft-kips}$$

Without knowing the frame member sizes, the alignment chart for the effective length factor cannot be used. Referring to Table C-C2.1 in the Commentary to the Specification, we see that case f corresponds most closely to the end conditions of the beam-column of this example, and $K_x = 2.0$ for the sidesway case.

FIGURE 6.20 ≡

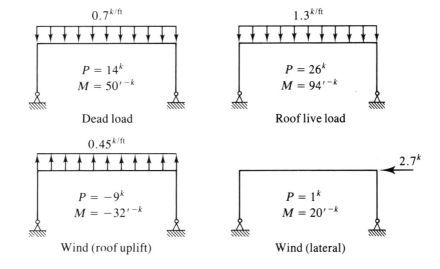

Dead load   Roof live load

Wind (roof uplift)   Wind (lateral)

For the braced condition, $K_x = 1.0$ will be used. Because the member is braced in the out-of-plane direction, $K_y = 1.0$ will be used.

A trial selection will be made using the method given in Part 3 of the *Manual*. From Table 3-2, the bending factor m is 1.55 for W12 shapes with $KL = 15$ feet.

$$P_{ueq} = P_u + M_{ux}m = 52.0 + 202.5(1.55) = 366 \text{ kips}$$

For $KL = K_yL = 15$ ft, a W12 × 58 has a design strength of $\phi_c P_n = 397$ kips. For the x-axis,

$$\frac{K_xL}{r_x/r_y} = \frac{2.0(15)}{2.10} = 14.3 \text{ ft} < 15 \text{ ft} \quad \therefore \quad KL = 15 \text{ ft controls}$$

Try a W12 × 58.

For the braced condition,

$$\frac{K_xL}{r_x} = \frac{1.0(15)(12)}{5.28} = 34.09$$

$$P_{e1x} = \frac{\pi^2 EA_g}{(K_xL/r_x)^2} = \frac{\pi^2(29,000)(17.0)}{(34.09)^2} = 4187 \text{ kips}$$

From AISC Equation C1-2,

$$B_1 = \frac{C_m}{1 - \dfrac{P_u}{P_{e1}}} = \frac{0.6}{1 - \dfrac{52.0}{4187}} = 0.608 < 1.0 \quad \therefore \quad \text{use } B_1 = 1.0.$$

Note that $B_1 = 1.0$ is the value originally assumed, and since B_2 will not change, the previously computed value of $M_u = 202.5$ ft-kips is unchanged.

From the Beam Design charts in Part 4 of the *Manual* with $L_b = 15$ ft, the design moment for a W12 × 58 with $C_b = 1.0$ is

$$\phi_b M_n = 220 \text{ ft-kips}$$

For a bending moment that varies linearly from zero at one end to a maximum at the other, the value of C_b is 1.67 (see Example 6.6). The corrected value of the design moment is therefore

$$\phi_b M_n = 1.67(220) = 367 \text{ ft-kips}$$

This moment, however, is larger than the plastic moment capacity of $\phi_b M_p =$

233 ft-kips, which can also be read from the charts. This means that the design strength must be limited to

$$\phi_b M_n = \phi_b M_p = 233 \text{ ft-kips}$$

From the Column Load Tables, we see that this shape is compact for any axial load, so local buckling limit states need not be investigated.

Determine the appropriate interaction formula:

$$\frac{P_u}{\phi_c P_n} = \frac{52.0}{397} = 0.1310 < 0.2 \quad \therefore \quad \text{use AISC Equation H1-1b.}$$

$$\frac{P_u}{2\phi_c P_n} + \left(\frac{M_{ux}}{\phi_b M_{nx}} + \frac{M_{uy}}{\phi_b M_{ny}} \right) = \frac{0.1310}{2} + \left(\frac{202.5}{233} + 0 \right) = 0.935 < 1.0 \quad \text{(OK)}$$

Answer   Use a W12 × 58.

In the preceding example, limiting the drift index was a design criterion, and there was no freedom to choose the method of computing the amplification factor B_2. If there had been no imposed drift index, a different value of B_2 could have been computed from AISC Equation C1-5 as follows:

$$\frac{K_x L}{r_x} = \frac{2.0(15)(12)}{5.28} = 68.18$$

$$P_{e2x} = \frac{\pi^2 E A_g}{(K_x L/r_x)^2} = \frac{\pi^2(29{,}000)(17.0)}{(68.18)^2} = 1047 \text{ kips}$$

$$B_2 = \frac{1}{1 - \dfrac{\sum P_u}{\sum P_{e2}}} = \frac{1}{1 - \dfrac{2(52.0)}{2(1047)}} = 1.052$$

6.9 ≡ TRUSSES WITH TOP CHORD LOADS BETWEEN JOINTS

If a compression member in a truss must support transverse loads between its ends, it will be subjected to bending as well as axial compression and is therefore a beam-column. This can occur in the top chord of a roof truss with purlins located between the joints. The truss can be modeled as an assembly of continuous chord members and pin-connected web members, and the axial loads and bending moments can be found by using a method of structural analysis such as the stiffness method. The magnitude of the moments involved, however, does

FIGURE 6.21 ≡

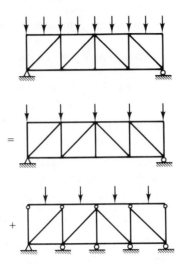

not usually warrant this degree of sophistication, and in most cases an approximate analysis will suffice. The following procedure is recommended:

1. Consider each member of the top chord as a fixed-end beam. Use the fixed-end moment as the maximum bending moment in the member. The top chord is actually one continuous member rather than a series of individual pin-connected members, so this approximation is more accurate than treating each member as a simple beam.

2. Add the reactions from this fixed-end beam to the actual joint loads to obtain total joint loads.

3. Analyze the truss with these total joint loads acting. The resulting axial load in the top chord member is the axial compressive load to be used in the design.

Alternatively, the bending moments and beam reactions can be found by treating the top chord as a continuous beam with supports at the panel points. This method is represented schematically in Figure 6.21.

≡ **EXAMPLE 6.11**

The parallel-chord roof truss in Figure 6.22 supports purlins at the top chord panel points and midway between the panel points. The *factored* loads transmitted by the purlins are as shown. Design the top chord. Use A36 steel and select a structural tee cut from a W-shape.

SOLUTION   The bending moments and panel point forces caused by the loads acting between the joints will be found by treating each top chord member as a fixed-

FIGURE 6.22 ≣

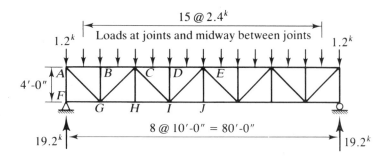

end beam. From Part 4 of the *Manual*, "Beam and Girder Design," the fixed-end moment for each top chord member is

$$M = M_{nt} = \frac{PL}{8} = \frac{2.4(10)}{8} = 3.0 \text{ ft-kips}$$

These end moments and the corresponding reactions are shown in Figure 6.23a. When the reactions are added to the loads that are directly applied to the joints, the loading condition shown in Figure 6.23b is obtained. The maximum axial compressive force will occur in member *DE* (and in the adjacent member, to the

FIGURE 6.23 ≣

(a)

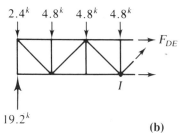

(b)

right of the center of the span) and can be found by considering the equilibrium of a free body of the portion of the truss to the left of section a–a:

$$\sum M_I = (19.2 - 2.4)(30) - 4.8(10 + 20) + F_{DE}(4) = 0$$

$$F_{DE} = -90 \text{ kips} \quad \text{(compression)}$$

Design for an axial load of 90 kips and a bending moment of 3.0 ft-kips.

Table 3-2 in Part 3 of the *Manual* does not provide bending factors for structural tees. Although Yura's method[41] was probably developed for I- and H-shaped members, it will be tried here. An examination of the Column Load Tables shows that a small shape will be needed, since the axial load is small and the moment is small relative to the axial load. If a 6-inch-deep tee is used,

$$P_{equiv} = P + \frac{2M_x}{d} + \frac{7.5M_y}{b} = 90 + \frac{2(3)(12)}{6} + 0 = 102 \text{ kips}$$

From the Column Load Tables, with $K_x L = 10$ ft and $K_y L = 5$ ft, **try a WT6 × 17.5** ($\phi_c P_n = 124$ kips).

Bending is about the x-axis, and the member is braced against sidesway:

$$M_{nt} = 3.0 \text{ ft-kips}, \quad M_{\ell t} = 0$$

Because there is a transverse load on the member, and the ends are restrained, $C_m = 0.85$ (the Commentary approach will not be used here).

Compute B_1:

$$\frac{KL}{r} = \frac{K_x L}{r_x} = \frac{10(12)}{1.76} = 68.18$$

$$P_{e1} = \frac{\pi^2 E A_g}{(KL/r)^2} = \frac{\pi^2(29,000)(5.17)}{(68.18)^2} = 318.3 \text{ kips}$$

$$B_1 = \frac{C_m}{1 - \dfrac{P_u}{P_{e1}}} = \frac{0.85}{1 - \dfrac{90}{318.3}} = 1.185$$

Amplified moment:

$$M_u = B_1 M_{nt} + B_2 M_{\ell t} = 1.185(3.0) + 0 = 3.555 \text{ ft-kips}$$

The nominal moment strength of a structural tee will be based on local buckling if the cross section is classified as slender; otherwise it will be based on lateral-torsional buckling (see AISC F1.2c and Section 5.14 of this book). For the flange,

$$\lambda = \frac{b_f}{2t_f} = \frac{6.560}{2(0.520)} = 6.308$$

$$\lambda_r = \frac{95}{\sqrt{F_y}} = \frac{95}{\sqrt{36}} = 15.83 > \lambda$$

For the web,

$$\lambda = \frac{d}{t_w} = \frac{6.25}{0.300} = 20.83$$

$$\lambda_r = \frac{127}{\sqrt{F_y}} = \frac{127}{\sqrt{36}} = 21.17 > \lambda$$

Since $\lambda < \lambda_r$ for both the flange and the web, the shape is not slender, and lateral-torsional buckling controls. From AISC Equation F1-15,

$$M_n = M_{cr} = \frac{\pi\sqrt{EI_y GJ}}{L_b}\left[B + \sqrt{1 + B^2}\right]$$

$$\leq 1.5M_y \text{ for stems in tension} \qquad \text{(AISC Equation F1-15)}$$

$$\leq 1.0M_y \text{ for stems in compression}$$

From AISC Equation F1-16,

$$B = \pm 2.3(d/L_b)\sqrt{I_y/J} = \pm 2.3\left[\frac{6.25}{5(12)}\right]\sqrt{\frac{12.2}{0.369}} = \pm 1.378$$

and the nominal strength from AISC Equation F1-15 is

$$M_n = \frac{\pi\sqrt{29{,}000(12.2)(11{,}200)(0.369)}}{5(12)}\left[\pm 1.378 + \sqrt{1 + (1.378)^2}\right]$$

$$= 2002[\pm 1.378 + 1.703] = 6168 \text{ in.-kips or } 650.5 \text{ in.-kips}$$

The positive value of B corresponds to tension in the stem of the tee, and the negative sign is used to obtain the strength when the stem is in compression. For the loading in this problem, the maximum moment occurs at both the fixed ends and at midspan, thus the strength is controlled by compression in the stem, and

$$M_n = 650.5 \text{ in.-kips} = 54.21 \text{ ft-kips}$$

Subject to a maximum of

$$1.0M_y = 1.0F_y S_x = \frac{1.0(36)(3.23)}{12} = 9.690 \text{ ft-kips} < 54.21 \text{ ft-kips}$$

$$\therefore \text{ use } M_n = 9.690 \text{ ft-kips}$$

$$\phi_b M_n = 0.90(9.690) = 8.721 \text{ ft-kips}$$

Determine which interaction formula to use:

$$\frac{P_u}{\phi_c P_n} = \frac{90}{124} = 0.7258 > 0.2 \quad \therefore \quad \text{use AISC Equation H1-1a.}$$

$$\frac{P_u}{\phi_c P_n} + \frac{8}{9}\left(\frac{M_{ux}}{\phi_b M_{nx}} + \frac{M_{uy}}{\phi_b M_{ny}}\right) = 0.7258 + \frac{8}{9}\left(\frac{3.555}{8.721} + 0\right)$$

$$= 0.7258 + 0.3623 = 1.09 > 1.0 \quad \text{(N.G.)}$$

The bending moment is small in this example, but so is the bending strength, and the bending term of the interaction formula is not insignificant. In searching for a better shape, we must be sure that the bending strength, as well as the axial compressive strength, is larger. The next shape in the Column Load Tables is a WT6 × 20, with an axial compressive design strength of 133 kips. An inspection of the Dimensions and Properties tables shows that we are entering a group of shapes in which the x-axis is the weak axis. This means that bending is now about the weak axis, and there is no lateral-torsional buckling limit state. Further, if the shape is nonslender, the nominal strength will be based on yielding and is equal to the plastic moment capacity, subject to a limit of $1.5M_y$.

∴   **Try a WT6 × 20** ($\phi_c P_n = 133$ kips).

Compute B_1:

$$\frac{KL}{r} = \frac{K_x L}{r_x} = \frac{10(12)}{1.57} = 76.43$$

$$P_{e1} = \frac{\pi^2 E A_g}{(KL/r)^2} = \frac{\pi^2(29{,}000)(5.89)}{(76.43)^2} = 288.6 \text{ kips}$$

$$B_1 = \frac{C_m}{1 - \dfrac{P_u}{P_{e1}}} = \frac{0.85}{1 - \dfrac{90}{288.6}} = 1.235$$

Amplified moment:

$$M_u = B_1 M_{nt} = 1.235(3.0) = 3.705 \text{ ft-kips}$$

Check the slenderness parameters:

Flange:

$$\lambda = \frac{b_f}{2t_f} = \frac{8.005}{2(0.515)} = 7.772$$

$$\lambda_r = \frac{95}{\sqrt{F_y}} = \frac{95}{\sqrt{36}} = 15.8 > \lambda$$

Web:

$$\lambda = \frac{d}{t_w} = \frac{5.970}{0.295} = 20.2 < \lambda_r = 21.17$$

Because bending is about the weak axis,

$$M_n = M_p = Z_x F_y = \frac{5.30(36)}{12} = 15.9 \text{ ft-kips}$$

subject to a maximum of

$$1.5M_y = 1.5F_y S_x = \frac{1.5(36)(2.95)}{12} = 13.28 \text{ ft-kips}$$

Since $M_p > 1.5 M_y$,

$$\phi_b M_n = \phi_b(1.5 M_y) = 0.90(13.28) = 11.95 \text{ ft-kips}$$

Determine which interaction formula to use:

$$\frac{P_u}{\phi_c P_n} = \frac{90}{133} = 0.6767 > 0.2 \quad \therefore \quad \text{use AISC Equation H1-1a.}$$

$$\frac{P_u}{\phi_c P_n} + \frac{8}{9}\left(\frac{M_{ux}}{\phi_b M_{nx}} + \frac{M_{uy}}{\phi_b M_{ny}}\right) = 0.6767 + \frac{8}{9}\left(\frac{3.705}{11.95} + 0\right)$$

$$= 0.952 < 1.0 \quad \text{(OK)}$$

Answer Use WT6 × 20.

PROBLEMS

6.1 A simply supported beam is subjected to the factored end couples (bending is about the strong axis) and the factored axial load shown in Figure P6.1. Lateral support is provided only at the ends. Neglect the weight of the beam and investigate this member as a beam-column. *Use the AISC Specification, but do not use a moment amplification factor for this analysis.* A36 steel is used.

$$30^k \; — \; \text{W6} \times 25 \; — \; 30^k$$

$$22'^{-k} \quad | \longleftarrow 10' \longrightarrow | \quad 15'^{-k}$$

FIGURE P6.1

6.2 The beam-column in Figure P6.2 is braced against sidesway, and lateral support of the compression flange is provided at the ends. Bending is about the

FIGURE P6.2

major axis, and A36 steel is used. The given loads are service loads, consisting of 25% dead load and 75% live load. Is this member adequate? *Use the AISC Specification, but do not use a moment amplification factor for this analysis.*

6.3   The member in Figure P6.3 has lateral support only at points A, B, and C. Bending is about the strong axis. Is the member adequate for the loading shown? Service loads are shown (the uniform load includes the weight of the member). A36 steel is used. *Use the AISC Specification, but do not use a moment amplification factor for this analysis.*

FIGURE P6.3

6.4   Rework Problem 6.1 to incorporate the moment amplification factor given by Expression 6.4.

6.5   Rework Problem 6.2 to incorporate the moment amplification factor given by Expression 6.4.

6.6   Rework Problem 6.3 to incorporate the moment amplification factor given by Expression 6.4.

6.7   Compute the moment amplification factor B_1 for the member shown in Figure P6.7. The axial force and bending moments are based on factored loads, and bending is about the x-axis.

FIGURE P6.7

6.8   Compute the moment amplification factor B_1 for the member shown in Figure P6.8. The axial force and bending moments are based on factored loads. Bending is about the x-axis.

560^k

$188'^{-k}$

$11'$

W14 × 132

$220'^{-k}$

FIGURE P6.8

6.9   The beam-column shown in Figure P6.9 is subjected to a factored axial compressive load of 25 kips and factored bending moments of 125 ft-kips about the x-axis and 20 ft-kips about the y-axis, both acting at one end.

   a. Determine the moment amplification factor B_1 for bending about the x-axis.

   b. Determine the moment amplification factor B_1 for bending about the y-axis.

25^k

$20'$

W10 × 68

FIGURE P6.9

6.10 The member in Figure P6.10 is fixed at both ends and subjected to the factored loads shown. Bending is about the x-axis.

   a. Compute the moment amplification factor B_1. Use a value of C_m based on the provisions of AISC C1, *not the Commentary.*

   b. Compute B_1 using a value of C_m based on the refinement given in the Commentary.

FIGURE P6.10

6.11 The loads in Figure P6.11 are factored. Compute the moment amplification factor B_1 in the following ways:

   a. Use a value of C_m based on the provisions of AISC C1, *not the Commentary.*

   b. Use a value of C_m based on the refinement given in the Commentary.

FIGURE P6.11

6.12 A horizontal member is subjected to the loading shown in Figure P6.12. The uniform load includes the weight of the member. Bending is about the major

FIGURE P6.12

axis, and lateral support of the compression flange is provided at the ends. If A36 steel is used, does this member satisfy the provisions of the AISC Specification? The loads are 50% dead load and 50% live load.

6.13   Investigate the member shown in Figure P6.13 for compliance with the AISC Specification. At the top, couples of 100 ft-kips about the strong axis and 15 ft-kips about the weak axis are applied. Both of these couples result from live load, and the axial load is 50% dead load and 50% live load. ASTM A242 steel is used. The member is braced at the ends only. $K_x L = K_y L = 20$ ft.

FIGURE P6.13

6.14   Check the beam column in Figure P6.14 for compliance with the AISC Specification. The end moments shown are applied about the major axis. The member

FIGURE P6.14

is braced against sidesway in both directions at each end and in the weak direction at midheight. A36 steel is used. The given moments and axial force are due to service loads consisting of 50% live load and 50% dead load.

6.15   The member shown in Figure P6.15 is laterally supported at its ends, and sidesway is prevented. Bending is about the major axis. Does this member satisfy the provisions of the AISC Specification? The loads are 50% live load and 50% dead load.

FIGURE P6.15

6.16   The member in Figure P6.16 is laterally supported only at its ends. If A572 Grade 50 steel is used, does the member satisfy the provisions of the AISC Specification? The loads are 50% live load and 50% dead load.

FIGURE P6.16

6.17   A W14 × 109 of A36 steel is to be investigated for use as a beam-column in an unbraced frame. The length is 14 ft. First-order analyses of the frame were performed for both the sway and the nonsway cases. The factored loads and moments corresponding to one of the load combinations to be investigated are given for this member in the following table:

Type of analysis	P_u (kips)	M_{top} (ft-kips)	M_{bot} (ft-kips)
Nonsway	200	45	25
Sway	—	40	95

The following values are also available from the results of a preliminary design:

$$\sum P_{e2} = 40,000 \text{ kps}, \quad \sum P_u = 6000 \text{ kips}$$

Use $K_x = 1.0$ (nonsway case), $K_x = 1.7$ (sway case), and $K_y = 1.0$. Determine whether this member satisfies the provisions of the AISC Specification.

6.18 A W14 × 82 of A36 steel, 16 feet long, is to be investigated for use as a column in an unbraced frame. The axial load and end moments obtained from a first-order analysis of the gravity loads (dead load and live load) are shown in Figure P6.18a. The frame is symmetric, and the gravity loads are symmetrically placed. Figure P6.18b shows the wind load moments obtained from a first-order analysis. All loads and moments are based on service loads, and all bending moments are about the strong axis. Effective length factors are $K_x = 0.85$ (braced case), $K_x = 1.2$ (unbraced case), and $K_y = 1.0$. Determine whether this member is in compliance with the AISC Specification.

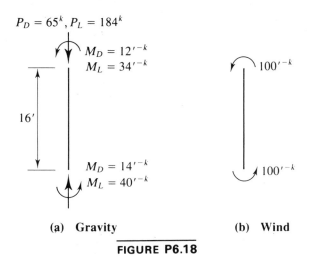

$P_D = 65^k, P_L = 184^k$

$M_D = 12'^{-k}$
$M_L = 34'^{-k}$

$100'^{-k}$

16'

$M_D = 14'^{-k}$
$M_L = 40'^{-k}$

$100'^{-k}$

(a) Gravity   (b) Wind

FIGURE P6.18

In each of Problems 6.19 through 6.21, use A36 steel and select the lightest W-shape for the given braced members. The axial load and end moments are based on factored loads. Bending is about the x-axis, and lateral support is provided only at the ends.

6.19

FIGURE P6.19

6.20

FIGURE P6.20

6.21

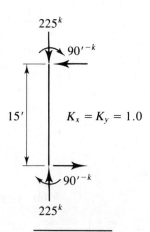

$K_x = K_y = 1.0$

FIGURE P6.21

6.22　Use Grade 50 steel and select the lightest W-shape for the member in Figure P6.22. The axial load and end moment are based on service loads consisting of 35% dead load and 65% live load. Lateral support is provided only at the ends.

FIGURE P6.22

6.23   The beam-column in Figure P6.23 is subjected to the factored axial load and
end moments shown. Select the lightest W-shape for each of the following
grades of steel: (a) A36, (b) $F_y = 50$ ksi, and (c) $F_y = 60$ ksi. In each case, lateral
support is provided only at the ends.

FIGURE P6.23

6.24   Use A36 steel and select the lightest W-shape for the member and loading of
Problem 6.16.

6.25   The uniform load in Figure P6.25 is a superimposed service live load. The axial load is a service load, half dead load and half live load. Use A36 steel and select the lightest W-shape. Lateral support is provided only at the ends.

FIGURE P6.25

6.26   The beam-column shown in Figure P6.26 is laterally braced at A, B, and C. Use A36 steel and select a W10 shape. All of the loads shown are factored.

FIGURE P6.26

6.27   Use A36 steel and select the lightest W12 shape. The loads in Figure P6.27 are service loads; the uniform load is a live load, and the axial load consists of 30 kips dead load and 70 kips live load.

FIGURE P6.27

6.28   Use A36 steel and select the lightest W14 shape to be used as a beam-column in an *unbraced* frame. The member length is 16 ft, and the effective length factors are $K_x = 1.0$ (nonsway), $K_x = 2.0$ (sway), and $K_y = 1.0$. Factored loads and moments based on first-order analyses are as follows: $P_u = 60$ kips, $M_{nt} = 200$ ft-kips, and $M_{\ell t} = 20$ ft-kips. Use $C_m = 0.6$.

6.29   The single-story unbraced frame shown in Figure P6.29 is subjected to dead load, roof live load, and wind. The results of an approximate analysis are summarized in the figure. The axial load and end moment are given separately for dead load, roof live load, wind uplift on the roof, and lateral wind load. All vertical loads are symmetrically placed and contribute only to the M_{nt} moments. The lateral load produces $M_{\ell t}$ moments. Use A36 steel and select a W10 shape for the columns (vertical members). Design for a drift index of $\frac{1}{400}$ based on service wind load. Bending is about the strong axis, and each column is laterally braced at the top and bottom.

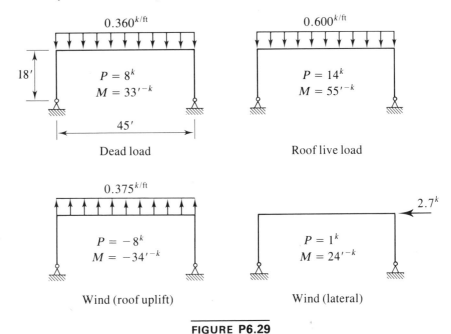

$0.360^{k/ft}$

$18'$

$P = 8^k$
$M = 33'^{-k}$

$45'$

Dead load

$0.600^{k/ft}$

$P = 14^k$
$M = 55'^{-k}$

Roof live load

$0.375^{k/ft}$

$P = -8^k$
$M = -34'^{-k}$

Wind (roof uplift)

2.7^k

$P = 1^k$
$M = 24'^{-k}$

Wind (lateral)

FIGURE P6.29

6.30 Use A36 steel and select a structural tee-shape for the top chord of the truss shown in Figure P6.30. The trusses are spaced at 20 ft and are subjected to the following loads:

Purlins: M6 × 4.4, located at joints and midway between joints
Snow: 20 lb per square ft of horizontal projection
Metal deck: 2 psf
Roofing: 4 psf
Insulation: 3 psf

$4'$

$8 \ @ \ 5'\text{-}0'' = 40'\text{-}0''$

FIGURE P6.30

6.31   Use A36 steel and select a structural tee shape for the top chord of the truss in Figure P6.31. This is the truss of Example 3.12, and the pertinent data are summarized as follows:

> The trusses are spaced at 20 ft on centers and support W6 × 12 purlins at the joints and midway between the joints.

Metal deck: 2 psf

Built-up roof: 5 psf

Snow: 18 psf of horizontal projection of roof surface

FIGURE P6.31

6.32   Write a computer program to analyze beam-columns in braced frames. The program should be restricted to the special case of no transverse loads and compact shapes with $L_b \le L_p$.

     Input: F_y (ksi), L (ft), L_p (ft), r_x (in.), r_y (in.), area (in.^{2}), Z_x (in.^{3}), K_x, K_y, P_u (kips), M_1 (ft-kips), and M_2 (ft-kips). (M_1 and M_2 are to be based on factored loads.)

     Output: Same as input, plus:

          1. A statement of the limitations of the program.
          2. The sum of the ratios from AISC Equation H1-1a or H1-1b.

   Restrictions and special instructions:

     1. No transverse loads on the member.

     2. Bending is about the x-axis.

     3. Use $M_n = M_p$;   that is, the shape must be compact, and the member length must not exceed L_p.

     4. If the program is written in FORTRAN, use a subroutine subprogram to compute the design compressive strength. Use four arguments: F_y, the two slenderness ratios, and $\phi_c P_n$.

6.33   Write a computer program to generate points for the construction of a curve of design compressive strength as a function of design moment for a given beam-column. (Such a curve is called an *interaction diagram.*)

        Input: Name of shape (e.g., W18 × 50), L (ft), F_y (ksi), L_p (ft), r_x (in.), r_y (in.), A (in.^{2}), Z_x (in.^{3}), K_x, K_y, and C_{mx}.

        Output: Same as input, plus a table of values of $\phi_c P_n$ in kips and $\phi_b M_n$ in ft-kips. Values of $\phi_c P_n$ should range from 0 to AF_y in increments of $AF_y/20$.

Restrictions:

    1. Bending is about the x-axis.

    2. The shape must be compact, and the member length must not exceed L_p. Use $M_n = M_p$.

7 SIMPLE CONNECTIONS

7.1 ≡ INTRODUCTION

Connections of structural steel members are of critical importance. An inadequate connection, which can be the "weak link" in a structure, has been the cause of numerous failures. Failure of structural *members* is rare; most structural failures are the result of poorly designed or detailed connections. The problem is compounded by the confusion that sometimes exists regarding responsibility for the design of connections. In many cases, the connections are not designed by the same engineer who designs the rest of the structure, but by someone associated with the steel fabricator who furnishes the material for the project. The structural engineer responsible for the production of the design drawings, however, is responsible for the complete design, including the connections. It is therefore incumbent upon the engineer to be proficient in connection design, if only for the purpose of validating a connection designed by someone else.

Modern steel structures are connected by welding or bolting (either high-strength or "common" bolts) or by a combination of both. Until fairly recently, connections were either welded or riveted. In 1947, the Research Council of Riveted and Bolted Structural Joints was formed, and its first specification was issued in 1951. This document authorized the substitution of high-strength bolts for rivets on a one-for-one basis. Since that time, high-strength bolting has rapidly gained in popularity, and today the widespread use of high-strength bolts has rendered the rivet obsolete in civil engineering structures. There are several reasons for this change. Two relatively unskilled workers can install high-strength bolts, whereas four skilled workers are required for riveting. In addition, the riveting operation is noisy and somewhat dangerous because of the practice of tossing the heated rivet from the point of heating to the point of installation. Riveted construction is still covered by the AISC Specification and the *Manual of Steel Construction*, however; many existing structures contain riveted joints, and an understanding of their behavior is essential for the strength evaluation and rehabilitation of older structures. The design and analysis of riveted connections is essentially the same as for connections with common bolts; only the material properties are different.

Welding has several advantages over bolting. A welded connection is often simpler in concept and requires few, if any, holes (sometimes erection bolts may be required to hold the members in position for the welding operation). Connections that are extremely complex with fasteners can become very simple

FIGURE 7.1 ☰

Riveted   Welded

when welds are used. A case in point is the plate girder shown in Figure 7.1. Before welding became widely used, this type of built-up shape was fabricated by riveting. To attach the flange plates to the web plate, angle shapes were used to transfer load between the two elements. If cover plates were added, the finished product became even more complicated. The welded version, however, is elegant in its simplicity. On the negative side, skilled workers are required for welding, and inspection can be difficult and costly. This last disadvantage can be partially overcome by using shop welding instead of field welding whenever possible. Quality welding can be more easily ensured under the controlled conditons of a fabricating shop. When a connection is made with a combination of welds and bolts, welding can be done in the shop and bolting in the field. In the single-plate beam-to-column connection in Figure 7.2, the plate is shop welded to the column flange and field bolted to the beam web.

In considering the behavior of different types of connections, it is convenient to categorize them according to the type of loading. The tension member lap splice shown in Figure 7.3(a) subjects the fasteners to forces that tend to shear the shank of the fastener. Similarly, the weld in (b) must resist shearing forces. The connection of a bracket to a column flange as in (c), whether by fasteners or welds, subjects the connection to shear when loaded as shown. The hanger connection in (d) puts the fasteners in tension, and the connection shown in (e) produces both shear and tension in the upper row of fasteners. The

FIGURE 7.2 ☰

FIGURE 7.3 ≡

(a)

(b)

Section

(c)

(d)

(e)

strength of a fastener depends on whether it is subjected to shear or tension, or both. Welds are weak in shear and are usually assumed to fail in shear, regardless of the direction of loading.

Once the force per fastener or force per unit length of weld has been determined, it is a simple matter to evaluate the adequacy of the connection. This determination is the basis for the two major categories of connections. If the line of action of the resultant force to be resisted passes through the center of gravity of the connection, each part of the connection is assumed to resist an equal share of the load, and the connection will be called a simple connection. In such connections, illustrated in Figure 7.3a and b, each fastener or each unit length of weld will resist an equal amount of force.* The load capacity of

---

* There is actually a small eccentricity in each of these two connections, but it is usually neglected.

the connection can then be found by multiplying the capacity of each fastener or inch of weld by the total number of fasteners or the total length of weld. The present chapter will be devoted to simple connections. Eccentrically loaded connections, to be covered in the following chapter, are those in which the line of action of the load does not act through the center of gravity of the connection. The connections shown in Figure 7.3c and e are of this type. In these cases, the load is not resisted equally by each fastener or each segment of weld, and the determination of the distribution of the load is the complicating factor in the design of this type of connection.

The AISC Specification deals with connections in Chapter J, "Connections, Joints and Fasteners." Bolts, rivets, and welds are covered, although rivets will not be considered in this textbook.

7.2 ≡ BOLTED SHEAR CONNECTIONS: FAILURE MODES

Before considering the strength of specific grades of bolts, the various modes of failure that are possible in connections with fasteners subjected to shear will be examined. There are two broad categories of failure: failure of the fastener and failure of the parts being connected. Consider the lap joint shown in Figure 7.4a. Failure of the fastener can be assumed to occur as shown. The average shearing stress in this case will be

$$f_v = \frac{P}{A} = \frac{P}{\pi d^2/4}$$

where P is the load acting on an individual fastener, A is the cross-sectional area of the fastener, and d is its diameter. The load can then be written as

$$P = f_v A$$

FIGURE 7.4 ≡

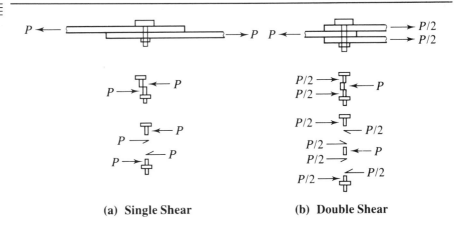

(a) Single Shear   (b) Double Shear

Although the loading in this case is not perfectly concentric, the eccentricity is small and can be neglected. The connection in Figure 7.4b is similar, but an analysis of free-body diagrams of portions of the fastener shank shows that each cross-sectional area is subjected to half of the total load, or, equivalently, two cross sections are effective in resisting the total load. From either point of view, the load is $P = 2f_v A$. This is a case of *double shear*. The first case, with only one shear plane, is called *single shear*. The addition of more thicknesses of material to the connection will increase the number of shear planes and further reduce the load on each plane, but this will also increase the length of the fastener and could subject it to bending.

Other modes of failure in shear connections involve failure of the parts being connected and fall into two general categories:

1. **Failure resulting from excessive tension, shear, or bending in the parts being connected.** If a tension member is being connected, tension on both the gross area and effective net area must be investigated. Depending on the configuration of the connection, block shear might also need to be considered. Block shear must also be examined in beam-to-column connections in which the top flange of the beam is coped. (Block shear has been covered in Chapters 3 and 5 of this book and in AISC J4.3.) Depending on the type of connection and loading, connection fittings such as gusset plates and framing angles may require an analysis for shear, tension, bending, or block shear. It should be obvious that the design of a tension member connection will usually be done in parallel with the design of the member itself, since the two processes are interdependent.

FIGURE 7.5 ≡

FIGURE 7.6 ≡

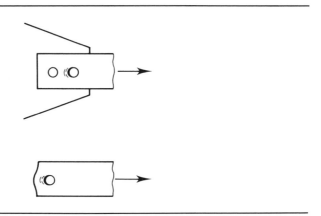

2. **Failure of the connected part because of bearing exerted by the fasteners.** If the hole is slightly larger than the fastener and the fastener is assumed to be placed loosely in the hole, contact between the fastener and the connected part will exist over approximately half of the circumference of the fastener when a load is applied. This is illustrated in Figure 7.5. The stress will vary from a maximum at point A to zero at point B; for simplicity, an average stress, computed as the applied force divided by the projected area of contact, is used.

Thus, the bearing stress would be computed as $f_p = P/(dt)$, where P is the force applied to the fastener, d is the fastener diameter, and t is the thickness of the part subjected to the bearing. The bearing load is therefore $P = f_p dt$.

The bearing problem can be complicated by the presence of a nearby bolt or the proximity of an edge in the direction of the load as in Figure 7.6. In addition to limiting the bearing force, a minimum spacing of fasteners and a minimum edge distance must be maintained.

7.3 ≡ BEARING STRENGTH AND SPACING REQUIREMENTS

Bearing strength is independent of the type of fastener, as the stress under consideration is on the part being connected rather than the fastener. For this reason, bearing strength, as well as spacing and edge-distance requirements, which are also independent of the type of fastener, will be considered before shear and tensile strength.

AISC Specification requirements for bearing strength are based on preventing excessive hole elongation. This will be a function of the applied load, the bearing area (bolt diameter times thickness of the connected part), how close the bolt holes are to each other in the direction of the bearing force (bolt spacing), and how close the bolt holes are to the edge of the connected part of the direction of the bearing force (edge distance). The design bearing strength is denoted by ϕR_n, where the resistance factor ϕ is 0.75. The nominal strength R_n will be given in the following paragraphs.

If the connection contains two or more bolts in the direction of the load, the center-to-center spacing s is at least 3 bolt diameters, and the edge distance L_e is at least $1\frac{1}{2}$ bolt diameters, the nominal strength is given by

$$R_n = 2.4dtF_u \qquad \text{(AISC Equation J3-1a)}$$

where

  d = bolt diameter

  t = thickness of connected part

  F_u = tensile strength of connected part

Figure 7.7a illustrates this case (two or more bolts in the direction of the load). The above expression, AISC Equation J3-1a, is valid for standard holes ($\frac{1}{16}$-inch larger than the bolt diameter) and "short-slotted" holes. This textbook will give only those AISC requirements that are applicable to standard holes.

If there are two or more bolts in the direction of the load and the edge distance L_e in that direction is less than $1\frac{1}{2}d$, or the spacing s is less than $3d$, the nominal bearing strength for *the bolt closest to the edge* is

$$R_n = F_u t L_e \leq 2.4dtF_u \qquad \text{(AISC Equation J3-2a)}$$

where L_e is the distance from the center of the hole to the edge. For the remaining bolts in the direction of the load,

$$R_n = F_u t(s - d/2) \leq 2.4dtF_u \qquad \text{(AISC Equation J3-2b)}$$

If there is only one bolt in the direction of the load (Figure 7.7b), the nominal strength is also given by AISC Equation J3-2a.

Additional Spacing and Edge-Distance Requirements

To maintain clearances between bolt nuts and to provide room for wrench sockets, AISC J3.3 requires that center-to-center spacing of fasteners (in any

FIGURE 7.7 ≡

(a)   (b)

direction) be no less than $2\frac{2}{3}d$ and preferably no less than $3d$, where d is the fastener diameter. Minimum edge distances (in any direction) are given in AISC Table J3.4 as a function of bolt size and type of edge — sheared, rolled, or gas cut.

Summary of Bearing Strength, Spacing, and Edge-Distance Requirements (Standard Holes Only)

A. Bearing strength
  1. Two or more bolts in line of force:

$s \geq 3d$ and $L_e \geq 1\frac{1}{2}d$: $\phi R_n = 0.75[(2.4dt)F_u]$ (AISC Equation J3-1a)

$L_e < 1\frac{1}{2}d$ or $s < 3d$:

For the bolt closest to the edge,

$\phi R_n = 0.75(F_u t L_e) \leq 0.75[2.4dtF_u]$ (AISC Equation J3-2a)

For the remaining bolts,

$\phi R_n = 0.75[F_u t(s - d/2)] \leq 0.75[2.4dtF_u]$ (AISC Equation J3-2b)

  2. One bolt in line of force:

$\phi R_n = 0.75(F_u t L_e) \leq 0.75[2.4dtF_u]$ (AISC Equation J3-2a)

B. Minimum spacing and edge distance

In any direction, both in the line of force and transverse to the line of force,

$s \geq 2\frac{2}{3}d$ (preferably 3d)

$L_e \geq$ value from Table J3.4.

EXAMPLE 7.1

Check bearing, bolt spacing, and edge distances for the connection shown in Figure 7.8.

SOLUTION

From AISC J3.3, the minimum spacing in any direction is

$2\frac{2}{3}d = 2.667(\frac{3}{4}) = 2.00$ in.

actual spacing $= 2.50$ in. > 2.00 in.   (OK)

The minimum edge distance in any direction is obtained from AISC Table J3.4. If we assume sheared edges (the worst case), the minimum edge distance is $1\frac{1}{4}$ in.

actual edge distance $= 1\frac{1}{4}$ in.   (OK)

FIGURE 7.8 ≡

Bearing strength:

Check spacing and edge distances in the direction of the load:

$$3d = 3(\tfrac{3}{4}) = 2.25 \text{ in.} < s = 2.5 \text{ in.}$$
$$1.5d = 1.5(\tfrac{3}{4}) = 1.125 \text{ in.} < L_e = 1.25 \text{ in.}$$

Since there are two bolts in the direction of the load, the bearing strength can be obtained from AISC Equation J3-1a. The bearing strength of the gusset plate will control because it is the thinner component in the connection.

$$\phi R_n = \phi(2.4dtF_u) = 0.75\left[2.4\left(\frac{3}{4}\right)\left(\frac{3}{8}\right)(58)\right] = 29.36 \text{ kips}$$

This is a simple connection, and each bolt resists an equal share of the load. For one bolt,

$$P_u = \frac{65}{4} = 16.25 \text{ kips} < 29.36 \text{ kips} \quad \text{(OK)}$$

Answer   Bearing strength, spacing, and edge-distance requirements are satisfied.

7.4 ≡ COMMON BOLTS

We begin our coverage of fastener strength with *common bolts*, which differ from high-strength bolts not only in material properties, but also in that we do not account for the clamping force from the tightening of the bolt. Common bolts, also known as *unfinished bolts*, are designated as ASTM A307.

The design shear strength of A307 bolts is ϕR_n, where the resistance factor ϕ is 0.75, and the nominal shear strength is

$$R_n = F_v A_b$$

where

F_v = ultimate shearing stress

A_b = cross-sectional area of the unthreaded part of the bolt (this is also known as the *nominal bolt area* or *nominal body area*)

The ultimate shearing stress is given in AISC Table J3.2 as 24 ksi, giving a nominal strength of

$$R_n = F_v A_b = 24A_b$$

EXAMPLE 7.2

Determine the design strength of the connection shown in Figure 7.9 based on shear and bearing.

SOLUTION

The given connection can be classified as a simple connection, and each fastener can be considered to resist an equal share of the load. It will be convenient in most cases to determine the strength corresponding to one fastener and then to multiply by the total number of fasteners.

Shear strength: This is a case of single shear, and the design strength of one bolt is

$$\phi R_n = 0.75(24A_b)$$

The nominal bolt area is

$$A_b = \frac{\pi d^2}{4} = \frac{\pi (3/4)^2}{4} = 0.4418 \text{ in.}^2$$

The design shear strength per bolt is therefore

$$\phi R_n = 0.75(24)(0.4418) = 7.952 \text{ kips}$$

Bearing strength: Although bearing load is a problem for the connected material and not the fasteners, bearing strength is a function of bolt size, and it is convenient to consider the bearing strength associated with each bolt. There are two bolts in the direction of the load, so the spacing and edge-distance limits will be checked.

$$1.5d = 1.5\left(\frac{3}{4}\right) = 1.125 \text{ in.} < L_e = 1.5 \text{ in.}$$

(for both the bar and the gusset plate edges)

$$3d = 3\left(\frac{3}{4}\right) = 2.25 \text{ in.} < s = 3 \text{ in.}$$

The conditions for the use of AISC Equation J3-1a are therefore satisfied. Since the gusset plate is thinner than the bar, the gusset plate has less bearing strength, and the gusset plate thickness of $\frac{3}{8}$ in. will be used.

$$\phi R_n = 0.75[2.4dtF_u] = 0.75\left[2.4\left(\frac{3}{4}\right)\left(\frac{3}{8}\right)(58)\right] = 29.36 \text{ kips}$$

This is larger than the shear strength of 7.952 kips, so the shear strength governs. For two bolts,

$$\phi R_n = 2(7.952) = 15.9 \text{ kips}$$

FIGURE 7.9 ≡

(Note that all spacing and edge-distance requirements are satisfied. For a sheared edge, the minimum edge distance required by AISC Table J3.4 is $1\frac{1}{4}$ in., and this is satisfied in both directions. Since $s > 3d$ and $L_e > 1.5d$ in the direction of the load, there are no other spacing and edge-distance requirements in this direction.)

Answer   Based on shear and bearing, the design strength of the connection is 15.9 kips. (Note that some other limit state that has not been checked, such as tension on the net area of the bar, may in fact govern the design strength.)

≡ **EXAMPLE 7.3**

A bar $4 \times \frac{3}{8}$ is used as a tension member to resist a service dead load of 8 kips and a service live load of 22 kips. This member was designed under the assumption that a single line of $\frac{3}{4}$-in.-diameter A307 bolts would be used to connect the member to a $\frac{3}{8}$-in. gusset plate. How many bolts are required?

SOLUTION   Factored load:

$$P_u = 1.2D + 1.6L = 1.2(8) + 1.6(22) = 44.80 \text{ kips}$$

Capacity of one bolt (from Example 7.2):
For shear,

$$\phi R_n = 7.952 \text{ kips per bolt}$$

For bearing,

$$\phi R_n = 29.36 \text{ kips per bolt}$$

Number of bolts required:

$$\frac{44.80 \text{ kips}}{7.952 \text{ kips/bolt}} = 5.63 \text{ bolts}$$

Answer   Use six $\frac{3}{4}$-in.-diameter A307 bolts.

7.5 ≡ HIGH-STRENGTH BOLTS

High-strength bolts for structural joints are available in two grades: ASTM A325 and ASTM A490. The AISC provisions for high-strength bolts are based in part on the provisions of the specification of the Research Council on Structural Connections of the Engineering Foundation.[23]

A490 bolts have a higher ultimate tensile strength than A325 bolts and are assigned a higher nominal strength. They were introduced long after A325 bolts had been in general use, primarily for use with high-strength steels.[44] A490 bolts are more expensive than A325 bolts, but usually fewer are required.

In certain cases, A325 and A490 bolts are installed to such a degree of tightness that they are subjected to extremely large tensile forces. For example, the initial tension in a $\frac{5}{8}$-in.-diameter A325 bolt can be as high as 19 kips. A complete list of minimum tension values, for those connections in which a minimum tension is required, is given in AISC Table J3.1, Minimum Bolt Tension. Each value is equal to 70% of the minimum tensile strength of the bolt. The purpose of such a large tensile force is to achieve the clamping force illustrated in Figure 7.10.

As a nut is turned and advanced along the threads of a bolt, the connected parts undergo compression and the bolt elongates. The free-body diagrams in Figure 7.10a show that the total compressive force acting on the connected part is numerically equal to the tension in the bolt. If an external load P is applied, a friction force will develop between the connected parts. The maximum possible value of this force is

$$F = \mu N$$

where μ is the coefficient of static friction between the connected parts, and N is the normal compressive force acting on the inner surfaces. The value of μ will depend on the surface condition of the steel — for example, whether it is painted or whether rust is present. Thus, each bolt in the connection is capable of resisting a load of $P = F$, even if the bolt shank does not bear on the connected part. As long as this frictional force is not exceeded, there is no bearing or shear. If P is greater than F and slippage occurs, shear and bearing will then exist and will affect the capacity of the connection.

Installation

How is this high tension accurately achieved? There are currently four authorized procedures for the installation of high-strength bolts:[23]

1. **The turn-of-the-nut method.** This procedure is based on the load-deformation characteristics of the fastener and the connected parts. One full turn of a nut corresponds to a fixed length of travel along the bolt threads, and this can be correlated to the elongation of the bolt. The stress–strain relationship for the bolt material can then be used to compute the tension in the bolt.

FIGURE 7.10 ≡

Elongation

Total = N

Connected parts and bolt

Connected parts only

(a) No external loads

Connected parts only

(b) External load applied

For a given size and type of bolt, therefore, the number of turns of the nut required to produce a given tensile force can be computed. Table 5 in the high-strength bolt specification[23] gives the required nut rotation for various bolt sizes in terms of the ratio of length to diameter. The specified rotation is from a snug position, with *snug* being defined as the tightness after a few impacts with an impact wrench or after the full effort of one worker with an ordinary spud wrench, with all parts of the connection in firm contact. In spite of all the uncertainties and variables involved, the turn-of-the-nut method has proved to be a reliable and surprisingly accurate method.

2. **Calibrated wrench tightening.** Torque wrenches are used for this purpose. The torque required to attain a specified tension in a bolt of a given size and grade is determined by tightening this bolt in a tension-indicating device. This calibration must be done daily during construction for bolts of each size and grade.

3. **Alternate design bolts.** These are specially designed A325 and A490 bolts whose tips twist off when the proper tension has been achieved. Special wrenches are required for their installation. Inspection of this type of bolt installation is particularly easy.

4. **Direct tension indicators.** The most common of these devices is a washer with protrusions on its surface. When the bolt is tightened, the protrusions are compressed in proportion to the tension in the bolt. A prescribed amount of deformation can be established for a given bolt, and when that amount has been achieved, the bolt will have the proper tension. This can be determined by measuring the gap between the nut or bolt head and the undeformed part of the washer surface. Inspection of the bolt installation is also simplified when this type of direct tension indicator is used, since only a feeler gage is required.

7.6 ≡ SHEAR STRENGTH OF HIGH-STRENGTH BOLTS

The design shear strength of both A325 and A490 bolts is ϕR_n, where the resistance factor ϕ is 0.75. As with common bolts, the nominal shear strength of high-strength bolts is given by the ultimate shearing stress times the nominal bolt area. Unlike A307 bolts, however, the shear strength of A325 and A490 bolts depends on whether the threads are in a plane of shear. Rather than use a reduced cross-sectional area when the threaded portion is subject to shear, the ultimate shearing stress is multiplied by a factor of 0.75, the approximate ratio of threaded area to unthreaded area. The strengths are given in AISC Table J3.2 and are summarized in Table 7.1. AISC Table J3.2 refers to threads in a plane of shear as "*not* excluded from shear planes" and refers to threads not in a plane of shear as "excluded from shear planes." The first category, threads included in the shear plane, is sometimes referred to as connection type "N," and an A325 bolt of this type can be denoted as an A325-N bolt. The designation "X" can be used to indicate that threads are excluded from the plane of shear — for example, an A325-X bolt.

TABLE 7.1 ≡

Fastener	Nominal Shear Strength $R_n = F_v A_b$
A325, threads in plane of shear	$48A_b$
A325, threads not in plane of shear	$60A_b$
A490, threads in plane of shear	$60A_b$
A490, threads not in plane of shear	$75A_b$

7.7 ≡ SLIP-CRITICAL CONNECTIONS

A connecton with high-strength bolts is classified as either a *slip-critical* connection or a *bearing-type* connection. A slip-critical connection is one in which no slippage is permitted — that is, the friction force must not be exceeded. In a bearing-type connection, slip is acceptable, and shear and bearing actually occur. In some types of structures, notably bridges, the load on connections can undergo many cycles of reversal. In such cases, fatigue of the fasteners can become critical if the connection is allowed to slip with each reversal, and a slip-critical connection is advisable. In most structures, however, slip is perfectly acceptable, and a bearing-type connection is adequate. (A307 bolts are only used in bearing-type connections.) Proper installation and achievement of the prescribed initial tension is necessary for slip-critical connections. Specific situations in which high-strength bolts must be fully tensioned are listed in AISC J1.11. In bearing-type connections, the only practical requirement for the installation of the bolts is that they be tensioned enough so that the surfaces of contact in the connection firmly bear on one another. This is the snug-tight condition referred to earlier in the discussion of the turn-of-the-nut method.

Although slip-critical connections theoretically are not subject to shear and bearing, they must have sufficient shear and bearing strength in the event of an overload that may cause slip to occur. To prevent slip, the *service load* "shear" on the fastener must not exceed ϕR_n, where $\phi = 1.0$ and

$$R_n = F_v A_b$$

Since resistance to slip is a serviceability requirement, the limiting load is a service load.[*] The allowable slip load per unit area of bolt, F_v, is a function of the coefficient of static friction between the connected parts, which in turn is a function of the condition of the surfaces. AISC Table J3.6 gives values of F_v for A325 and A490 bolts for various types of holes and Class A surfaces of contact. A Class A surface is one with clean mill scale (mill scale is an iron oxide that forms on the steel when it is reheated for rolling). For this surface condition and standard size holes, $F_v = 17$ ksi for A325 bolts, and $F_v = 21$ ksi for A490 bolts. In this book, Class A surfaces will normally be assumed. The design slip-critical shear strength can therefore be written as

$$\phi R_n = 17 A_b \text{ for A325 bolts}$$
$$= 21 A_b \text{ for A490 bolts}$$

where $\phi = 1.0$. If the loading includes wind loads, the total load may be reduced by 25% — that is, use a load of 0.75 × the sum of the service loads (AISC J3.8).

---

[*] AISC also permits slip-critical connections to be designed for factored loads; this is covered in AISC Appendix J3.8b.

≡ **EXAMPLE 7.4**

The connection shown in Figure 7.11a uses $\frac{3}{4}$-in.-diameter A325 bolts with the threads in the shear plane. No slip is permitted. Both the tension member and the gusset plate are of A36 steel.

   a. Determine the design strength.

   b. Determine the maximum total service load (dead plus live only). Assume Class A surfaces.

SOLUTION   **a.** Shear strength of one bolt:

$$A_b = \frac{\pi(3/4)^2}{4} = 0.4418 \text{ in.}^2$$

$$\phi R_n = \phi F_v A_b = 0.75(48)(0.4418) = 15.90 \text{ kips}$$

Bearing strength of one bolt:

$$1.5d = 1.5\left(\frac{3}{4}\right) = 1.125 \text{ in.} < L_e = 1.5 \text{ in.}$$

                  (for both the tension member and gusset-plate edges)

$$3d = 3\left(\frac{3}{4}\right) = 2.25 \text{ in.} < s = 3 \text{ in.}$$

Since there are two bolts in the direction of the load, the bearing strength is

$$\phi R_n = \phi(2.4dtF_u) = 0.75(2.4)\left(\frac{3}{4}\right)\left(\frac{3}{8}\right)(58) = 29.36 \text{ kips}$$

(The thickness of the gusset plate controls.)

The strength is governed by shear: For four bolts,

$$\phi R_n = 4(15.90) = 63.6 \text{ kips}$$

FIGURE 7.11 ≡

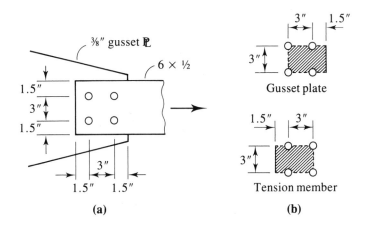

(a)   (b)

Strength based on tension on the gross area of the tension member:

$$\phi_t P_n = \phi_t F_y A_g = 0.90(36)\left(6 \times \frac{1}{2}\right) = 97.2 \text{ kips}$$

Tension on the net area: All elements of the cross section are connected, so shear lag is not a factor and $A_e = A_n$:

$$\phi_t P_n = \phi_t F_u A_e = \phi_t F_u t(w_g - \sum d) = 0.75(58)\left(\frac{1}{2}\right)\left[6 - 2\left(\frac{7}{8}\right)\right]$$

$$= 92.4 \text{ kips}$$

Block shear strength: The failure block for the gusset plate has the same dimensions as the block for the tension member except for the thickness (Figure 7.11b). The gusset plate, which is the thinner element, will control.

Shear areas: There are two potential shear-failure planes:

$$A_{gv} = 2 \times \frac{3}{8}(3 + 1.5) = 3.375 \text{ in.}^2$$

and since there are 1.5 hole diameters per horizontal line of bolts,

$$A_{nv} = 2 \times \frac{3}{8}\left[(3 + 1.5) - (1.5)\left(\frac{7}{8}\right)\right] = 2.391 \text{ in.}^2$$

Tension areas:

$$A_{gt} = \frac{3}{8}(3) = 1.125 \text{ in.}^2$$

$$A_{nt} = \frac{3}{8}\left(3 - \frac{7}{8}\right) = 0.7969 \text{ in.}^2$$

AISC Equation J4-3a:

$$\phi R_n = \phi[0.6F_y A_{gv} + F_u A_{nt}] = 0.75[0.6(36)(3.375) + 58(0.7969)]$$
$$= 0.75[72.90 + 46.22] = 89.3 \text{ kips}$$

AISC Equation J4-3b:

$$\phi R_n = \phi[0.6F_u A_{nv} + F_y A_{gt}] = 0.75[0.6(58)(2.391) + 36(1.125)]$$
$$= 0.75[83.21 + 40.50] = 92.8 \text{ kips}$$

The fracture term (the one involving F_u) in the second equation is larger than the one in the first equation, therefore AISC Equation J4-3b governs.

design strength for block shear = 92.8 kips

Of all the limit states investigated, the strength corresponding to bolt shear is the smallest.

Answer   The design strength of the connection is 63.6 kips.

b. Because of the constraint that no slippage of the connection is permitted, this connection is classified as slip-critical and the total service load must not exceed the allowable slip load. For one A325 bolt,

$$\phi R_n = 17A_b = 17(0.4418) = 7.511 \text{ kips}$$

For four bolts,

$$\phi R_n = 4(7.511) = 30.0 \text{ kips}$$

Answer   To prevent slip, the total service load must not exceed 30.0 kips.

EXAMPLE 7.5

A $\frac{5}{8}$-in.-thick tension member is connected to two $\frac{1}{4}$-in. splice plates as shown in Figure 7.12. The given loads are service loads. A36 steel and $\frac{5}{8}$-in.-diameter A325 bolts will be used. If slip *is* permissible, how many bolts are required? Each bolt centerline shown represents a row of bolts in the direction of the width of the plates.

SOLUTION   Design strength for one bolt:

Shear:

$$A_b = \frac{\pi \left(\frac{5}{8}\right)^2}{4} = 0.3068 \text{ in.}^2$$

Assume that the bolt threads are in the planes of shear:

$$\phi R_n = \phi F_v A_b \times 2 \text{ planes of shear} = 0.75(48)(0.3068)(2) = 22.09 \text{ kips}$$

Bearing: The bolt spacing in the direction of the load is greater than $3d$, and the edge distance is greater than $1.5d$, so AISC Equation J3-1a will be used for the nominal bearing strength. The bearing force on the $\frac{5}{8}$-in.-thick tension member will be twice as large as the bearing force on each of the $\frac{1}{4}$-in. splice plates. Since the total load on the splice plates is the same as the load on the tension member, for the splice plates to be critical the total splice plate thickness must be less than the thickness of the tension member, and it is. This will be demonstrated in the following calculations.

Tension member:

$$\phi R_n = \phi(2.4dtF_u) = 0.75\left[2.4\left(\frac{5}{8}\right)\left(\frac{5}{8}\right)(58)\right] = 40.78 \text{ kips}$$

Splice plates:

$$\phi R_n = \phi(2.4dtF_u) = 0.75\left[2.4\left(\frac{5}{8}\right)\left(\frac{1}{4}+\frac{1}{4}\right)(58)\right] = 32.62 \text{ kips}$$

FIGURE 7.12 ≡

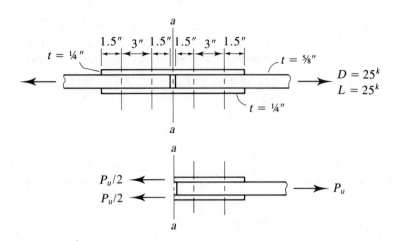

Bearing is controlled by the splice plates, but the overall design strength per bolt is controlled by shear, and $\phi R_n = 22.09$ kips.

Factored load:

$$P_u = 1.2D + 1.6L = 1.2(25) + 1.6(25) = 70 \text{ kips}$$

$$\text{number of bolts required} = \frac{\text{total load}}{\text{load per bolt}}$$

$$= \frac{70}{22.09} = 3.17 \text{ bolts}$$

Answer   Use four bolts, two per line, on each side of the splice. A total of eight bolts will be required for the connection.

≡ **EXAMPLE 7.6**

The C6 × 13 shown in Figure 7.13 has been selected to resist a factored tensile load of 108 kips. It is to be attached to a $\frac{3}{8}$-in. gusset plate with $\frac{7}{8}$-in.-diameter A325 bolts. Assume that the threads are in the plane of shear and that slip of the connection is permissible. Determine the number and required layout of bolts such that the length of connection h is a minimum. A36 steel is used.

SOLUTION   Determine the capacity of a single bolt:

Shear:

$$A_b = \frac{\pi(7/8)^2}{4} = 0.6013 \text{ in.}^2$$

$$\phi R_n = \phi F_v A_b = 0.75(48)(0.6013) = 21.65 \text{ kips}$$

FIGURE 7.13 ≡

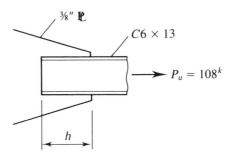

Bearing: The gusset plate is thinner than the web of the channel and will control the bearing strength. Assuming that along a line parallel to the force there are two or more bolts, the spacing is at least $3d$, and the edge distance is at least $1\frac{1}{2}d$,

$$\phi R_n = \phi(2.4dtF_u) = 0.75(2.4)\left(\frac{7}{8}\right)\left(\frac{3}{8}\right)(58) = 34.26 \text{ kips}$$

Shear controls: number of bolts required $= \dfrac{108}{21.65} = 4.99$

Although five bolts will furnish enough capacity, try six bolts so that a symmetrical layout with two gage lines of three bolts each can be used as in Figure 7.14. (Two gage lines are used to minimize the length of the connection.) It is unknown whether the design of this tension member was based on the assumption of one line or two lines of fasteners; the tensile capacity of the channel with two lines of bolts must be checked before proceeding.

Tension on the gross area:

$$\phi_t P_n = 0.90F_y A_g = 0.90(36)(3.83) = 124 \text{ kips}$$

Tension on the effective net area:

$$A_n = 3.83 - 2(1.0)(0.437) = 2.96 \text{ in.}^2$$

Since the exact length of the connection is not yet known, use the average value of U from the Commentary:

$$A_e = U A_n = 0.85(2.96) = 2.51 \text{ in.}^2$$
$$\phi_t P_n = 0.75F_u A_e = 0.75(58)(2.51) = 109 \text{ kips} \quad \text{(controls)}$$

FIGURE 7.14 ≡

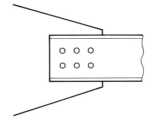

The member capacity is therefore based on two gage lines of bolts.

Check spacing and edge-distance transverse to the load:

From AISC J3.3,

$$\text{minimum spacing} = 2.667\left(\frac{7}{8}\right) = 2.33 \text{ in.}$$

From AISC Table J3.4,

$$\text{minimum edge distance} = 1\tfrac{1}{8} \text{ in.}$$

A spacing of 3 in. and an edge distance of $1\frac{1}{2}$ in. will be used transverse to the load.

The minimum length of the connection can be established by using the minimum permissible spacing and edge distances in the longitudinal direction. The minimum spacing in any direction is

$$2\tfrac{2}{3}d = 2.33 \text{ in.} \quad \text{Try } 2\tfrac{1}{2} \text{ in.}$$

The minimim edge distance in any direction is $1\frac{1}{8}$ in. If these minimum distances are used in the longitudinal direction, the capacity of the bolts nearest the edge is

$$\phi R_n = \phi(F_u t L_e) = 0.75(58)\left(\frac{3}{8}\right)(1.125) = 18.35 \text{ kips}$$

This is less than the maximum permitted value of $\phi(2.4dtF_u) = 34.26$ kips.

The capacity of each of the remaining bolts is

$$\phi R_n = \phi\left[F_u t\left(s - \frac{d}{2}\right)\right] = 0.75\left[58\left(\frac{3}{8}\right)\left(2.5 - \frac{7/8}{2}\right)\right]$$
$$= 33.64 \text{ kips} < 34.26 \text{ kips}$$

The total bearing strength is

$$2(18.35) + 4(33.64) = 171.3 \text{ kips} > P_u = 108 \text{ kips} \quad \text{(OK)}$$

The tentative connection design is shown in Figure 7.15 and will now be checked for block shear in the gusset plate (the geometry of the failure block in the channel is identical, but the gusset plate is thinner).

Shear areas:

$$A_{gv} = \frac{3}{8}(2.5 + 2.5 + 1.125)(2) = 4.594 \text{ in.}^2$$

and since there are $2\frac{1}{2}$ hole diameters along each of the shear surfaces,

$$A_{nv} = \frac{3}{8}[6.125 - 2.5(1.0)](2) = 2.719 \text{ in.}^2$$

FIGURE 7.15 ≡

Tension areas:

$$A_{gt} = \frac{3}{8}(3) = 1.125 \text{ in.}^2$$

$$A_{nt} = \frac{3}{8}(3 - 1.0) = 0.75 \text{ in.}^2$$

Check for tension yield and shear fracture with AISC Equation J4-3a:

$$\phi R_n = \phi[0.6F_y A_{gv} + F_u A_{nt}]$$
$$= 0.75[0.6(36)(4.594) + 58(0.75)] = 0.75[99.23 + 43.50] = 107.0 \text{ kips}$$

Check for tensile fracture and shear yield with AISC Equation J4-3b:

$$\phi R_n = \phi[0.6F_u A_{nv} + F_y A_{gt}]$$
$$= 0.75[0.6(58)(2.719) + 36(1.125)] = 0.75[94.62 + 40.50] = 101.3 \text{ kips}$$

The fracture term (the one involving F_u) in the second equation is larger than the one in the first equation; therefore, the second equation governs.

design strength for block shear = 101.3 kips < 108 kips   (N.G.)

The simplest way to increase the block shear strength for this connection is to increase the shear areas by increasing the bolt spacing. If this is done, AISC Equation J4-3b will still control. Although the required spacing can be determined by trial and error, it can be solved for directly, which will be done here. From Equation J4-3b, let

$$0.75[0.6(58)A_{nv} + 40.50] = 108 \text{ kips}$$
$$\text{required } A_{nv} = 2.974 \text{ in.}^2$$

$$A_{nv} = \frac{3}{8}(2s + 1.125 - 2.5)(2) = 2.974 \text{ in.}^2$$

required $s = 2.67$ in.   Use $s = 3.0$ in.

With a spacing of 3.0 in., the net shear area is

$$A_{nv} = \frac{3}{8}(3 + 3 + 1.125 - 2.5)(2) = 3.469 \text{ in.}^2$$

FIGURE 7.16 ≡

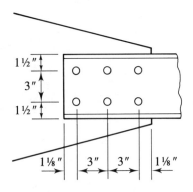

and the block shear strength from AISC Equation J4-3b is

$$\phi R_n = \phi[0.6F_u A_{nv} + F_y A_{gt}]$$
$$= 0.75[0.6(58)(3.469) + 36(1.125)] = 0.75[120.7 + 40.50] = 120.9 \text{ kips}$$

Using the spacing and edge distances determined above, the minimum length is therefore

$h = 1.125$ in. at the end of the channel

+ 2 spaces @ 3.0 in.

+ 1.125 in. at the end of the gusset plate

= 8.25 in. total

Answer   Use the connection detail as shown in Figure 7.16.

The bolt layout in the above example is symmetrical with respect to the longitudinal centroidal axis of the member. Consequently, the resultant resisting force provided by the fasteners also acts along this line, and the geometry is consistent with the definition of a simple connection. If an odd number of bolts had been required and two rows used, the symmetry would not exist and the connection would be eccentric. In such cases, the designer has several choices: (1) ignore the eccentricity, assuming the effects are negligible; (2) account for the eccentricity; (3) use a staggered pattern of fasteners that would preserve the symmetry; or (4) add an extra bolt and remove the eccentricity. Most engineers would probably choose the last alternative.

7.8 ≡ HIGH-STRENGTH BOLTS IN TENSION

When a tensile load is applied to a bolt with no initial tension, the tensile force in the bolt is equal to the applied load. If the bolt is pretensioned, however, a large part of the applied load is used to relieve the compression, or clamping

FIGURE 7.17 ≡

forces, on the connected parts. This will be demonstrated in the following development.[45] Figure 7.17 shows a hanger connection consisting of a structural tee shape bolted to the bottom flange of a W-shape and subjected to a tensile load. A single bolt and a portion of the connected parts will be isolated and examined both before and after loading.

Before loading, all forces are internal, and a free-body diagram of the assembly is as shown in Figure 7.18a. For simplicity, all forces will be assumed

FIGURE 7.18 ≡

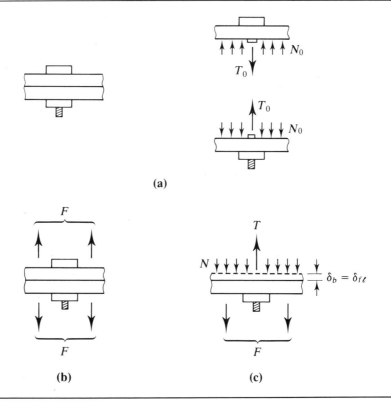

to be symmetric with respect to the axis of the bolt, and any eccentricity will be neglected. If the connected parts are considered as separate free bodies, the forces consist of the bolt tension T_0 and the normal clamping force N_0, shown here as a uniformly distributed force. Equilibrium requires that T_0 be equal to N_0. When the external tensile load is applied, the forces on the assembly are as shown in part b of the figure, with F representing the total tensile force applied to one bolt (again, the actual distribution of the applied force per bolt has been idealized for simplicity). Figure 7.18c shows the forces acting on a free-body diagram of the segment of the structural tee flange and the corresponding segment of the bolt. Summing forces in the direction of the bolt axis gives

$$T = F + N$$

The application of the force F will increase the bolt tension and cause it to elongate by an amount δ_b. Compression in the flange of the structural tee will be reduced, resulting in a distortion $\delta_{f\ell}$ in the same sense as δ_b. An approximate relationship between the applied force and the change in bolt tension follows.

From elementary mechanics of materials, the axial deformation of an axially loaded uniform member is given by

$$\delta = \frac{PL}{AE} \tag{7.1}$$

where

$\qquad P =$ applied axial force

$\qquad L =$ original, undeformed length

$\qquad A =$ cross-sectional area

$\qquad E =$ modulus of elasticity

Equation 7.1 can be solved for the load:

$$P = \frac{AE\delta}{L} \tag{7.2}$$

The change in bolt force corresponding to a given axial displacement δ_b can therefore be given by

$$\Delta T = \frac{A_b E_b \delta_b}{L_b} \tag{7.3}$$

where the subscript indicates a property or dimension of the bolt. Application of Equation 7.2 to the compression flange requires a somewhat more liberal interpretation of the load distribution in that N must be treated as if it were uniformly applied over a given surface area. $A_{f\ell}$. If this is done, the change in the force N is obtained from Equation 7.2 as

$$\Delta N = \frac{A_{f\ell} E_{f\ell} \delta_{f\ell}}{L_{f\ell}} \tag{7.4}$$

where $L_{f\ell}$ is the flange thickness. As long as the connected parts (the two flanges) remain in contact, the bolt deformation δ_b and the flange deformation $\delta_{f\ell}$ will be equal. Since $E_{f\ell}$ is approximately equal to E_b,[46] and $A_{f\ell}$ is much larger than A_b,

$$\frac{A_{f\ell} E_{f\ell} \delta_{f\ell}}{L_{f\ell}} \gg \frac{A_b E_b \delta_b}{L_b}$$

and therefore

$$\Delta N \gg \Delta T$$

The ratio of ΔT to ΔN is in the range of 0.05 to 0.1.[45] As a consequence, ΔT will be no greater than $0.1\Delta N$, demonstrating that most of the applied load is devoted to relieving the compression of the connected parts. To estimate the magnitude of the load required to completely overcome the clamping effect and cause the parts to separate, consider the free-body diagram in Figure 7.19. When the parts have separated,

$$T = F$$

or

$$T_0 + \Delta T = F \tag{7.5}$$

At the point of impending separation, the bolt elongation and plate decompression are equal, and

$$\Delta T = \frac{A_b E_b}{L_b} \delta_b = \frac{A_b E_b}{L_b} \delta_{f\ell} \tag{7.6}$$

where $\delta_{f\ell}$ is the elongation corresponding to the release of the initial compression force N_0. From Equation 7.1,

$$\delta_{f\ell} = \frac{N_0 L_{f\ell}}{A_{f\ell} E_{f\ell}}$$

FIGURE 7.19 ≡

Substituting into Equation 7.6,

$$\Delta T = \left(\frac{A_b E_b}{L_b}\right)\left(\frac{N_0 L_{f\ell}}{A_{f\ell} E_{f\ell}}\right) = \left(\frac{A_b E_b/L_b}{A_{f\ell} E_{f\ell}/L_{f\ell}}\right)N_0 = \left(\frac{A_b E_b/L_b}{A_{f\ell} E_{f\ell}/L_{f\ell}}\right)T_0 \approx 0.1 T_0$$

From Equation 7.5,

$$T_0 + 0.1 T_0 = F, \qquad F = 1.1 T_0$$

Therefore, at the instant of separation, the bolt tension is approximately 10% larger than its initial value at installation. Once the connected parts separate, however, any increase in external load will be resisted entirely by a corresponding increase in bolt tension. If the bolt tension is assumed to be equal to the externally applied force (just as if there were no initial tension) and the connection is loaded until the connected parts separate, the bolt tension will be underestimated by less than 10%. This has been accounted for in the nominal strength values given in AISC Table J3.2. These values depend on the connected parts not separating, so the initial tension must not be overcome. For this reason, high-strength bolts subject to direct tension must be pretensioned to the values prescribed in AISC Table J3.1, whether the connection is slip-critical or not.

In summary, the tensile force in the bolt should be computed without considering the initial tension.

Prying Action

In most connections in which fasteners are subjected to tension forces, the flexibility of the connected parts can lead to deformations that increase the tension applied to the fasteners. A hanger connection of the type used in the previous discussion is subject to this type of behavior. The additional tension is called a *prying force* and is illustrated in Figure 7.20, which shows the forces on a free body of the hanger. Before the external load is applied, the normal compressive force N_0 is centered around the bolt. As the load is applied, if the flange is flexible enough to deform as shown, the compressive forces will migrate toward the edges of the flange. This redistribution will change the relationship between all forces, and the bolt tension will increase. If the connected parts are sufficiently rigid, however, this shifting of forces will not occur, and there will be no prying action. The maximum value of the prying force will be reached when only the corners of the flange remain in contact with the other connected part.

In connections of this type, bending caused by the prying action will usually control the design of the connected part. AISC J3.6 requires that prying action be included in the computation of tensile loads applied to fasteners.

A procedure for the determination of prying forces, based on research reported in *Guide to Design Criteria for Bolted and Riveted Joints*,[45] is given in the *Manual* in Part 11, "Connections For Tension and Compression" (Volume II). The specific case examined is the connection of a segment of a structural tee

FIGURE 7.20 ≡

Before external load After external load

Maximum prying force

shape, but a pair of back-to-back angles would be treated in exactly the same way. The method is presented here in a somewhat different form but gives the same results.

The method used is based on the model shown in Figure 7.21. All forces are for one fastener. Thus, T is the external factored tension force applied to one bolt, Q is the prying force corresponding to one bolt, and B_c is the total bolt force. The prying force has shifted to the tip of the flange and is at its maximum value.

The equations that follow are derived from a consideration of equilibrium using the free-body diagrams in Figure 7.21. From the summation of moments at b–b in Figure 7.21b,

$$Tb - M_{a-a} = Qa \tag{7.7}$$

From Figure 7.21c,

$$M_{b-b} = Qa \tag{7.8}$$

Finally, equilibrium of forces requires that

$$B_c = T + Q \tag{7.9}$$

These three equilibrium equations can be combined to obtain a single equation for the total bolt force, which includes the effects of prying action. We first define the variable α as the ratio of the moment per unit length along the bolt

FIGURE 7.21 ≡

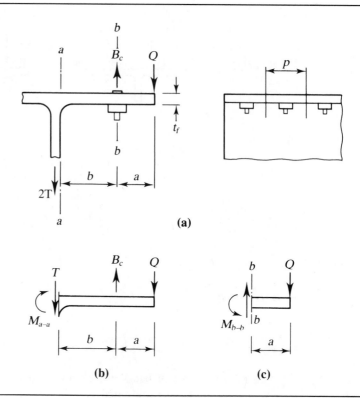

(a)

(b)   **(c)**

line to the moment per unit length at the face of the stem. For the bolt line, the length is a net length.

$$\alpha = \frac{M_{b-b}/(p - d')}{M_{a-a}/p} = \frac{M_{b-b}}{M_{a-a}}\left(\frac{1}{1 - d'/p}\right) = \frac{M_{b-b}}{\delta M_{a-a}} \tag{7.10}$$

where

p = length of flange tributary to one bolt

d' = diameter of bolt hole

$$\delta = 1 - \frac{d'}{p} = \frac{\text{net area at bolt line}}{\text{gross area at web face}}$$

M_{a-a} = design strength at $a-a$ = $\phi_b M_p = \phi_b(pt_f^2 F_y/4)$

(The numerical evaluation of α will require the use of another equation, which will be developed subsequently.) With this notation we can combine the three equilibrium equations 7.7–7.9 to obtain the total bolt force, B_c:

$$B_c = T\left[1 + \frac{\delta\alpha}{(1 + \delta\alpha)}\frac{b}{a}\right] \tag{7.11}$$

At this level of loading, deformations are so large that the resultant of tensile stresses in the bolt does not coincide with the axis of the bolt. As a consequence, the bolt force predicted by Equation 7.11 is conservative and does not quite agree with test results. Much better agreement is obtained if the force B_c is shifted toward the stem of the tee by an amount $d/2$, where d is the bolt diameter. The modified values of b and a are therefore defined as

$$b' = b - \frac{d}{2}$$

and

$$a' = a + \frac{d}{2}$$

(For best agreement with test results, the value of a should be no greater than $1.25b$.)

With this modification we can write Equation 7.11 as

$$B_c = T\left[1 + \frac{\delta\alpha}{(1 + \delta\alpha)}\frac{b'}{a'}\right] \tag{7.12}$$

We can evaluate α from Equation 7.12 by setting the bolt force B_c equal to the bolt design tensile strength, which we will denote by B. This results in

$$\alpha = \frac{\left(\dfrac{B}{T} - 1\right)\dfrac{a'}{b'}}{\delta\left[1 - \left(\dfrac{B}{T} - 1\right)\dfrac{a'}{b'}\right]} \tag{7.13}$$

Two limit states are possible: tensile failure of the bolts and bending failure of the tee. Failure of the tee is assumed to occur when plastic hinges form at section a–a, the face of the stem of the tee, and at b–b, the bolt line, thereby creating a beam mechanism. The moment of each of these locations will be equal to M_p, the plastic moment capacity of a length of flange tributary to one bolt. If the absolute value of α, obtained from Equation 7.13, is less than 1.0, the moment at the bolt line is less than the moment at the face of the stem, indicating that the beam mechanism has not formed, and the controlling limit state will be tensile failure of the bolt. The bolt force B_c in this case will be equal to the design strength B. If the absolute value of α is equal to or greater than 1.0, plastic hinges have formed at both a–a and b–b, and the controlling limit state is flexural failure of the tee flange. Since the moments at these two locations are limited to the plastic moment M_p, α should be set equal to 1.0.

The three equilibrium Equations 7.7–7.9 can also be combined into a single equation for the required flange thickness, t_f. From Equations 7.7 and 7.8, we can write

$$Tb' - M_{a-a} = M_{b-b}$$

where b' has been substituted for b. From Equation 7.10,

$$Tb' - M_{a-a} = \delta\alpha M_{a-a} \tag{7.14}$$

Setting M_{a-a} equal to the design strength,

$$M_{a-a} = \phi_b M_p = \phi_b \frac{p t_f^2 F_y}{4}$$

where t_f is the required flange thickness. Substituting into Equation 7.14, we obtain

$$t_f = \sqrt{\frac{4Tb'}{\phi_b p F_y (1 + \delta\alpha)}}$$

With $\phi_b = 0.90$,

$$\text{Required } t_f = \sqrt{\frac{4.444 T b'}{p F_y (1 + \delta\alpha)}} \tag{7.15}$$

The design of connections subjected to prying is essentially a trial-and-error process. When selecting the size or number of bolts, an allowance for the prying force must be made. The selection of the tee flange thickness is more difficult in that it is a function of the bolt selection and tee dimensions. The Preliminary Hanger Connection Selection Table provided in Part 11 of the *Manual* can be helpful when choosing a trial shape. Once the trial shape has been selected and the number of bolts and their layout estimated, Equations 7.12 and 7.15 can be used to verify or disprove the choices.

If the actual flange thickness is different from the required value, the actual values of α and B_c will be different from those previously calculated. If the actual bolt force, which includes the prying force Q, is desired, α will need to be recomputed as follows.

First, combine Equations 7.7 and 7.8, using b' instead of b:

$$M_{b-b} = Tb' - M_{a-a}$$

From Equation 7.10,

$$\alpha = \frac{M_{b-b}}{\delta M_{a-a}}$$

$$= \frac{Tb' - M_{a-a}}{\delta M_{a-a}} = \frac{Tb'/M_{a-a} - 1}{\delta}$$

Setting M_{a-a} equal to the design moment,

$$M_{a-a} = \phi_b M_p = 0.90 \left(\frac{p t_f^2 F_y}{4} \right)$$

then

$$\alpha = \frac{\dfrac{Tb'}{0.90pt_f^2 F_y/4} - 1}{\delta} = \frac{1}{\delta}\left(\frac{4.444Tb'}{pt_f^2 F_y} - 1\right) \tag{7.16}$$

The total bolt force can then be found from Equation 7.12.

EXAMPLE 7.7

An 8-in.-long WT10.5 × 66 is attached to the bottom flange of a beam as shown in Figure 7.22. This hanger must support a factored load of 90 kips. Determine the number of $\frac{7}{8}$-in.-diameter A325 bolts required and investigate the adequacy of the tee. A36 steel is used.

SOLUTION   Bolt area:

$$A_b = \frac{\pi(7/8)^2}{4} = 0.6013 \text{ in.}^2$$

Design strength of one bolt:

$$B = \phi R_n = \phi F_t A_b = 0.75(90.0)(0.6013) = 40.59 \text{ kips}$$

FIGURE 7.22 ≣

WT10.5 × 66

5½″

12.44″

5.5″

$t_f = 1.035''$

b | a

0.650″

Number of bolts required $= \frac{90}{40.59} = 2.22$. A minimum of four bolts will be needed to maintain symmetry:

Try four bolts.

From the dimensions shown in Figure 7.22,

$$b = \frac{(5.5 - 0.650)}{2} = 2.245 \text{ in.}$$

$$a = \frac{(12.44 - 5.5)}{2} = 3.470 \text{ in.}$$

$1.25b = 1.25(2.425) = 3.031 \text{ in.} < 3.470 \text{ in.}$ Use $a = 3.031$ in.

$$b' = b - \frac{d}{2} = 2.425 - \frac{7/8}{2} = 1.988 \text{ in.}$$

$$a' = a + \frac{d}{2} = 3.031 + \frac{7/8}{2} = 3.468 \text{ in.}$$

Factored external load per bolt, excluding prying force $= T = \frac{90}{4} = 22.5$ kips.

$$d' = d + \frac{1}{8} = \frac{7}{8} + \frac{1}{8} = 1 \text{ in.}$$

$$p = \frac{8}{2} = 4 \text{ in.}$$

$$\delta = 1 - \frac{d'}{p} = 1 - \frac{1}{4} = 0.75$$

Compute α:

$$\frac{B}{T} - 1 = \frac{40.59}{22.5} - 1 = 0.8040$$

$$\frac{a'}{b'} = \frac{3.468}{1.988} = 1.744$$

From Equation 7.13,

$$\alpha = \frac{\left(\dfrac{B}{T} - 1\right)\dfrac{a'}{b'}}{\delta\left[1 - \left(\dfrac{B}{T} - 1\right)\dfrac{a'}{b'}\right]} = \frac{0.8040(1.744)}{0.75[1 - 0.8040(1.744)]} = -4.65$$

$|\alpha| > 1.0$ ∴   Use $\alpha = 1.0$.

From Equation 7.15,

$$\text{Required } t_f = \sqrt{\frac{4.444 T b'}{p F_y (1 + \delta\alpha)}} = \sqrt{\frac{4.444(22.5)(1.988)}{4(36)(1 + 0.75)}}$$

$$= 0.888 \text{ in.} < 1.035 \text{ in.}   \text{(OK)}$$

Both the number of bolts selected and the flange thickness are adequate, and no further computations are required. To illustrate the procedure, however, the prying force will be computed using Equations 7.16 and 7.12. From Equation 7.16,

$$\alpha = \frac{1}{\delta}\left(\frac{4.444Tb'}{pt_f^2 F_y} - 1\right) = \frac{1}{0.75}\left[\frac{4.444(22.5)(1.988)}{4(1.035)^2(36)} - 1\right] = 0.3848$$

From Equation 7.12, the total bolt force, including prying, is

$$B_c = T\left[1 + \frac{\delta\alpha}{(1 + \delta\alpha)}\frac{b'}{a'}\right]$$

$$= 22.5\left\{1 + \frac{0.75(0.3848)}{[1 + 0.75(0.3848)]}\left(\frac{1.988}{3.4868}\right)\right\} = 25.39 \text{ kips}$$

The prying force is

$$Q = B_c - T = 25.39 - 22.5 = 2.89 \text{ kips}$$

Answer   A WT10.5 × 66 is satisfactory. Use four $\frac{7}{8}$-in.-diameter A325 bolts.

If the flange thickness had proved to be inadequate, the alternatives would include trying a larger tee-shape or using more bolts to reduce T, the external load per bolt. The prying force in the above example adds approximately 13% to the externally applied load. Neglect of this additional tension could have serious consequences. The designer should be alert for situations in which it may arise and, when feasible, minimize it by avoiding overly flexible connection elements.

7.9 ≡ COMBINED SHEAR AND TENSION IN FASTENERS

In most of the situations in which a bolt is subjected to both shear and tension, the connection is loaded eccentrically and falls within the realm of Chapter 8. There are, however, simple connections in which the fasteners are in a state of combined loading. Figure 7.23 shows a structural tee segment connected to the flange of a column for the purpose of attaching a bracing member. This bracing member is oriented in such a way that the line of action of the member force passes through the center of gravity of the connection. The vertical component of the load will put the fasteners in shear, and the horizontal component will cause tension (with the possible inclusion of prying forces). Since the line of action of the load acts through the center of gravity of the connection, each fastener can be assumed to take an equal share of each component.

FIGURE 7.23 ≡

As in other cases of combined loading, AISC uses an interaction formula approach. Bearing-type connections and slip-critical connections are treated differently.

The shear and tensile strengths for bearing-type bolts are based on test results and can be taken from the elliptical interaction curve shown in Figure 7.24. The equation of this curve, solved for the tensile stress limit, is

$$\text{limiting } f_t = \frac{F_t}{F_v} \sqrt{F_v^2 - f_v^2}$$

where F_t and F_v are the usual limits on tensile and shearing stresses in the absence of interaction, and f_v is the shearing stress resulting from the factored loads. AISC approximates this curve with three straight-line segments as shown. The general form of the equation of the sloping line is

$$\text{nominal tension stress limit} = F_t = C - R f_v$$

where C and R are constants that depend on the type of fastener and the location of the threads. This tension stress limit cannot exceed the value of F_t from Table J3.2, represented by the horizontal line segment in Figure 7.24.

FIGURE 7.24 ≡

AISC J3.7 and Table J3.5 give values of F_t for bearing-type connections with rivets, common bolts, and high-strength bolts. Values for the usual high-strength bolts are as follows:

A325 bolts:

$$F_t = 117 - 1.9 f_v \leq 90 \quad \text{when threads are in the shear plane}$$
$$F_t = 117 - 1.5 f_v \leq 90 \quad \text{when threads are not in the shear plane}$$

A490 bolts:

$$F_t = 147 - 1.9 f_v \leq 113 \quad \text{when threads are in the shear plane}$$
$$F_t = 147 - 1.5 f_v \leq 113 \quad \text{when threads are not in the shear plane}$$

The shearing strengths are the same as for the case of shear only.

In slip-critical connections in which the bolts are subject to both shear and tension, the effect of the applied tensile force is to relieve some of the clamping force, thereby reducing the available friction force. The AISC Specification reduces the slip-critical shear strength for this case. From AISC J3.9, the slip-critical shear strength is to be multiplied by the factor

$$1 - \frac{T}{T_b}$$

where T is the service tensile load on the bolt, and T_b is the prescribed initial bolt tension from AISC Table J3.1.

The approach to the analysis of bolted connections loaded in both shear and tension can be summarized as follows:

Bearing-Type Connections

1. Check the shear and bearing resulting from factored loads against the usual design strengths.

2. Check the factored load tension against the reduced design tension strength.

Slip-Critical Connections

1. Check the factored load tension, shear, and bearing against the usual design strengths.

2. Check the applied service load shear against the reduced slip-critical shear strength.

EXAMPLE 7.8

A WT10.5 × 31 is used as a bracket to transmit a 60 kip service load to a W14 × 90 column as shown in Figure 7.23. The load consists of 15 kips dead load and 45 kips live load. Four $\frac{7}{8}$-in.-diameter A325 bolts are used. Both the column and bracket are of A36 steel. Assume all spacing and edge-distance requirements are satisfied, including those necessary for the use of the full design strength in bearing, and determine the adequacy of the bolts for the following types of connections: (a) bearing-type connection with the threads in shear and (b) slip-critical connection with the threads in shear.

SOLUTION   Factored load $= 1.2D + 1.6L = 1.2(15) + 1.6(45) = 90$ kips

a. Bearing-type connection, threads in shear plane:

Total shear force:

$$\frac{3}{5}(90) = 54 \text{ kips}$$

Shear force per bolt $= \dfrac{54}{4} = 13.5$ kips

$$A_b = \frac{\pi(7/8)^2}{4} = 0.6013 \text{ in.}^2$$

$$\phi R_n = \phi F_v A_b = 0.75(48)(0.6013) = 21.65 \text{ kips} > 13.5 \text{ kips} \quad \text{(OK)}$$

Bearing (flange of tee controls):

$$\phi R_n = \phi(2.4dtF_u)$$
$$= 0.75(2.4)\left(\frac{7}{8}\right)(0.615)(58) = 56.18 \text{ kips} > 13.5 \text{ kips} \quad \text{(OK)}$$

Total tensile load:

$$\frac{4}{5}(90) = 72 \text{ kips}$$

Tensile load per bolt $= \dfrac{72}{4} = 18$ kips.

From AISC Table J3.5,

$$F_t = 117 - 1.9f_v \leq 90$$
$$= 117 - 1.9\left(\frac{13.5}{0.6013}\right) = 74.34 \text{ ksi}$$

$$\phi R_n = \phi F_t A_b = 0.75(74.34)(0.6013) = 33.53 \text{ kips} > 18 \text{ kips} \quad \text{(OK)}$$

Answer   The connection is adequate as a bearing-type connection. (In order not to obscure the combined loading features of this problem, prying action has not been included in the analysis shown here.)

b. Slip-critical connection, threads in shear plane:

From part a, shear and bearing are satisfactory. Also from part a, the reduced tensile strength is adequate, so the full tensile strength will be adequate.

Without tension, the allowable slip load (assuming Class A surfaces) is

$$\phi R_n = 17A_b = 17(0.6013) = 10.22 \text{ kips}$$

In the presence of tension, this must be multiplied by the reduction factor $1 - T/T_b$.

T = service load per bolt

$$= \frac{4}{5}\left(\frac{60}{4}\right) = 12 \text{ kips}$$

From AISC Table J3.1, the prescribed initial tension for a $\frac{7}{8}$-in. A325 bolt is

$$T_b = 39 \text{ kips}$$

The reduced allowable slip load per bolt is therefore

$$10.22\left(1 - \frac{12}{39}\right) = 7.08 \text{ kips}$$

The service load shear per bolt is

$$\frac{3}{5}\left(\frac{60}{4}\right) = 9 \text{ kips} > 7.08 \text{ kips} \quad \text{(N.G.)}$$

Answer   The connection is inadequate as a slip-critical connection.

Connections with fasteners in shear and tension can be designed by trial, but a more direct procedure can be used if one assumes that the design is controlled by the strength that is reduced. If the assumption turns out to be correct, no iteration is required. This technique will be illustrated in the following example.

EXAMPLE 7.9

A concentrically loaded connection is subjected to a service load shear force of 50 kips and a service load tensile force of 100 kips. The loads are 25% dead load and 75% live load. The fasteners will be in single shear, and bearing strength will be controlled by a $\frac{5}{16}$-in.-thick connected part. Assume that all spacing and edge distances are satisfactory, including those that permit the full bearing strength to be used. Determine the required number of $\frac{3}{4}$-in.-diameter A325 bolts for the following cases: (a) bearing-type connection with threads in the plane of shear and (b) slip-critical connection with threads in the plane of shear. All contact surfaces have clean mill scale.

Consider this to be a preliminary design so that no consideraton of prying action is necessary.

SOLUTION

Factored load shear = $1.2[0.25(50)] + 1.6[0.75(50)] = 75$ kips

Factored load tension = $1.2[0.25(100)] + 1.6[0.75(100)] = 150$ kips

a. Bearing-type connection, threads in shear:

Assume that tension controls:

$$F_t = 117 - 1.9f_v \le 90$$

$$\phi F_t = 0.75(117 - 1.9f_v) \le 0.75(90)$$

$$= 87.8 - 1.42f_v \le 67.5$$

Let $\phi F_t = \dfrac{150}{\sum A_b}$ and $f_v = \dfrac{75}{\sum A_b}$

where $\sum A_b$ is the total bolt area. Solving for $\sum A_b$,

$$\frac{150}{\sum A_b} = 87.8 - 1.42\left(\frac{75}{\sum A_b}\right)$$

$$150 = 87.8 \sum A_b - 1.42(75)$$

$$\sum A_b = 2.921 \text{ in.}^2$$

Area of one bolt:

$$A_b = \frac{\pi(3/4)^2}{4} = 0.4418 \text{ in.}^2$$

number required $= \dfrac{\sum A_b}{A_b} = \dfrac{2.921}{0.4418} = 6.6$

Try 7 bolts.

Check shear:

$$\phi R_n = \phi(48A_b) \times 7 \text{ bolts}$$

$$= 0.75(48)(0.4418)(7) = 111 \text{ kips} > 75 \text{ kips}   \text{ (OK)}$$

Check bearing:

$$\phi R_n = \phi(2.4dtF_u) \times 7 \text{ bolts}$$

$$= 0.75(2.4)\left(\frac{3}{4}\right)\left(\frac{5}{16}\right)(58)(7) = 171 \text{ kips} > 75 \text{ kips}   \text{ (OK)}$$

(Note: Spacing and edge-distance requirements must be checked in the final design.)

Answer   Use seven bolts. (If the bolts are placed in two rows, use eight bolts for symmetry.)

b. Slip-critical connection:

Assume that the slip load controls. The reduced slip-critical shear strength is

$$\phi R_n\left(1 - \frac{T}{T_b}\right)$$

For one bolt,

$$\phi R_n = 17A_b$$
$$T_b = 28 \text{ kips} \quad \text{(from AISC Table J3.1)}$$
$$T = \frac{100}{n}$$

where n is the number of bolts. Setting the total service load shear force equal to the reduced slip-critical shear strength for n bolts,

$$50 = n(17A_b)\left(1 - \frac{100/n}{28}\right) = n(17)(0.4418)\left(1 - \frac{100/n}{28}\right)$$
$$50 = 7.511n - 26.82$$
$$n = 10.33$$

Since seven bolts are adequate for shear, bearing, and tension (with a *reduced* tensile strength), these limit states will not have to be checked.

Answer   Use 11 $\frac{3}{4}$-in.-diameter A325 bolts. (If the bolts are placed in two or more rows, use 12 bolts for symmetry.)

7.10 ≡ WELDED CONNECTIONS

Structural welding is a process wherein the parts to be connected are heated and fused together, with supplementary molten metal added to the joint. For example, the tension member lap joint shown in Figure 7.25a can be constructed by welding across the ends of both connected parts. A relatively small depth of material will become molten, and upon cooling, the structural steel and the weld metal will act as one continuous part where they are joined. The additional metal is deposited from a special electrode, which is part of an electrical circuit that includes the connected part, or base metal. In the shielded metal arc welding (SMAW) process, shown schematically in Figure 7.26, current arcs across a gap between the electrode and base metal, heating the connected parts

FIGURE 7.25 ≡

(a)   (b)

FIGURE 7.26 ≡

and depositing part of the electrode into the molten base metal. A special coating on the electrode vaporizes and forms a protective gaseous shield, preventing the molten weld metal from oxidizing before it solidifies. The electrode is moved across the joint, and a weld bead is deposited, its size depending on the rate of travel of the electrode. As the weld cools, impurities rise to the surface, forming a coating called *slag* that must be removed before painting or making another pass with the electrode.

Shielded metal arc welding is normally done manually and is the process universally used for field welds. For shop welding, an automatic or semiautomatic process is usually used. Foremost among these is submerged arc welding (SAW). In this process, the end of the electrode and the arc are submerged in a granular flux that melts and forms a gaseous shield. There is more penetration

FIGURE 7.27 ≡

(a) Complete penetration groove welds

(b) Partial penetration groove welds

FIGURE 7.28 ≡

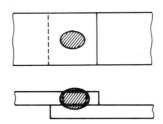

into the base metal than with shielded metal arc welding, and a higher strength is assigned. Other commonly used processes for shop welding include gas shielded metal arc, flux cored arc, and electroslag welding.

Quality control of welded connections is particularly difficult, since defects below the surface, or even minor flaws at the surface, will escape visual detection. Welders must be properly certified, and for critical work, special inspection techniques such as radiography or ultrasonic testing can be used.

The two most common types of welds are the *fillet weld* and the *groove weld*. The lap joint illustrated in Figure 7.25a is made with fillet welds, which are defined as those placed in a corner formed by two parts in contact. Fillet welds can also be used in a tee joint, shown in Figure 7.25b. Groove welds are those deposited in a gap, or groove, between two parts to be connected. They are most frequently used for butt, tee, and corner joints. In most cases, one or both of the connected parts will have beveled edges, called *prepared edges*, as shown in Figure 7.27a, although relatively thin material can be groove welded with no edge preparation. The welds shown in Figure 7.27a are complete penetration welds and can be made from one side, sometimes with the aid of a backing bar. Partial penetration groove welds can be made from one or both sides, with or without edge preparation (Figure 7.27b).

Figure 7.28 shows the plug or slot weld, which sometimes is used when more weld is needed than there is length of edge available. A circular or slotted hole is cut in one of the parts to be connected and is filled with the weld metal.

Of the two major types of welds, fillet welds are the most common and will be considered here in some detail. The design of complete penetration groove welds is a trivial exercise in that the weld will have the same strength as the base metal and the connected parts can be treated as completely continuous at the joint. The strength of a partial penetration groove weld will depend on the amount of penetration; once that has been determined, the design procedure will be essentially the same as that for a fillet weld.

7.11 ≡ FILLET WELDS

The design and analysis of fillet welds is based on the assumption that the cross section of the weld is a 45° right triangle as shown in Figure 7.29. Any reinforce-

FIGURE 7.29 ≡

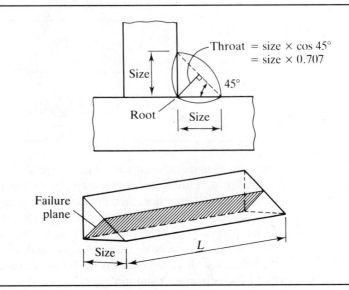

ment (buildup outside the hypotenuse of the triangle) or penetration is neglected. The size of a fillet weld is designated by the length of one of the two equal sides of this idealized cross section. Standard weld sizes are specified in increments of $\frac{1}{16}$ inch. Although a given length of weld can be loaded in any direction in shear, compression, or tension, a fillet weld is weakest in shear and is always assumed to fail in this mode. Specifically, failure is assumed to occur in shear on a plane through the throat of the weld. For fillet welds made with the shielded metal arc process, the throat is the perpendicular distance from the corner, or root, of the weld to the hypotenuse and is equal to 0.707 times the size of the weld. (The effective throat thickness for a weld made with the submerged arc welding process is larger. In this book, the shielded metal arc welding process will be conservatively assumed.) Thus, for a given length of weld L subjected to a load of P, the critical shearing stress is

$$f_v = \frac{P}{0.707 \times \text{size} \times L}$$

If the weld ultimate shearing stress F_W is used in this equation, the nominal load capacity of the weld can be written as

$$R_n = 0.707 \times \text{size} \times L \times F_W$$

and the nominal design strength is

$$\phi R_n = 0.707 \times \text{size} \times L \times \phi F_w \tag{7.17}$$

The strength of a fillet weld depends on the weld metal used — that is, it is a function of the type of electrode. The strength of the electrode is defined as its ultimate tensile strength, with strengths of 60, 70, 80, 90, 100, 110, and 120 kips per square inch available for the shielded metal arc welding process. The standard notation for specifying an electrode is the letter E followed by two or three digits indicating the tensile strength in kips per square inch and two digits specifying the type of coating. Since strength is the property of primary concern to the design engineer, the last two digits are usually represented by XX, and a typical designation would be E70XX or just E70, indicating an electrode with an ultimate tensile strength of 70 ksi. Electrodes should be selected to match the base metal. For the commonly used grades of steel, only the following two electrodes need be considered:

Use E70XX electrodes with steels that have a yield stress less than 60 ksi.

Use E80XX electrodes with steels that have a yield stress of 60 ksi or 65 ksi.

The notation for designating electrodes, as well as all provisions of the AISC Specification dealing with welds, have been taken from the *Structural Welding Code* of the American Welding Society.[47] Criteria not covered in the AISC Specification can be found in the AWS Code.

The design strengths of welds are given in AISC Table J2.5. The ultimate shearing stress F_W in a fillet weld is 0.6 times the tensile strength of the weld metal, denoted by F_{EXX}. The design stress is therefore ϕF_W, where $\phi = 0.75$ and $F_W = 0.60F_{EXX}$. For the usual two electrodes, the design strengths (stresses) are as follows:

E70XX: $\phi F_W = 0.75[0.60(70)] = 31.5$ ksi

E80XX: $\phi F_W = 0.75[0.60(80)] = 36$ ksi

An additional requirement is that the factored load shear on the base metal shall not produce a stress in excess of ϕF_{BM}, where F_{BM} is the nominal shear strength of the connected material. Thus, the factored load on the connection is subject to a limit of

$$\phi R_n = \phi F_{BM} \times \text{area of base metal subject to shear} \tag{7.18}$$

AISC J5, "Connecting Elements," gives the shear yielding strength as ϕR_n, where

$$\phi = 0.90$$
$$R_n = 0.60 A_g F_y \tag{AISC Equation J5-3}$$

and A_g is the area subjected to shear. The shear strength of the base metal can therefore be taken as

$$\phi F_{BM} = 0.90(0.60)F_y = 0.54F_y$$

As a consequence, when the load is in the same direction as the axis of the weld, the base metal must also be investigated by using the relationship of Equation 7.18. This can be explained by examining the welded bracket connection in

FIGURE 7.30 ≡

Figure 7.30. Assume that the load is close enough to the welded end so that any eccentricity is negligible. If the two welds are the same size, the design strength of *each* weld per inch of length can be found from Equation 7.17 as

$$0.707 \times \text{size} \times \phi F_W$$

whereas from Equation 7.18, the capacity of the bracket plate in shear per inch of length is

$$t \times \phi F_{BM}$$

Obviously, more load cannot be resisted by the weld than the base metal (the bracket in this case) is capable of carrying. This investigation must always be made when the base metal is subjected to shear.

In most welded connection problems, whether analysis or design, it is advantageous to work with the strength per unit length of weld, which will be either the strength of the weld itself or the strength of the base metal, whichever is smaller. This approach will be illustrated in the following examples.

≡ **EXAMPLE 7.10**

A flat bar used as a tension member is connected to a gusset plate as shown in Figure 7.31. The welds are $\frac{3}{16}$-in. fillet welds made with E70XX electrodes. The connected parts are of A36 steel. Assume that the tensile strength of the member is adequate and determine the design strength of the welded connection.

FIGURE 7.31 ≡

SOLUTION   Since the welds are placed symmetrically about the axis of the member, this connection qualifies as a simple connection, and each part of the weld will resist an equal share of the load. Only the total length of weld matters, and as long as the resultant applied load passes through the center of gravity of the weld (ignoring slight eccentricities), the location and orientation of the individual weld segments are irrelevant.

Strength of weld metal:
For E70 electrodes,

$$\phi F_W = 31.5 \text{ ksi}$$

$$0.707 \times \text{size} \times \phi F_W = 0.707\left(\frac{3}{16}\right)(31.5) = 4.176 \text{ kips/in.}$$

Check the strength of the base metal (the smallest thickness controls). From Equation 7.18,

$$\phi R_n = \phi F_{BM} \times \text{area subject to shear}$$

$$= \phi F_{BM} \times t = 0.54 F_y t = 0.54(36)\left(\frac{1}{4}\right)$$

$$= 4.86 \text{ kips/in.}$$

The weld strength controls. For the connection,

$$\phi R_n = 4.176 \text{ kips/in.} \times (4 + 4) \text{ in.} = 33.4 \text{ kips}$$

Answer   Design strength of the weld is 33.4 kips.

≡ **EXAMPLE 7.11**

A connection of the type used in Example 7.10 must resist a factored load of 40 kips. What total length of $\frac{3}{16}$-in. fillet weld, E70XX electrodes, is required?

SOLUTION   Capacity per inch of weld:
From Example 7.10,

$$\phi R_n/\text{in.} = 4.176 \text{ kips/in.}$$

Total length required:

$$\frac{40 \text{ kips}}{4.176 \text{ kips/in.}} = 9.58 \text{ in.}$$

Answer   Use 10 in. total, 5 in. each side.

FIGURE 7.32 ≡

Practical design of welded connections requires a consideration of such details as maximum and minimum weld sizes and lengths. The requirements for fillet welds are found in AISC J2.2b and are summarized here.

Minimum Size

The minimum size permitted is a function of the thickness of the thicker connected part and is given in AISC Table J2.4.

Maximum Size

Along the edge of a part less than $\frac{1}{4}$-in. thick, the maximum fillet weld size is equal to the thickness of the part. For thicker parts, the maximum size is $t - \frac{1}{16}$ in., where t is the thickness of the part.

Minimum Length

The minimum permissible length of a fillet weld is four times its size. This is certainly not a severe limitation, but if this length is not available, a shorter length can be used if the effective size of the weld is taken as one-fourth its length. Connections of the type shown in Figure 7.32, similar to those in the preceding examples, are in the special case category for shear lag in welded connections, covered in Chapter 3 of this book. AISC B3 states that the length of the welds in this case may not be less than the distance between them — that is, $\ell \geq w$.

End Returns

When a weld extends to the corner of a member, it must be continued around the corner as shown in Figure 7.33. This continuation, called an *end return*, is

FIGURE 7.33 ≡

for the purpose of avoiding stress concentrations and must have a length of at least two times the weld size. The length of end returns may be included in the computation of load capacity.

Small welds are generally cheaper than large welds. The maximum size that can be made with a single pass of the electrode is approximately $\frac{5}{16}$ inch, and multiple passes will add to the cost. In addition, for a given load capacity, although a small weld must be made longer, a larger and shorter weld will require more volume of weld metal.

EXAMPLE 7.12

A bar $4 \times \frac{1}{2}$ of A36 steel is used as a tension member to carry a service dead load of 6 kips and a service live load of 18 kips. It is to be attached to a $\frac{3}{8}$-in. gusset plate as shown in Figure 7.34. Design a welded connection.

SOLUTION

Since the base metal in this connection is A36 steel, E70XX electrodes will be used. No restrictions on connection length have been stipulated, so weld length will not be limited, and the smallest permissible size will be used.

$$\text{minimum size} = \frac{3}{16} \text{ in.} \quad \text{(AISC Table J2.4)}$$

Try a $\frac{3}{16}$-in. fillet weld, E70XX electrodes:

$$\text{capacity per inch} = 0.707\left(\frac{3}{16}\right)(31.5) = 4.176 \text{ kips/in.}$$

$$\text{capacity of base metal} = 0.54F_y t = 0.54(36)\left(\frac{3}{8}\right) = 7.29 \text{ kips/in.}$$

The weld strength of 4.176 kips/in. governs.

Factored load:

$$P_u = 1.2D + 1.6L = 1.2(6) + 1.6(18) = 36 \text{ kips}$$

$$\text{required length} = \frac{36}{4.176} = 8.62 \text{ in.}$$

$$\text{minimum length} = 4\left(\frac{3}{16}\right) = 0.75 \text{ in.} < 8.62 \text{ in.} \quad \text{(OK)}$$

FIGURE 7.34 ≡

FIGURE 7.35 ≣

End returns:

$$\text{minimum length} = 2\left(\frac{3}{16}\right) = 0.375 \text{ in.} \qquad \text{Use 1 in.}$$

For this type of connection, the length of the side welds must be at least equal to the transverse distance between them, or 4 inches in this case. The total length of weld provided, including end returns, will be

$$2(4 + 1) = 10 \text{ in.} > 8.62 \text{ in. required}$$

Answer   Use a $\frac{3}{16}$-in. fillet weld, E70XX electrodes, length = 10 in. (Figure 7.35).

Weld Symbols

Welds are specified on design drawings by standard symbols, which provide a convenient method for describing the required weld configuration. Details are given in Part 8 of the *Manual*, "Bolts, Welds, and Connected Elements" (Volume II), and will not be fully covered here. It is not the intent of this book to give a complete description of all symbols and all options but to provide a brief introduction to the standard symbols for fillet welds. The following discussion refers to the symbols shown in Figure 7.36.

The basic symbol is a horizontal line (reference line) containing information on the type, size, and length of weld and an inclined arrow pointing to the weld. A right triangle with the vertical leg on the left side is used to indicate a fillet weld. If the symbol for the type of weld is below the reference line, the weld is on the arrow side of the joint; that is, the part of the joint that the arrow is touching. If the symbol is above the line, the weld is on the other side of the joint, which may or may not be hidden from view in the drawing. The numbers on the reference line, from left to right, are the weld size and length. They should always be shown in this order. If both the front and back sides of the joint are to be welded, all information should be shown on each side of the reference line. A circle at the bend in the reference line is an instruction to weld all around the joint. To specify the process used or to furnish other information, a tail can be placed at the end of the reference line and the desired notation placed beside it. If no such reference is to be provided, the tail is omitted. Finally, a flag placed at the bend in the reference line indicates a field weld.

FIGURE 7.36 ≡

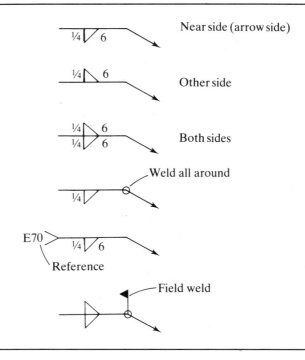

EXAMPLE 7.13

A plate $\frac{3}{4} \times 8$ of A36 steel is used as a tension member and is to be connected to a $\frac{3}{8}$-in.-thick gusset plate as shown in Figure 7.37. The length of the connection cannot exceed eight inches, and all welding must be done on the near side. Design a weld to develop the full tensile capacity of the member.

SOLUTION   Design strength of member based on gross area:

$$\phi_t P_n = 0.90 F_y A_g = 0.90(36)\left(\frac{3}{4}\right)(8) = 194.4 \text{ kips}$$

FIGURE 7.37 ≡

FIGURE 7.38 ≡

Design strength of member based on net area: For a flat bar connection, if the welds are along the sides only, $A_e = UA_g$. If there is also a transverse weld at the end, then $A_e = A_g$. Assuming the latter,

$$\phi_t P_n = 0.75 F_u A_e = 0.75(58)\left(\frac{3}{4}\right)(8) = 261 \text{ kips}$$

Design for a factored load of 194.4 kips

Use E70 electrodes:

$$\phi F_W = 31.5 \text{ ksi}$$

From AISC Table J2.4, minimum weld size $= \frac{1}{4}$ in.

Try $\frac{1}{4}$-in. E70 fillet weld.

design strength per inch of weld $= 0.707\left(\frac{1}{4}\right)(31.5) = 5.568 \text{ kips/in.}$

strength of base metal $= 0.54 F_y t = 0.54(36)\left(\frac{3}{8}\right) = 7.290 \text{ kips/in.}$

required length of weld $= \dfrac{194.4}{5.568} = 34.9 \text{ in.}$

With the given constraints, the available length is $8 + 8 + 8 = 24$ inches, so a $\frac{1}{4}$-in. weld cannot be used. If the length is set at 24 inches, the required weld size from Equation 7.17 is

$$\frac{\phi R_n}{0.707 \times L \times \phi F_w} = \frac{194.4}{0.707(24)(31.5)} = 0.364 \text{ in.}; \quad \text{Try } \frac{3}{8} \text{ in.}$$

maximum weld size $= \dfrac{3}{4} - \dfrac{1}{16} = \dfrac{11}{16}$ in. $> \dfrac{3}{8}$ in.   (OK)

minimum length $= 4\left(\dfrac{3}{8}\right) = 1.5$ in. < 24 in.   (OK)

Answer   Use the weld shown in Figure 7.38.

≡ PROBLEMS ≡

NOTE   In all problems, the given loads are static, and fatigue need not be considered.

In Problems 7.1 through 7.3, compute the design strength of the given connection.

7.1

FIGURE P7.1

7.2

FIGURE P7.2

7.3

FIGURE P7.3

7.4   Determine the number of $\frac{3}{4}$-in.-diameter A307 bolts required along line ab in Figure P7.4. The given loads are service loads. A36 steel is used.

$D = 7^k$
$L = 14^k$

$\frac{3}{8}''$ ℙ

$2L4 \times 4 \times \frac{1}{4}$

Section

FIGURE P7.4

7.5   The splice plates shown in Figure P7.5 are $\frac{1}{4}$-in. thick. How many $\frac{7}{8}$-in.-diameter A307 bolts are required? The given load is a factored load. A36 steel is used.

90^k $\frac{3}{8}''$ 90^k

FIGURE P7.5

7.6   **a.** How many $\frac{7}{8}$-in.-diameter A307 bolts are required for the connection shown in Figure P7.6?

b. What is the minimum width w?

c. What is the minimum splice length L?

L

w

A36 steel

$P_u = 120^k$
(factored)

$\frac{1}{2}''$ $\frac{1}{2}''$

FIGURE P7.6

7.7   A 10-ft-long tension member must resist a service dead load of 12 kips and a service live load of 36 kips. The connection is to be made with $\frac{7}{8}$-in.-diameter A307 bolts. Select a single equal-leg angle for the member, and design the connection. Use A36 steel.

7.8   Determine the design strength of the lap joint in Figure P7.8 for each of the
following cases. Assume the threads are in the shear plane in each case.

a. $\frac{7}{8}$-in.-diameter A307 bolts

b. $\frac{7}{8}$-in.-diameter A325 bolts, slip-critical connection. In addition to the design
strength, compute the maximum permissible total service load.

c. $\frac{7}{8}$-in.-diameter A490 bolts, bearing-type connection

FIGURE P7.8

7.9   Determine the required number of fasteners for the splice in Figure P7.9 for
each of the following cases. In each case, the threads are not in the shear plane.

a. $\frac{7}{8}$-in. A307 bolts

b. $\frac{7}{8}$-in. A325 bolts, bearing-type connection

FIGURE P7.9

7.10   Rework Problem 7.1 with A325 slip-critical bolts, threads in the plane of shear. In addition, compute the maximum permissible total service load.

7.11   Rework Problem 7.2 with A325 bolts, threads not in the shear plane, and slippage permitted.

7.12   Rework Problem 7.3 with A325 bolts and slippage permitted. It is not known whether the threads are in the plane of shear.

7.13   Rework Problem 7.4 with A325 bolts. No slip can be tolerated.

7.14   Rework Problem 7.7 with $\frac{1}{2}$-in. A490 slip-critical bolts. Threads are in the plane of shear.

7.15   Investigate both the tee and the bolts in the hanger connection in Figure P7.15. Include the effects of prying. The 150 kip load is a factored load.

FIGURE P7.15

7.16   Determine the adequacy of the hanger connection in Figure P7.16. Account for prying action. The 30 kip load is a factored load.

FIGURE P7.16

In Problems 7.17 through 7.20, assume that all spacings and edge distances meet all AISC Specification requirements (including those necessary for full bearing strength) and neglect prying forces. In each case, the line of action of the applied load passes through the center of gravity of the connection.

7.17 A structural tee bracket is attached to a column flange with six bolts as shown in Figure P7.17. Check this connection for compliance with the AISC Specification. The given load is a factored load.

¾″ ϕ A325 bolts
Bearing-type bolts
with threads in shear

FIGURE P7.17

7.18 If the connection shown in Figure P7.18 is not permitted to slip, how many $\frac{7}{8}$-in.-diameter A325 bolts are needed? The 100 kip load is a service load, consisting of 25 kips dead load and 75 kips live load. Assume that the threads are in the plane of shear.

FIGURE P7.18

7.19 The structural tee bracket shown in Figure P7.19 is attached to the column flange with six $\frac{7}{8}$-in.-diameter A325 bearing-type bolts. The threads are not in

FIGURE P7.19

the plane of shear. The given load is a factored load. Check these bolts for compliance with the AISC Specification.

7.20 A double-angle tension member is attached to a $\frac{7}{8}$-in. gusset plate, which in turn is connected to a column flange via another pair of angles as shown in Figure P7.20. The given load is a service load consisting of 25% dead load and 75% live load. All connections are to be made with $\frac{7}{8}$-in.-diameter A490 slip-critical bolts. Assume that the threads are in shear. Determine the required number of bolts and show their location on a sketch. A36 steel is used throughout.

FIGURE P7.20

In Problems 7.21 through 7.24, determine the design strength of the given connection.

7.21

FIGURE P7.21

7.22

FIGURE P7.22

7.23

¼″ fillet welds
E70 electrodes
A36 steel

⁵⁄₁₆″ ℞

8″

Bar 7 × ⁷⁄₁₆

FIGURE P7.23

7.24

³⁄₈″ ℞

¼″ fillet welds
E70 electrodes
A36 steel

P

8″

℞ ¼ × 8

FIGURE P7.24

7.25 A flat bar tension member is connected to a gusset plate as shown in Figure P7.25. The 75 kip load is a factored load. Determine the required length L.

³⁄₈″ ℞

Bar 5 × ½

1″

75^k

³⁄₁₆″ fillet welds
E70 electrodes

A36

L

FIGURE P7.25

7.26 Use a $\frac{1}{4}$-in. fillet weld with E70XX electrodes and find the total required length of weld (Figure P7.26). Use A572 Grade 50 steel.

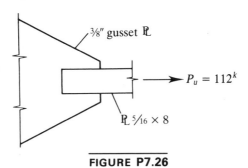

³⁄₈″ gusset ℞

$P_u = 112^k$

℞ ⁵⁄₁₆ × 8

FIGURE P7.26

7.27 Each weld shown in Figure P7.27 is a $\frac{3}{16}$-in. E70 fillet weld. What is the required length L?

FIGURE P7.27

7.28 Three sides of a tension member are welded with E70 fillet welds as shown in Figure P7.28. What size weld is required?

FIGURE P7.28

7.29 A single-angle tension member is to be connected to a gusset plate as shown in Figure P7.29. If a fillet weld and E70 electrode is used, what is the *minimum* permissible value of L?

FIGURE P7.29

7.30 Use the minimum permissible weld size, and compute the total length of fillet weld required for the angle-to-plate connection in Figure P7.30. E70XX electrodes and A36 steel are used.

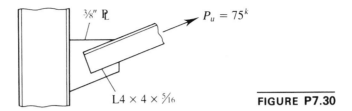

FIGURE P7.30

7.31 Design a weld for each of the connections in Figure P7.31. The loads are service loads, with equal parts dead and live load.

A36 steel

A572, grade 65 steel

(a) **(b)**

FIGURE P7.31

7.32 Design a welded connection to resist the full design strength of the tension member in Figure P7.32.

A36 steel **FIGURE P7.32**

7.33 Design a welded connection to resist 50% of the design strength of the tension member shown in Figure P7.33.

A242 steel **FIGURE P7.33**

7.34   Write a computer program to compute the various design strengths of a given high-strength bolt.

Input: Bolt diameter (in.), ASTM designation of bolt (325 or 490), F_u of the connected parts (ksi), and the thicknesses of the connected parts (in.).

Output: Same as input, plus the design tensile strength, design single shear strength (both with threads in shear and with threads not in shear), design bearing strength, and slip-critical service load, all in kips. Use AISC Equation J3-1a for the bearing strength and assume Class A surfaces for the slip-critical load.

7.35   Write a computer program to select the number of bolts required in a shear connection to resist given loads.

Input: Service dead load (kips), service live load (kips), ASTM designation of bolt (307, 325, or 490), the thicknesses of the connected parts (in.), F_u of the connected parts (ksi), and the following code for the type of connection:
SX = slip critical with threads excluded from the shear plane
SN = slip critical with threads not excluded from the shear plane
BX = bearing-type with threads excluded from the shear plane
BN = bearing-type with threads not excluded from the shear plane

Output: Same as input, plus the number of bolts required for diameters of $\frac{1}{2}$ in. to $1\frac{1}{2}$ in. in increments of $\frac{1}{8}$ in. Present your results in tabular form. Use AISC Equation J3-1a for the bearing strength and assume Class A surfaces for the slip-critical load.

7.36   Write a computer program to investigate high-strength bolts for compliance with the AISC Specification when they are subjected to both shear and tension. The program should be specialized for bearing-type connections with threads in the plane of shear.

Input: Bolt diameter (in.), type of bolt (325 or 490), number of bolts, F_u of connected parts (ksi), thicknesses of connected parts (in.), total factored shear load (kips), and total factored tension load (kips).

Output: Same as input, plus the shear and tension design strengths. Use AISC Equation J3-1a for the bearing strength.

7.37   Write a computer program to compute the length of various size fillet welds required to resist a given load.

Input: Applied load (kips), F_y of base metal (36 ksi to 65 ksi inclusive), and the thicknesses of the connected parts (in.).

Output: Same as input, plus the required electrode and a table of lengths for weld sizes that vary from the minimum to the maximum in increments of $\frac{1}{16}$ in. Present your output in the following form:

P = 40 KIPS

FY = 36 KSI

T1 = 0.50 IN.

T2 = 0.375 IN.

E70XX ELECTRODES

Size (in.)	Length (in.)	End Returns (in.)
0.1875	14.4	0.375
.	.	.
.	.	.
.	.	.

ECCENTRIC CONNECTIONS

EXAMPLES OF ECCENTRIC CONNECTIONS

An eccentric connection is one in which the resultant of the applied loads does not pass through the center of gravity of the fasteners or welds. If the connection has a plane of symmetry, the centroid of the shear area of the fasteners or welds may be used as the reference point, and the perpendicular distance from the line of action of the load to the centroid is called the *eccentricity*. Although a majority of connections are probably loaded eccentrically, in many cases the eccentricity is small and may be neglected.

The *framed beam* connection shown in Figure 8.1a is a typical eccentric connection. This connection, in either bolted or welded form, is commonly used to connect beams to columns. Although the eccentricities in this type of connection are usually negligible, they do exist and will be used here for illustration.

FIGURE 8.1 ≡

(a)

(b)

(c)

Two different connections are actually involved: the attachment of the beam to the framing angles and the attachment of the angles to the column. These connections illustrate the two basic categories of eccentric connections: those causing only shear in the fasteners or welds and those causing both shear and tension.

If the beam and angles are considered separately from the column (Figure 8.1b), the reaction *R* is seen to act at an eccentricity *e* from the centroid of the areas of the fasteners in the beam web. These fasteners are thus subjected to both a shearing force and a couple that lies in the plane of the connection and causes torsional shearing stress.

If the column and the angles are isolated from the beam as in Figure 8.1c, the fasteners in the column flange are seen to be subjected to the reaction *R* acting at an eccentricity *e* from the plane of the fasteners, producing the same couple as before. In this case, however, since the load is not in the plane of the fasteners, the couple will tend to put the upper part of the connection in tension and compress the lower part. The fasteners at the top of the connection will therefore be subjected to both shear and tension.

Although a bolted connection was used here for illustration, welded connections can be similarly categorized as either shear only or shear plus tension.

Maximum factored load reactions for various framed beam connections are given in Tables 9-2 through 9-12 in Part 9 of the *Manual*, "Simple Shear and PR Moment Connections" (Volume II). The relatively small eccentricity in these connections is neglected, and only the direct shear is considered.

8.2 ≡ ECCENTRIC BOLTED CONNECTIONS: SHEAR ONLY

The column bracket connection shown in Figure 8.2 is an example of a bolted connection subjected to eccentric shear. Two approaches exist for the solution of this problem: the traditional elastic analysis and the more accurate (but more complex) ultimate strength analysis. Both will be illustrated.

FIGURE 8.2 ≡

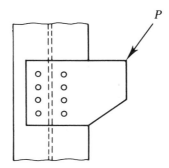

Elastic Analysis

In Figure 8.3a, the fastener shear areas and the load are shown separate from the column and bracket plate. The eccentric load P can be replaced by the same load acting at the centroid plus the couple $M = Pe$, where e is the eccentricity. If this is done, the load will be concentric, and each fastener can be assumed to resist an equal share of the load, given by $p_c = P/n$, where n is the number of fasteners. The fastener forces resulting from the couple can be found by considering the shearing stress in the fasteners to be the result of torsion of a cross section made up of the cross-sectional areas of the fasteners. If such an assumption is made, the shearing stress in each fastener can be found from the torsion formula

$$f_v = \frac{Md}{J} \tag{8.1}$$

where

   d = distance from the centroid of the area to the point where the stress is being computed

   J = polar moment of inertia of the area about the centroid

and the stress f_v is perpendicular to d. Although the torsion formula is applicable only to the right circular cylinders, its use here is conservative, yielding stresses that are somewhat larger than the actual stresses.

FIGURE 8.3 ≡

(a)   (b)

If the parallel-axis theorem is used, and the polar moment of inertia of each circular area about its own centroid is neglected, J for the total area can be approximated as

$$J = \sum Ad^2 = A \sum d^2$$

provided all fasteners have the same area A. Equation 8.1 can then be written as

$$f_v = \frac{Md}{A \sum d^2}$$

and the shear force in each fastener caused by the couple is

$$p_m = Af_v = A \frac{Md}{A \sum d^2} = \frac{Md}{\sum d^2}$$

The two components of shear force thus determined can be added vectorially to obtain the resultant force p as shown in Figure 8.3b, where the lower right-hand fastener is used as an example. When the largest resultant is determined, the fastener size is selected so as to resist this force. The critical fastener cannot always be found by inspection, and several force calculations may be necessary.

It is generally more convenient to work with rectangular components of forces. For the typical fastener shown in Figure 8.4, the horizontal and vertical components of force resulting from direct shear are

$$p_{cx} = \frac{P_x}{n} \quad \text{and} \quad p_{cy} = \frac{P_y}{n}$$

In terms of the x- and y-coordinates of the centers of the fastener areas,

$$\sum d^2 = \sum (x^2 + y^2)$$

where the origin of the coordinate system is at the centroid of the total fastener shear area. The x-component of p_m is given by

$$p_{mx} = \frac{y}{d} p_m = \frac{y}{d} \frac{Md}{\sum d^2} = \frac{y}{d} \frac{Md}{\sum (x^2 + y^2)} = \frac{My}{\sum (x^2 + y^2)}$$

Similarly,

$$p_{my} = \frac{Mx}{\sum (x^2 + y^2)}$$

and the total fastener force is

$$p = \sqrt{(\sum p_x)^2 + (\sum p_y)^2}$$

where

$$\sum p_x = p_{cx} + p_{mx}$$
$$\sum p_y = p_{cy} + p_{my}$$

FIGURE 8.4 ≡

If P, the load applied to the connection, is a factored load, then the force p on the fastener is the factored load to be resisted in shear and bearing — that is, the required design strength.

≡ **EXAMPLE 8.1**

Determine the critical fastener force in the bracket connection of Figure 8.5.

SOLUTION   The centroid of the fastener group can be found by using a horizontal axis through the lower row and applying the principal of moments:

$$\bar{y} = \frac{2(5) + 2(8) + 2(11)}{8} = 6 \text{ in.}$$

FIGURE 8.5 ≡

The horizontal and vertical components of the load are

$$P_x = \frac{1}{\sqrt{5}} (50) = 22.36 \text{ kips} \leftarrow$$

$$P_y = \frac{2}{\sqrt{5}} (50) = 44.72 \text{ kips} \downarrow$$

Referring to Figure 8.6a, we can compute the moment of the load about the centroid as follows:

$$M = 44.72(12 + 2.75) - 22.36(14 - 6) = 480.7 \text{ in.-kips (clockwise)}$$

Figure 8.6b shows the directions of all component bolt forces and the relative magnitudes of the components caused by the couple. Using these directions and relative magnitudes as a guide and bearing in mind that forces add by the parallelogram law, one can conclude that the lower right-hand fastener will have the largest resultant force.

FIGURE 8.6 ≡

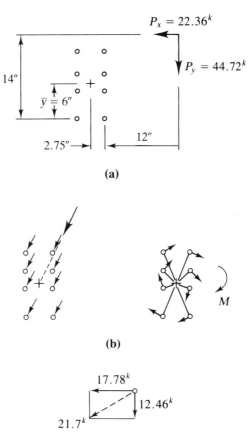

(a)

(b)

(c)

The horizontal and vertical components of force in each bolt resulting from the concentric load are

$$p_{cx} = \frac{22.36}{8} = 2.795 \text{ kips} \leftarrow$$

$$p_{cy} = \frac{44.72}{8} = 5.590 \text{ kips} \downarrow$$

For the couple,

$$\sum (x^2 + y^2) = 8(2.75)^2 + 2[(6)^2 + (1)^2 + (2)^2 + (5)^2] = 192.5 \text{ in.}^2$$

$$p_{mx} = \frac{My}{\sum (x^2 + y^2)} = \frac{480.7(6)}{192.5} = 14.98 \text{ kips} \leftarrow$$

$$p_{my} = \frac{Mx}{\sum (x^2 + y^2)} = \frac{480.7(2.75)}{192.5} = 6.867 \text{ kips} \downarrow$$

$$\sum p_x = 2.795 + 14.98 = 17.78 \text{ kips} \leftarrow$$

$$\sum p_y = 5.590 + 6.867 = 12.46 \text{ kips} \downarrow$$

$$p = \sqrt{(17.78)^2 + (12.46)^2} = 21.7 \text{ kips} \qquad \text{(see Figure 8.6c)}$$

Answer The critical fastener force is 21.7 kips. Inspection of the magnitudes and directions of the horizontal and vertical components of the forces confirms the earlier conclusion that the fastener selected is indeed the critical one.

Ultimate Strength Analysis

The foregoing procedure is relatively easy to apply but is inaccurate, albeit on the conservative side. The major flaw in the analysis is the implied assumption that the fastener load-deformation relationship is linear and that the yield stress is not exceeded. Experimental evidence shows that this is not the case and that individual fasteners do not have a well-defined shear yield stress. The procedure to be described here determines the ultimate strength of the connection by using an experimentally determined nonlinear load-deformation relationship for the individual fasteners.

The experimental study reported in Crawford and Kulak[48] used $\frac{3}{4}$-in.-diameter A325 bearing-type bolts and A36 steel plates, but the results can be used with little error for A325 bolts of different sizes and steels of other grades. The procedure gives conservative results when used with slip-critical bolts and with A490 bolts.[13]

The bolt force R corresponding to a deformation Δ is given by

$$R = R_{ult}(1 - e^{-\mu\Delta})^\lambda$$
$$= 74(1 - e^{-10\Delta})^{0.55} \tag{8.2}$$

where

$$R_{ult} = \text{bolt shear force at failure} = 74 \text{ kips}$$
$$e = \text{base of natural logarithms}$$
$$\mu = \text{a regression coefficient} = 10$$
$$\lambda = \text{a regression coefficient} = 0.55$$

The ultimate strength of the connection is based on the following assumptions:

1. At failure, the fastener group rotates about an instantaneous center (I.C.).
2. The deformation of each fastener is proportional to its distance from the instantaneous center and acts perpendicularly to the radius of rotation.
3. The capacity of the connection is reached when the ultimate strength of the fastener farthest from the instantaneous center is reached. (Figure 8.7 shows the bolt forces as resisting forces acting to oppose the applied load.)
4. The connected parts remain rigid.

As a consequence of the second assumption, the deformation of an individual fastener is

$$\Delta = \frac{r}{r_{max}} \Delta_{max} = \frac{r}{r_{max}} \quad (0.34)$$

where

$$r = \text{distance from the instantaneous center to the fastener}$$
$$r_{max} = \text{distance to the farthest fastener}$$
$$\Delta_{max} = \text{deformation of the farthest fastener at ultimate} = 0.34 \text{ in.}$$
$$\text{(determined experimentally)}$$

FIGURE 8.7 ≡

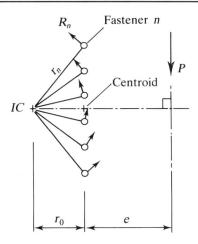

As with the elastic analysis, it is more convenient to work with rectangular components of forces. These are given by

$$R_y = \frac{x}{r} R \quad \text{and} \quad R_x = \frac{y}{r} R$$

where x and y are the horizontal and vertical distances from the instantaneous center to the fastener. At the instant of failure, equilibrium must be maintained, and the following three equations of equilibrium will be applied to the fastener group (refer to Figure 8.7):

$$\sum F_x = \sum_{n=1}^{m} (R_x)_n - P_x = 0 \tag{8.3}$$

$$M_{IC} = P(r_0 + e) - \sum_{n=1}^{m} (r_n \times R_n) = 0 \tag{8.4}$$

and

$$\sum F_y = \sum_{n=1}^{m} (R_y)_n - P_y = 0 \tag{8.5}$$

where the subscript n identifies an individual fastener and m is the total number of fasteners. The general procedure is to assume the location of the instantaneous center, then determine if the corresponding value of P satisfies the equilibrium equations. If so, this location is correct, and P is the capacity of the connection. The specific procedure is as follows:

1. Assume a value for r_0.
2. Solve for P from Equation 8.4.
3. Substitute r_0 and P into Equations 8.3 and 8.5.
4. If these equations are satisfied within an acceptable tolerance, the analysis is complete. Otherwise, a new trial value of r_0 must be selected and the process repeated.

For the usual case of vertical loading, Equation 8.3 will automatically be satisfied. For simplicity and without loss of generality, we will consider only this case. Even with this assumption, however, the computations for even the most trivial problems are overwhelming, and computer assistance is needed. Part b of the following example was worked with the aid of a standard spreadsheet program for personal computers.

≡ **EXAMPLE 8.2**

The bracket connection shown in Figure 8.8 must support an eccentric factored load of 53 kips. The connection was designed to have two vertical rows of four bolts, but one bolt was inadvertently omitted. If $\frac{7}{8}$-in.-diameter A325 bearing-type bolts are used, is the connection adequate? Assume that the bolt threads

FIGURE 8.8 ≡

are in the plane of shear. Use A36 steel and perform the following analyses: (a) elastic analysis; (b) ultimate strength analysis.

SOLUTION **a.** Elastic analysis:

For an $x–y$ coordinate system with the origin at the center of the lower left bolt (Figure 8.9),

$$\bar{y} = \frac{2(3) + 2(6) + 1(9)}{7} = 3.857 \text{ in.}$$

$$\bar{x} = \frac{3(3)}{7} = 1.286 \text{ in.}$$

$$\sum (x^2 + y^2) = 4(1.286)^2 + 3(1.714)^2 + 2(3.856)^2 + 2(0.857)^2$$
$$+ 2(2.143)^2 + 1(5.143)^2 = 82.27 \text{ in.}^2$$

$$e = 3 + 5 - 1.286 = 6.714 \text{ in.}$$

$$M = Pe = 53(6.714) = 355.8 \text{ in.-kips (clockwise)}$$

$$p_{cy} = \frac{53}{7} = 7.571 \text{ kips} \downarrow \qquad p_{cx} = 0$$

From the directions and relative magnitudes shown in Figure 8.9, the lower right bolt is judged to be critical:

$$p_{mx} = \frac{My}{\sum (x^2 + y^2)} = \frac{355.8(3.857)}{82.27} = 16.68 \text{ kips} \leftarrow$$

$$p_{my} = \frac{Mx}{\sum (x^2 + y^2)} = \frac{355.8(1.714)}{82.27} = 7.413 \text{ kips} \downarrow$$

$$\sum p_x = 16.68 \text{ kips}$$

$$\sum p_y = 7.571 + 7.413 = 14.98 \text{ kips}$$

$$p = \sqrt{(16.68)^2 + (14.98)^2} = 22.4 \text{ kips}$$

FIGURE 8.9 ≡

Determine the bolt design strength.

Bearing:

Bolt spacing: $3d = 3\left(\dfrac{7}{8}\right) = 2.62$ in. < 3 in. furnished   (OK)

Edge distance: $1.5d = 1.5\left(\dfrac{7}{8}\right) = 1.31$ in. < 2 in.   (OK)

Since there are more than two bolts in the direction of the load, AISC Equation J3-1a may be used:

$R_n = 2.4dtF_u$

$= 2.4\left(\dfrac{7}{8}\right)(0.455)(58) = 55.42$ kips

and the design strength is

$\phi R_n = 0.75(55.42) = 41.6$ kips > 22.4 kips   (OK)

Shear:

$A_b = \dfrac{\pi(7/8)^2}{4} = 0.6013$ in.^{2}

$\phi R_n = \phi F_v A_b = 0.75(48)(0.6013) = 21.6$ kips < 22.4 kips   (NG)

Answer   The connection is unsatisfactory by elastic analysis.

b. Ultimate strength analysis:

This analysis will be performed with the aid of standard spreadsheet software. The results of the final trial value of $r_0 = 1.57104$ are given in Table 8.1. The coordinate system and bolt numbering scheme are shown in Figure 8.10.

TABLE 8.1 ≡

Fastener	x	y	r	Δ	R	rR	R_y
1	0	0	3.868	0.255	70.774	273.731	5.221
2	3	0	5.067	0.334	72.553	367.598	47.045
3	0	3	0.903	0.060	47.649	43.046	15.050
4	3	3	3.395	0.224	69.563	236.188	67.310
5	0	6	2.162	0.143	63.631	137.555	8.398
6	3	6	3.922	0.259	70.891	278.061	59.377
7	0	9	5.151	0.340	72.631	374.107	4.023
Sum						1710.287	206.424

From Equation 8.4,

$$P(r_0 + e) = \sum rR$$

$$P = \frac{\sum rR}{r_0 + e} = \frac{1710.29}{1.57104 + 6.71429} = 206.424 \text{ kips}$$

where e has been expressed to six significant figures for consistency.
From Equation 8.5,

$$\sum F_y = \sum R_y - P = 206.424 - 206.424 = 0.000$$

Since the applied load has no horizontal component, Equation 8.3 is automatically satisfied.

The load of 206.424 kips just computed is the failure load for the connection and is based on the critical fastener reaching its ultimate load capacity. If the connection failure load is multiplied by the ratio of fastener design strength to the fastener ultimate strength in Crawford and Kulak[48] (74 kips), we obtain the connection capacity.

FIGURE 8.10 ≡

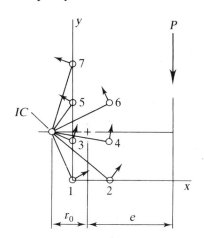

From part a, the design strength of one bolt (based on shear) is 21.6 kips.

Maximum factored load = 206(21.6/74) = 60.1 kips > 53 kips   (OK)

Answer   The connection is satisfactory by ultimate strength analysis.

Tables 8-18 through 8-25 in Part 8 of the *Manual* give coefficients for the design or analysis of common configurations of fastener groups subjected to eccentric vertical loads. For each arrangement of fasteners considered, the tables give a value for C, the ratio of connection failure load to fastener ultimate strength. To obtain a safe connection load, this constant must be multiplied by the design strength of the particular fastener used. For eccentric connections not included in the tables, the elastic method, which is conservative, may be used. Of course, a computer program or spreadsheet software may also be used to perform an ultimate strength analysis.

8.3 ≡ ECCENTRIC BOLTED CONNECTIONS: ≡ SHEAR PLUS TENSION

In a connection such as the one for the tee-stub bracket of Figure 8.11, an eccentric load creates a couple that will increase the tension in the upper row of fasteners and decrease it in the lower row. If the fasteners are bolts with no initial tension, the upper bolts will be put into tension, and the lower ones will not be affected. Regardless of the type of fastener, each one will receive an equal share of the *shear* load.

If the fasteners are pretensioned high-strength bolts, the contact surface between the column flange and the bracket flange will be uniformly compressed before the external load is applied. The bearing pressure will be equal to the total bolt tension divided by the area of contact. As the load P is gradually

FIGURE 8.11 ≡

FIGURE 8.12 ≡

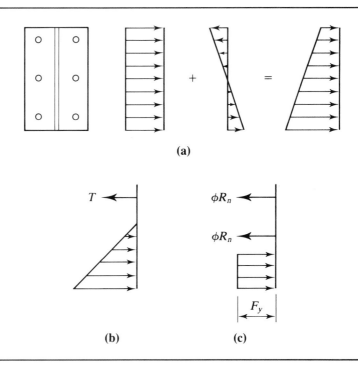

(a)

(b)   (c)

applied, the compression at the top will be relieved and the compression at the bottom will increase (Figure 8.12a). When the compression at the top has been completely overcome, the components will separate, and the couple *Pe* will be resisted by tensile bolt forces and compression on the remaining surface of contact (Figure 8.12b). As the ultimate load is approached, the forces in the bolts will approach their ultimate tensile strengths (which can be taken as the design strength in tension, reduced because of the shear), and the bearing pressure can be taken as equal to the yield stress of the bracket material. This condition is shown in Figure 8.12c.

Although this model is a simplification of a very complex problem, it is a reasonable one and will be used here. The location of the "neutral axis" in Figure 8.12c is determined from a consideration of force equilibrium.

≡ **EXAMPLE 8.3**

A beam-to-column connection is made with a structural tee as shown in Figure 8.13. Eight $\frac{3}{4}$-in.-diameter A325 fully tightened bearing-type bolts are used to attach the flange of the tee to the column flange. Investigate the adequacy of this connection (the tee-to-column connection) if it is subjected to a factored load of 88 kips at an eccentricity of 3 inches. Assume that the bolt threads are in the plane of shear. All structural steel is A36.

FIGURE 8.13 ≡

W10 × 49

SOLUTION   Shear load per bolt $= \dfrac{88}{8} = 11$ kips

$$A_b = \frac{\pi(3/4)^2}{4} = 0.4418 \text{ in.}^2$$

Design strength in shear:

$$\phi R_n = \phi F_v A_b = 0.75(48)(0.4418) = 15.90 \text{ kips} > 11 \text{ kips} \quad \text{(OK)}$$

Design strength in bearing:

$$\phi R_n = \phi(2.4 d t F_u) = 0.75(2.4)\left(\frac{3}{4}\right)(0.560)(58)$$

$$= 43.8 \text{ kips} > 11 \text{ kips} \quad \text{(OK)}$$

The connection is satisfactory in shear and bearing. To check tension, assume the limit state of Figure 8.12c and compute the corresponding design moment (moment strength). From AISC Table J3.5,

$$F_t = 117 - 1.9 f_v \leq 90 \text{ ksi}$$

$$= 117 - 1.9\left(\frac{11}{0.4418}\right) = 69.69 \text{ ksi} < 90 \text{ ksi}$$

and the design strength is

$$\phi R_n = \phi F_t A_b = 0.75(69.69)(0.4418) = 23.09 \text{ kips}$$

or

$$2(23.09) = 46.18 \text{ kips for two bolts}$$

Assume that the neutral axis at failure is below the bottom row of bolts as shown in Figure 8.14. The axis can be located by summing forces:

$$\sum F = 0: \quad 8a(36) = 4(46.18)$$

$$a = 0.6414 \text{ in.}$$

FIGURE 8.14 ≡

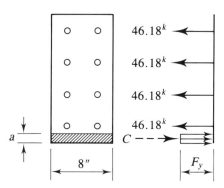

This places the axis below the bottom row of bolts as assumed. The design strength can be obtained by computing the moment of the couple formed by the compressive and tensile forces. Taking moments about the resultant compressive force C,

$$\phi M_n = \sum M_c = 46.18\left[\left(10.5 - \frac{0.6414}{2}\right) + \left(7.5 - \frac{0.6414}{2}\right)\right.$$
$$\left. + \left(4.5 - \frac{0.6414}{2}\right) + \left(1.5 - \frac{0.6414}{2}\right)\right] = 1049 \text{ in.-kips}$$

The factored load moment is

$$M_u = P_u e = 88(3) = 264 \text{ in.-kips} < 1049 \text{ in.-kips} \quad \text{(OK)}$$

Answer   The connection is satisfactory.

8.4 ≡ ECCENTRIC WELDED CONNECTIONS: SHEAR ONLY

Eccentric welded connections are analyzed in much the same way as bolted connections, except that unit lengths of weld replace individual fasteners in the computations. As in the case of eccentric bolted connections loaded in shear, welded shear connections can be investigated by either elastic or ultimate strength methods.

Elastic Analysis

The load on the bracket in Figure 8.15a may be considered to act in the plane of the weld — that is, the plane of the throat. If this slight approximation is made, the load will be resisted by the area of weld shown in Figure 8.15b. Computations are simplified, however, if a unit throat dimension is used. The

FIGURE 8.15 ≡

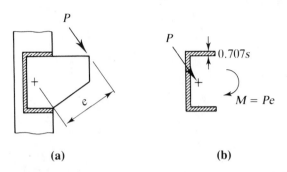

(a)   (b)

calculated load can then be multiplied by 0.707 times the weld size to obtain the actual load.

An eccentric load in the plane of the weld subjects the weld to both direct shear and torsional shear. Since all elements of the weld resist an equal portion of the direct shear, the direct shear stress is

$$f_1 = \frac{P}{1(L)}$$

where L is the total length of the weld and is numerically equal to the shear area, since a unit throat size has been assumed. If rectangular components are used,

$$f_{1x} = \frac{P_x}{L} \quad \text{and} \quad f_{1y} = \frac{P_y}{L}$$

where P_x and P_y are the x- and y-components of the applied load. The shearing stress caused by the couple is found with the torsion formula

$$f_2 = \frac{Md}{J}$$

where

d = distance from the centroid of the shear area to the point where the stress is being computed

J = polar moment of inertia of that area.

Figure 8.16 shows this stress at the upper right-hand corner of the given weld. In terms of rectangular components,

$$f_{2x} = \frac{My}{J} \quad \text{and} \quad f_{2y} = \frac{Mx}{J}$$

FIGURE 8.16 ≡

Also,

$$J = \int_A r^2 \, dA = \int_A (x^2 + y^2) \, dA = \int_A x^2 \, dA + \int_A y^2 \, dA = I_y + I_x$$

where I_x and I_y are the rectangular moments of inertia of the shear area. Once the rectangular components of f_1 and f_2 have been found, they can be added vectorially to obtain the resultant shearing stress at the point of interest:

$$f_v = \sqrt{(\textstyle\sum f_x)^2 + (\textstyle\sum f_y)^2}$$

As with bolted connections, the critical location for this resultant stress can usually be determined from an inspection of the relative magnitudes and directions of the direct and torsional shearing stress components.

EXAMPLE 8.4

Determine the size of weld required for the bracket connection in Figure 8.17. The 60 kip load is a factored load. A36 steel is used for both the bracket and the column.

SOLUTION

The eccentric load may be replaced by a concentric load and a couple, as shown in Figure 8.17. The direct shearing stress, in kips per inch, is the same for all segments of the weld and is equal to

$$f_{1y} = \frac{60}{8 + 12 + 8} = \frac{60}{28} = 2.143 \text{ kips/in.}$$

Before computing the torsional component of shearing stress, the location of the centroid of the weld shear area must be determined. From the principle of moments with summation of moments about the y-axis,

$$\bar{x}(28) = 8(4)(2) \qquad \bar{x} = 2.286 \text{ in.}$$

The eccentricity e is $10 + 8 - 2.286 = 15.71$ in., and the torsional moment is

$$M = Pe = 60(15.71) = 942.6 \text{ in.-kips}$$

FIGURE 8.17 ≡

If the moment of inertia of each horizontal weld about its own centroidal axis is neglected, the moment of inertia of the total weld area about its horizontal centroidal axis is

$$I_x = \frac{1}{12}(1)(12)^3 + 2(8)(6)^2 = 720.0 \text{ in.}^4$$

Similarly,

$$I_y = 2\left[\frac{1}{12}(1)(8)^3 + 8(4 - 2.286)^2\right] + 12(2.286)^2 = 195.0 \text{ in.}^4$$

and

$$J = I_x + I_y = 720.0 + 195.0 = 915.0 \text{ in.}^4$$

Figure 8.17 shows the directions of both components of stress at each corner of the connection. By inspection, either the upper right-hand corner or the lower right-hand corner may be taken as the critical location. If the lower right-hand corner is selected,

$$f_{2x} = \frac{My}{J} = \frac{942.6(6)}{915.0} = 6.181 \text{ kips/in.}$$

$$f_{2y} = \frac{Mx}{J} = \frac{942.6(8 - 2.286)}{915.0} = 5.886 \text{ kips/in.}$$

$$f_v = \sqrt{(6.181)^2 + (2.143 + 5.886)^2} = 10.13 \text{ kips/in.}$$

Check the strength of the base metal. From Equation 7.18,

$$\phi R_n = \phi F_{BM} \times \text{area subject to shear}$$
$$= \phi F_{BM} \times t = 0.54 F_y t = 0.54(36)(\tfrac{9}{16})$$
$$= 10.94 \text{ kips/in.} > 10.13 \text{ kips/in.} \quad (\text{OK})$$

From Equation 7.17, the weld strength is

$$\phi R_n = 0.707 \times s \times L \times \phi F_W$$

The matching electrode for A36 steel is E70, with $\phi F_W = 31.5$ ksi. The required weld size is therefore

$$s = \frac{\phi R_n}{0.707 L \phi F_W} = \frac{10.13}{0.707(1.0)(31.5)} = 0.455 \text{ in.}$$

Answer   Use a $\frac{1}{2}$-in. fillet weld, E70 electrode.

Ultimate Strength Analysis

Eccentric welded shear connections may be safely designed by elastic methods, but the factor of safety against failure will be larger than necessary and will vary from connection to connection.[49] This type of analysis suffers from some of the same shortcomings as the elastic method for eccentric bolted connections, including the assumption of a linear load-deformation relationship for the weld. Another source of error is the assumption that the strength of the weld is independent of the direction of the applied load. An ultimate strength procedure is presented in Part 8 in Volume II of the *Manual* and will be summarized here. It is based on the research by Butler *et al.*[49] and Kulak and Timler[50] and closely parallels the method used for eccentric bolted connections.[48]

Instead of considering individual fasteners, we treat the continuous weld as an essembly of discrete segments. At failure, the applied connection load is resisted by forces in each element, with each force acting perpendicular to the radius constructed from an instantaneous center of rotation to the centroid of the segment (Figure 8.18). This concept is essentially the same as that used for the fasteners. However, determining which element has the maximum deformation and computing the force in each element at failure is much more difficult.

FIGURE 8.18 ≡

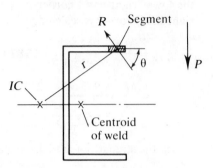

To determine the critical element, the ratio Δ_{max}/r is computed for each element, where

$$\Delta_{max} = 1.087w(\theta + 6)^{-0.65} \leq 0.17w$$

θ = the angle the resisting force makes with the axis of the weld segment (Figure 8.18)

w = weld leg size

r = the distance from the I.C. to the centroid of the segment

The element with the smallest ratio is the one that reaches its ultimate capacity first. The deformation of the other elements can then be computed as

$$\Delta = \frac{r}{r_{max}} \Delta_{max}$$

where

r = radius for the element

$\Delta_{max}/r_{max} = \Delta_{max}/r$ for the critical element.

The resisting force for each element can be found from the following expression:

$$R = 0.60F_{EXX}(1.0 + 0.50 \sin^{1.5} \theta)[p(1.9 - 0.9p)]^{0.3}$$

where

F_{EXX} = weld electrode strength

$$p = \frac{\Delta}{\Delta_{max}}$$

(Unlike the bolted case, R is a function of θ.) The preceding computations are based on an assumed location of the instantaneous center of rotation. If it is the actual location, the equations of equilibrium will be satisfied. The remaining details are the same as for a bolted connection:

1. Solve for the load capacity from the equation

$$\sum M_{IC} = 0$$

where IC refers to the instantaneous center.

2. If the two force equilibrium equations are satisfied, the assumed location of the instantaneous center and the load found in Step 1 are correct; otherwise, assume a new location and repeat the entire process.

The absolute necessity for the use of a computer is obvious. Computer solutions for various common configurations of eccentric welded shear connections are given in tabular form in Part 8 of the *Manual*. Tables 8-38 through 8-45 give factored load capacities for various common combinations of horizontal and vertical weld segments based on an ultimate strength analysis. These tables may be used for either design or analysis and will cover almost any situation the designer is likely to encounter. For those connections not covered by the tables, the more conservative elastic method may be used.

EXAMPLE 8.5

Determine the weld size required for the connection of Example 8.4 based on ultimate strength considerations. Use the tables for Eccentrically Loaded Weld Groups given in Part 8 of the *Manual*.

SOLUTION

The weld of Example 8.4 is the same type as the one shown in Table 8-42 (Angle = 0°), and the loading is similar. The following geometric constants are required for entry into the table:

$$a = \frac{a\ell}{\ell} = \frac{e}{\ell} = \frac{15.7}{12} = 1.3$$

$$k = \frac{k\ell}{\ell} = \frac{8}{12} = 0.67$$

By interpolation in Table 8-42 for $a = 1.3$,

$$C = 1.14 \text{ for } k = 0.6$$

and

$$C = 1.30 \text{ for } k = 0.7$$

Interpolating between these two values for $k = 0.67$,

$$C = 1.25$$

For E70XX electrodes, $C_1 = 1.0$.

The required value of D is

$$D = \frac{P_u}{CC_1\ell} = \frac{60}{1.25(1.0)(12)} = 4.0$$

The required weld size is therefore

$$\frac{1}{16}(4.0) = 0.25 \quad \text{(versus 0.455 in. required in Example 8.4)}$$

Answer   Use a $\frac{1}{4}$-in. fillet weld, E70 electrode.

Special Provision for Axially Loaded Members

When a structural member is axially loaded, the stress is uniform over the cross section, and the resultant force may be considered to act along the *gravity axis*, which is a longitudinal axis through the centroid. For the member to be concentrically loaded at its ends, the resultant resisting force furnished by the connection must also act along this axis. If the member has a symmetrical cross section, this can be accomplished by placing the welds or bolts symmetrically. If

FIGURE 8.19 ≡

FIGURE 8.20 ≡

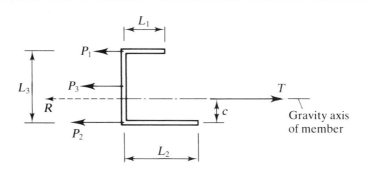

the member is one with an unsymmetrical cross section, such as the double-angle section in Figure 8.19, a symmetrical placement of welds or bolts will result in an eccentrically loaded connection, with a couple of Te (Figure 8.19b).

AISC J1.8 permits this eccentricity to be neglected in statically loaded members. When the member is subjected to fatigue caused by repeated loading or reversal of stress, the eccentricity must be either accounted for or eliminated by an appropriate placement of the welds or bolts. (Of course, this may be done even if the member is subjected to static loads only.) The correct placement can be determined by applying the force and moment equilibrium equations. For the welded connection of Figure 8.20, the first equation can be obtained by summing moments about the lower longitudinal weld:

$$\sum M_{L_2} = Tc - P_3 \frac{L_3}{2} - P_1 L_3 = 0$$

This equation can be solved for P_1, the required resisting force in the upper longitudinal weld. This value can then be substituted into the force equilibrium equation:

$$\sum F = T - P_1 - P_2 - P_3 = 0$$

This equation can be solved for P_2, the required resisting force in the lower longitudinal weld. For a given size weld, the lengths L_1 and L_2 can then be determined. This procedure, known as *balancing the welds*, will be illustrated in the following example.

≡ **EXAMPLE 8.6**

A tension member consists of a double-angle section, 2L5 × 3 × $\frac{1}{2}$, with the long legs placed back-to-back. The angles are attached to a $\frac{3}{8}$-in.-thick gusset plate. All steel is A36. Design a welded connection, balanced to eliminate eccentricity, that will resist the full tensile capacity of the member.

SOLUTION Load capacity of member based on the gross section:

$$\phi_t P_n = 0.90 F_y A_g = 0.90(36)(7.5) = 243.0 \text{ kips}$$

Capacity based on the net section:

Since the length of the connection is not yet known, U cannot be computed from AISC Equation B3-2. Assuming a value of 0.85,

$$A_e = U A_g = 0.85(7.5) = 6.375 \text{ in.}^2$$
$$\phi_t P_n = 0.75 F_u A_e = 0.75(58)(6.375) = 277.3 \text{ kips} > 243.0 \text{ kips}$$

Based on the assumed value of U, yielding of the gross section is the controlling limit state, and $\phi_t P_n = 243.0$ kips. For one angle, the load to be resisted is

$$\frac{243.0}{2} = 121.5 \text{ kips}$$

Since A36 steel is used, the approximate electrode is E70XX.

$$\text{Minimum weld size} = \frac{3}{16} \text{ in.} \quad \text{(AISC Table J2.4)}$$

$$\text{Maximum size} = \frac{1}{2} - \frac{1}{16} = \frac{7}{16} \text{ in.} \quad \text{(AISC J2.2b)}$$

Try a $\frac{5}{16}$-in. fillet weld, E70 electrode.

$$\begin{aligned}
\text{Capacity per inch of length} &= 0.707 s (\phi F_W) \\
&= 0.707 \left(\frac{5}{16} \right)(31.5) \\
&= 6.960 \text{ kips/in.}
\end{aligned}$$

$$\begin{aligned}
\text{Capacity of base metal in shear} &= t(\phi F_{BM}) = t(0.54 F_y) \\
&= \left(\frac{3}{8} \right)(0.54)(36) \\
&= 7.29 \text{ kips/in.}
\end{aligned}$$

The weld strength controls, $\therefore$ use 6.960 kips/in.

FIGURE 8.21 ≡

Referring to Figure 8.21, the capacity of the weld across the end of the angle is

$$P_3 = 6.960(5) = 34.80 \text{ kips.}$$

$$\sum M_{L_2} = 121.5(3.25) - 34.80\left(\frac{5}{2}\right) - P_1(5) = 0$$

$$P_1 = 61.58 \text{ kips}$$

$$\sum F = 121.5 - 61.58 - 34.80 - P_2 = 0, \qquad P_2 = 25.12 \text{ kips}$$

$$L_1 = \frac{P_1}{6.960} = \frac{61.58}{6.960} = 8.85 \text{ in.} \qquad Use\ 9\ in.$$

$$L_2 = \frac{25.12}{6.960} = 3.61 \text{ in.} \qquad Use\ 4\ in.$$

Answer   Use the weld shown in Figure 8.22.

FIGURE 8.22 ≡

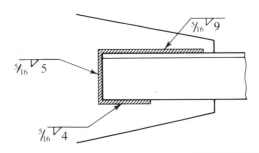

8.5 ≡ ECCENTRIC WELDED CONNECTIONS: SHEAR PLUS TENSION

Many eccentric connections, particularly beam-to-column connections, place the welds in tension as well as shear. Two such connections are illustrated in Figure 8.23.

The seated beam connection consists primarily of a short length of angle that serves as a "shelf" to support the beam. The welds attaching this angle to the column must resist the moment caused by the eccentricity of the reaction as well as the beam reaction in direct shear. The angle connecting the top flange provides torsional stability to the beam at its end and does not assist in supporting the reaction. It may be attached to the beam web instead of the top flange. The beam-to-angle connections can be made with either welds or bolts and will not carry any calculated load.

The framed beam connection is very common and subjects the vertical angle-to-column welds to the same type of load as the seated beam connection. The beam-to-angle part of the connection is also eccentric, but the load is in the

FIGURE 8.23 ≡

(a) Seated beam connection

(b) Framed beam connection

plane of shear and there is no tension. Both the seated and the framed connections have their bolted counterparts.

In each of the connections discussed, the vertical welds on the column flange are loaded as shown in Figure 8.24. As with the bolted connection in Section 8.3, the eccentric load P can be replaced by a concentric load P and a couple $M = Pe$. The shearing stress is

$$f_v = \frac{P}{A}$$

FIGURE 8.24 ≡

where A is the total throat area of the weld. The maximum tensile stress can be computed from the flexure formula

$$f_t = \frac{Mc}{I}$$

where I is the moment of inertia about the centroidal axis of the area consisting of the total throat area of the weld, and c is the distance from the centroidal axis to the farthest point on the tension side. The maximum resultant stress can be found by adding these two components vectorially:

$$f_r = \sqrt{f_v^2 + f_t^2}$$

For units of kips and inches, this stress will be in kips per square inch. If a unit throat size is used in the computations, the same numerical value can also be expressed as kips per linear inch. If f_r is derived from factored loads, it may be compared with the design strength of a unit length of weld. Although this procedure assumes elastic behavior, it will be conservative when used in an LRFD context.

EXAMPLE 8.7

An L6 × 4 × $\frac{1}{2}$ is used in a seated beam connection as shown in Figure 8.25. It must support a factored load reaction of 22 kips. All structural steel is A36, and E70XX electrodes will be used. What size fillet welds are required for the connection to the column flange?

SOLUTION

As in previous design examples, a unit throat size will be used in the calculations. Although an end return is needed for a weld of this type, for simplicity it will be conservatively neglected in the following calculations. In any event, its length could only be estimated at this point, since the weld size has not yet been determined.

FIGURE 8.25 ≡

FIGURE 8.26 ≡

With the given beam setback of $\frac{3}{4}$ inch, the beam is supported by 3.25 inches of the 4-in. outstanding leg of the angle. If the reaction is assumed to act through the center of this contact length, the eccentricity of the reaction with respect to the weld is

$$e = 0.75 + \frac{3.25}{2} = 2.375 \text{ in.}$$

and the moment is

$$M = Pe = 22(2.375) = 52.25 \text{ in.-kips}$$

For the assumed weld configuration in Figure 8.26,

$$I = \frac{2(1)(6)^3}{12} = 36 \text{ in.}^4, \qquad c = \frac{6}{2} = 3 \text{ in.}$$

$$f_t = \frac{Mc}{I} = \frac{52.25(3)}{36} = 4.354 \text{ kips/in.}$$

$$f_v = \frac{P}{A} = \frac{22}{2(1)(6)} = 1.833 \text{ kips/in.}$$

$$f_r = \sqrt{f_t^2 + f_v^2} = \sqrt{(4.354)^2 + (1.833)^2} = 4.724 \text{ kips/in.}$$

The required weld size s can be found by equating f_r to the weld capacity per inch of length:

$$f_r = 0.707s(\phi F_w)$$
$$4.724 = 0.707s(31.5), \qquad s = 0.212 \text{ in.}$$

From AISC Table J2.4,

Minimum weld size $= \frac{1}{4}$ in.   (based on column flange thickness)

From AISC J2.2b,

Maximum size $= \frac{1}{2} - \frac{1}{16} = \frac{7}{16}$ in.

Check shear capacity of base metal:

Applied direct shear $= f_v = 1.833$ kips/in.

$$\text{Shear capacity of angle leg} = t(\phi F_{BM}) = t(0.54F_y) = \frac{1}{2}(0.54)(36)$$

$$= 9.72 \text{ kips/in.} > 1.833 \text{ kips/in.} \text{(OK)}$$

Answer   Use a $\frac{1}{4}$-in. fillet weld, E70XX electrodes.

The end returns were neglected in the preceding example, but they could have been included by performing a second iteration with an end return length of twice the weld size found in the first iteration. (Of course, this is not warranted here, since the minimum weld size governed the design.) End returns will be accounted for in the following example.

≡ **EXAMPLE 8.8**

A welded framed beam connection is shown in Figure 8.27. The framing angles are $4 \times 3 \times \frac{1}{2}$, and the column is a W12 × 72. A36 steel is used throughout, and the welds are $\frac{3}{8}$-in. fillet welds made with E70XX electrodes. Determine the maximum factored load beam reaction as limited by the welds at the column flange.

SOLUTION

The beam reaction will be assumed to act through the center of gravity of the connection to the framing angles. The eccentricity of the load with respect to the welds at the column flange will therefore be the distance from this center of gravity to the column flange. For a unit throat size and the weld shown in Figure 8.28a,

$$\bar{x} = \frac{2(2.5)(1.25)}{32 + 2(2.5)} = 0.1689 \text{ in.}$$

$$e = 3 - 0.1689 = 2.831 \text{ in.}$$

FIGURE 8.27 ≡

W12 × 72

L4 × 3 × ½

FIGURE 8.28 ≡

(a)   **(b)**

and the moment on the column flange welds is

$$M = Re = 2.831R \text{ in.-kips}$$

where R is the beam reaction in kips.

From the dimensions given in Figure 8.28b, the properties of the column flange welds can be computed as follows:

$$\bar{y} = \frac{32(16)}{32 + 0.75} = 15.63 \text{ in.}$$

$$I = \frac{1(32)^3}{12} + 32(16 - 15.63)^2 + 0.75(15.63)^2 = 2918 \text{ in.}^4$$

For the two welds,

$$I = 2(2918) = 5836 \text{ in.}^4$$

$$f_t = \frac{Mc}{I} = \frac{2.831R(15.63)}{5836} = 0.007582R \text{ kips/in.}$$

$$f_v = \frac{R}{A} = \frac{R}{2(32 + 0.75)} = 0.01527R \text{ kips/in.}$$

$$f_r = \sqrt{(0.007582R)^2 + (0.01527R)^2} = 0.01705R \text{ kips/in.}$$

Let $0.01705R = 0.707s(\phi F_w)$ and solve for R:

$$0.01705R = 0.707\left(\frac{3}{8}\right)(31.5), \qquad R = 489.8 \text{ kips}$$

Check the shear capacity of the base metal:

$$t(\phi F_{BM}) = t(0.54F_y) = 0.5(0.54)(36) = 9.72 \text{ kips/in.}$$

Direct shear to be resisted:

$$\frac{489.8}{2(32.75)} = 7.48 \text{ kips/in.} < 9.72 \text{ kips/in.} \quad (\text{OK})$$

Answer   Maximum factored load beam reaction $= 490$ kips.

8.6 ☰ MOMENT-RESISTING CONNECTIONS

In all beam-to-column or beam-to-beam connections, there is some degree of moment restraint, even in connections designed to be simple, or moment-free. It is impossible to construct a perfectly frictionless pin or hinge, and most connections designed to be moment-free go far beyond that condition. On the other hand, it is also very difficult to fabricate a perfectly rigid joint that is capable of transferring 100% of the moment capacity of one member to another. Thus, although one might think that the framed and seated beam connections shown in Figure 8.23 would be somewhat rigid, they will actually transmit very little moment if the connecting angles are sufficiently flexible. As noted earlier, the eccentricity of the load with respect to the bolts or welds is very small and is usually neglected.

The AISC Specification defines two types of construction in Section A2.2, "Types of Construction":

Type FR — Fully Restrained (Rigid, or Continuous, Framing). This type of structure has moment-resisting joints capable of transmitting any moment the member can resist, with no relative rotation of the members that meet at the joint. This type of construction is authorized without any restrictions. If the frame is designed as a rigid frame, the connections must be designed accordingly — that is, as moment connections.

Type PR — Partially Restrained (Semirigid Framing). In this type of construction, the frame is designed on the basis of a known amount of restraint, intermediate between simple and rigid, at each joint. In general, moment restraint will be on the order of 20% to 90% of the member moment capacity. The chief drawback to this type of framing is the requirement for a rigorous structural analysis that incorporates this partial joint restraint. Implicit in this requirement is the need for moment-rotation curves for the connections.

If the partial restraint is neglected, beams can be treated as simply supported with no moment resistance at the joints. The aforementioned framed and seated beam connections are in this category. In general, connections that transfer less than 20% of the member capacity are considered simple. Beam supports designed in this manner are sometimes called *shear connections* since only the reaction, or end shear, is transmitted.

Frames with shear connections must be braced in the plane of the frame since there is no "frame action" to provide lateral stability. This bracing can take several forms: diagonal bracing members, shear walls, or lateral support from an adjacent structure. The moments resulting from lateral loads (usually wind) can also be provided for in the design of selected beam-to-column connections. In this approach, the connection is assumed to act as a simple connection in resisting the dead and live loads (gravity loads) and as a moment connection with the limited capability of resisting the wind moments. Of course, the connection does not know this and will actually be a partially restrained connection regardless of the loading. If the beam is designed as if it were simply supported, the maximum gravity load moment will be overestimated, and the beam will be somewhat overdesigned. In many cases, however, the wind moments will be small and the overconservatism will be slight. If this concept of simple framing is used, the Specification requires that the following conditions be met:

1. Although the beams (girders) are not simply supported, they must be capable of supporting the gravity loads as if they were.

2. The connections and the connected members (beams and columns) must be capable of resisting the wind moments.

3. The connections must have enough inelastic rotational capacity so that the fasteners or welds will not be overloaded under the combined action of gravity loads and wind.

In this book, we will consider only two types of connections: *simple connections* designed for gravity loads (with lateral frame stability provided by a positive bracing system) and *rigid connections* designed for more than 90% of the moment capacity of the beam. Simple connections have already been considered in the discussion of framed and seated beam connections, and we will now turn our attention to rigid connections.

Several examples of commonly used moment connections are illustrated in Figure 8.29. As a general rule, most of the moment transfer is through the beam flanges, and most of the moment capacity is developed there. The connection in Figure 8.29a typifies this concept. The plate connecting the beam web to the column is shop welded to the column and field bolted to the beam. With this arrangement, the beam is conveniently held in position so that the flanges can be field welded to the column. The plate connection is designed to resist only shear and takes care of the beam reaction. Complete penetration groove welds connect the beam flanges to the column and can transfer a moment equal to the moment capacity of the beam flanges. This will constitute most of the moment capacity of the beam, but a small amount of restraint will also be provided by the plate connection. (Because of strain hardening, the full plastic moment capacity of the beam can actually be developed through the flanges.) In making the flange connection, a small portion of the beam web is removed and a "backing bar" is used at each flange to permit all welding to be done from the top. When the flange welds cool, they will shrink, typically about $\frac{1}{8}$ inch. The

FIGURE 8.29 ≡

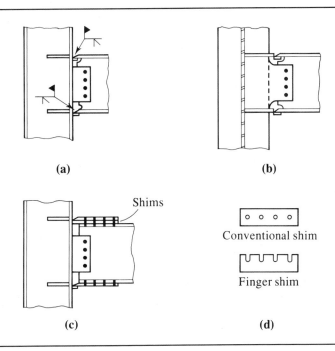

(a)   (b)

Shims

Conventional shim

Finger shim

(c)   (d)

resulting longitudinal displacement can be accounted for by using slotted bolt holes and tightening the bolts after the welds have cooled. This connection also uses column stiffeners, which are not always required. (See Section 8.7.)

The moment connection of Figure 8.29a also illustrates a recommended connection design practice: Whenever possible, welding should be done in the fabricating shop, and bolting should be done in the field. Shop welding is less expensive and can be more closely controlled.

In most beam-to-column moment connections, the members are part of a plane frame and are oriented as shown in Figure 8.29a — that is, with the webs in the plane of the frame so that bending of each member is about its major axis. In cases where a beam must frame into the web of a column rather than its flange (for example, in a space frame), the connection of Figure 8.29b can be used. This connection is similar to the one in Figure 8.29a but requires the use of column stiffeners to make the connections to the beam flanges.

Although the connection of Figure 8.29a is simple in concept, its execution requires close tolerances. If the beam is shorter than anticipated, the gap between the column and the beam flange may cause difficulties in welding, even when a backing bar is used. The three-plate connection in Figure 8.29c does not suffer from this handicap, and it has the additional advantage of being completely field bolted. The flange plates and the web plate are all shop welded to the column flange and field bolted to the beam. To provide for variation in the beam depth, the distance between flange plates is made larger than the nominal depth of the beam, usually by about $\frac{3}{8}$ inch. This gap is filled at the top flange during erection with *shims*, which are thin strips of steel used for adjusting the fit at joints. Shims may be one of two types: either conventional or "finger" shims, which can be inserted after the bolts are in place (Figure 8.29d).

The following example illustrates the design of a three-plate moment connection, including the requirements for connecting elements, which are covered by AISC J5.

≣ **EXAMPLE 8.9**

Design a three-plate moment connection of the type shown in Figure 8.30 for the connection of a W21 × 50 beam to the flange of a W14 × 99 column. Assume a beam setback of $\frac{1}{2}$ in. A frame analysis shows that the connection must transfer a factored load moment of 210 ft-kips and a factored load shear of 33 kips. All plates are to be shop welded to the column with E70XX electrodes and field bolted to the beam with A325 bearing-type bolts. A36 steel is used throughout.

SOLUTION

Web plate (neglect eccentricity):

Try $\frac{3}{4}$-in. diameter bolts. Assume that threads are in the plane of shear.

Shear capacity of one bolt:

$$\phi F_v A_b = 0.75(48)(0.4418) = 15.90 \text{ kips}$$

$$\text{Number of bolts required} = \frac{33}{15.90} = 2.08; \quad \textit{Try 3.}$$

Determine plate thickness required for bearing:

Assume $\phi R_n = \phi(2.4dtF_u)$. This will be the case if there are two or more fasteners parallel to the line of force, a bolt spacing of at least 3 bolt diameters, and an edge distance of at least 1.5 bolt diameters. The spacing and edge distances shown in Figure 8.31a satisfy these requirements and will be used.

FIGURE 8.30 ≣

W14 × 99

Load resisted by each bolt $= \dfrac{33}{3} = 11$ kips. Let $\phi R_n = 11$ kips:

$$\phi(2.4dtF_u) = 0.75(2.4)\left(\frac{3}{4}\right)t(58) = 11$$

Solving for the thickness,

$$t = 0.140 \text{ in.}$$

Bearing on beam web:

$$t_w = 0.380 \text{ in.} > 0.140 \text{ in. required} \quad \text{(OK)}$$

Determine plate thickness required for shear:
Consider a vertical section through the plate. From AISC J5, "Connecting Elements,"

$$\phi R_n = 0.90[0.60A_g F_y] \qquad \text{(AISC Equation J5-3)}$$
$$33 = 0.90[0.60(9t)(36)]$$
$$t = 0.189 \text{ in.} \quad \text{(controls)}$$

Use $t = \frac{1}{4}$ in.

Connection of shear plate to column flange:
Minimum fillet weld size $= \frac{1}{4}$ in. (Based on the thicker connected part, the minimum is $\frac{5}{16}$ in., but it need not be larger than the thickness of the thinner part.)

$$\text{Capacity per inch of length} = 0.707s(\phi F_W) = 0.707\left(\frac{1}{4}\right)(31.5)$$
$$= 5.568 \text{ kips/in.}$$

Shear capacity of base metal:

$$t\phi F_{BM} = t(0.54F_y) = \frac{1}{4}(0.54)(36) = 4.86 \text{ kips/in.} \quad \text{(controls)}$$

FIGURE 8.31 ≡

(a) (b)

The required length of $\frac{1}{4}$-in. fillet weld is therefore

$$\frac{33}{4.86} = 6.79 \text{ in.}$$

A continuous weld on one side of the plate would suffice, but both sides are usually welded, and that will be done here.

The minimum width of the plate can be determined from a consideration of edge distances. The load being resisted (the beam reaction) is vertical, so the horizontal edge distance need only conform to the clearance requirements of AISC Table J3.4. Assuming a sheared edge, the minimum edge distance is $1\frac{1}{4}$ inch.

With a beam setback of $\frac{1}{2}$ inch and edge distances of $1\frac{1}{2}$ inches (see Figure 8.31b), the width of the plate is

$$0.5 + 2(1.5) = 3.5 \text{ in.}$$

Use a bar $3\frac{1}{2} \times \frac{1}{4}$.

Flange plates:

Find the force at interface between beam flange and plate. From Figure 8.32,

$$M = Hd \quad \text{and} \quad H = \frac{M}{d} = \frac{210(12)}{20.83} = 121.0 \text{ kips}$$

Try $\frac{3}{4}$-in. A325 bolts. (Since $\frac{3}{4}$-in.-diameter bolts were selected for the shear connection, try the same size here.) If bolt shear controls, the number of bolts required is

$$\frac{121.0}{15.90} = 7.61 \qquad \text{Use 8 (4 pairs)}$$

Use edge distances of $1\frac{1}{2}$ inches and spacing of 3 inches so as to qualify for a nominal bearing strength of $2.4dtF_u$:

$$\phi R_n = 0.75(2.4)t(58) = 104.4t$$

Determine the minimum plate thickness required for bearing. The load per bolt is

$$\phi R_n = \frac{121.0}{8} = 15.12 \text{ kips}$$

Equating this to the bearing strength,

$$104.4t = 15.12 \text{ kips}, \qquad t = 0.145 \text{ in.}$$

Both flange plates will be designed as tension connecting elements. (Although one of the plates will be in compression, the connection details preclude any

FIGURE 8.32 ≡

stability problems.) The minimum cross section required for tension on the gross and net areas will now be determined. From AISC Equation J5-1,

$$\phi R_n = 0.90(A_g F_y)$$

$$\text{Required } A_g = \frac{\phi R_n}{0.90 F_y} = \frac{H}{0.90 F_y} = \frac{121.0}{0.90(36)} = 3.735 \text{ in.}^2$$

From AISC Equation J5-2,

$$\phi R_n = 0.75 A_n F_u$$

$$\text{Required } A_n = \frac{\phi R_n}{0.75 F_u} = \frac{H}{0.75 F_u} = \frac{121.0}{0.75(58)} = 2.782 \text{ in.}^2$$

Try a plate width of $w_g = 6.5$ in. (This is equal to the beam flange width.)

Thickness needed to satisfy gross area requirement:

$$A_g = 6.5t = 3.735 \text{ in.}^2 \qquad t = 0.575 \text{ in.}$$

Thickness needed to satisfy net area requirement:

$$A_n = tw_n = t(w_g - \sum d_{\text{hole}}) = t\left[6.5 - 2\left(\frac{7}{8}\right)\right] = 4.750t$$

Let $4.750t = 2.782 \text{ in.}^2$; $t = 0.586 \text{ in.}$ (controls)

This thickness is also greater than that required for bearing, so it will be the minimum acceptable thickness.

Try a bar $6\frac{1}{2} \times \frac{5}{8}$.

Since this plate is a tension connecting element, its net area cannot exceed $0.85A_g$ in the computations (AISC J5.2):

$$A_n = \frac{5}{8}\left[6.5 - 2\left(\frac{7}{8}\right)\right] = 2.969 \text{ in.}^2$$

$$0.85A_g = 0.85(0.625)(6.5) = 3.453 \text{ in.}^2 > 2.969 \text{ in.}^2 \quad \text{(OK)}$$

Use a bar $6\frac{1}{2} \times \frac{5}{8}$.

Part of the flange area of the beam will be lost because of the bolt holes, and the moment capacity will be reduced. AISC B10 permits this reduction to be neglected when

$$0.75F_u A_{fn} \geq 0.9F_y A_{fg}$$ (AISC Equation B10-1)

where

A_{fg} = gross flange area

$\quad = b_f t_f = 6.530(0.535) = 3.494 \text{ in.}^2$

A_{fn} = net flange area

$$\quad = t_f(b_f - \sum d_h) = 0.535\left[6.530 - 2\left(\frac{7}{8}\right)\right] = 2.557 \text{ in.}^2$$

Evaluating AISC Equation B10-1,

$0.75F_u A_{fn} = 0.75(58)(2.557) = 111.2 \text{ kips}$

$0.9F_y A_{fg} = 0.9(36)(3.494) = 113.2 \text{ kips} > 111.2 \text{ kips}$

Since AISC Equation B10-1 is not satisfied, the flexural strength should be based on an effective flange area of

$$A_{fe} = \frac{5}{6}\frac{F_u}{F_y}A_{fn}$$ (AISC Equation B10-3)

$$\quad = \frac{5}{6}\left(\frac{58}{36}\right)(2.557) = 3.433 \text{ in.}^2$$

FIGURE 8.33 ≡

2"   3 @ 3"   1½"

¾" φ A325 bolts
E70XX electrodes

shim as required

³⁄₁₆
³⁄₁₆

Bar 3½ × ¼ × 0'-9"

Bar 6½ × ⅝ × 1'-0½"

This is an insignificant difference from the actual gross flange area of 3.494 in.^{2}, so the flexural strength will not be modified.

Answer   Use the connection shown in Figure 8.33 (column stiffener requirements will be considered in the next section).*

---

* Figure 8.33 also shows the symbol for a bevel groove weld, used here for the beam flange plate-to-column connection.

8.7 ≣ COLUMN STIFFENERS AND OTHER REINFORCEMENT

Most of the moment transferred from the beam to the column in a rigid connection is in the form of a couple consisting of the tensile and compressive forces in the beam flanges. As a consequence of the application of these relatively large concentrated forces, reinforcement of the column is sometimes necessary. For negative moment, as would be the case with gravity loading, these forces are directed as shown in Figure 8.34, with the top flange of the beam delivering a tensile force to the column, and the bottom flange delivering a compressive force.

Both of these forces are transmitted to the column web, with compression being more critical because of the stability problem. The tensile load at the top

FIGURE 8.34 ≣

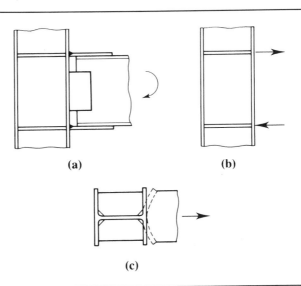

(a)

(b)

(c)

can distort the column flange, exaggerated in Figure 8.34c, creating additional load on the weld connecting the beam flange to the column flange. A stiffener of the type shown can provide anchorage for the column flange. Obviously, this stiffener must be welded to both the web and the flange. If the applied moment never changes direction, the stiffener resisting the compressive load (the bottom stiffener in this illustration) can be fit to bear on the flange and need not be welded to it.

AISC Specification Requirements

The AISC requirements for column web reinforcement are covered in Chapter K, "Strength Design Considerations." For the most part, these provisions are based on theoretical analyses that have been modified to fit test results. If the applied factored load transmitted by the beam flange or flange plate exceeds the design strength ϕR_n for any of the limit states considered, stiffeners must be used.

To avoid a local bending failure of the column flange, the tensile load from the beam flange must not exceed

$$\phi R_n = \phi(6.25 t_f^2 F_{yf}) \qquad \text{(AISC Equation K1-1)}$$

where

$\phi = 0.90$

t_f = thickness of the column flange

F_{yf} = yield stress of the column flange

For the limit state of local web yielding in compression,

$$\phi R_n = \phi[(5k + N)F_{yw} t_w] \qquad \text{(AISC Equation K1-2)}$$

or, when the load is applied within a distance from the end of the member equal to the depth of the member,

$$\phi R_n = \phi[(2.5k + N)F_{yw} t_w] \qquad \text{(AISC Equation K1-3)}$$

where

$\phi = 1.0$

k = distance from the outer flange surface of the column to the toe of the fillet in the web

N = length of applied load = thickness of beam flange or flange plate

F_{yw} = yield stress of the column web

t_w = thickness of the column web

AISC Equations K1-2 and K1-3 were also used in Section 5.12 of this book to investigate web yielding in beams subjected to concentrated loads.

To prevent web crippling when the compressive load is delivered to one flange only, as in the case of an exterior column with a beam connected to one side, the applied load must not exceed the design strength given by one of the following equations. (Web crippling has also been addressed in Section 5.12.) When the load is applied at a distance of at least $d/2$ from the end of the column,

$$\phi R_n = \phi 135 t_w^2 \left[1 + 3\left(\frac{N}{d}\right)\left(\frac{t_w}{t_f}\right)^{1.5} \right] \sqrt{\frac{F_{yw} t_f}{t_w}} \qquad \text{(AISC Equation K1-4)}$$

where

$\phi = 0.75$

d = total column depth

If the load is applied at the end of the column,

$$\phi R_n = \phi 68 t_w^2 \left[1 + 3\left(\frac{N}{d}\right)\left(\frac{t_w}{t_f}\right)^{1.5} \right] \sqrt{\frac{F_{yw} t_f}{t_w}} \quad \text{for } \frac{N}{d} \leq 0.2$$

$$\text{(AISC Equation K1-5a)}$$

or

$$\phi R_n = \phi 68 t_w^2 \left[1 + \left(4\frac{N}{d} - 0.2\right)\left(\frac{t_w}{t_f}\right)^{1.5} \right] \sqrt{\frac{F_{yw} t_f}{t_w}} \quad \text{for } \frac{N}{d} > 0.2$$

$$\text{(AISC Equation K1-5b)}$$

Compression buckling of the web must be investigated when loads are delivered to *both* column flanges. This would occur at an interior column with beams connected to both sides. The design strength for this limit state is

$$\phi R_n = \phi \left[\frac{4100 t_w^3 \sqrt{F_{yw}}}{h} \right] \qquad \text{(AISC Equation K1-8)}$$

where

$\phi = 0.90$

h = column web depth from toe of fillet to toe of fillet (Figure 8.35)

If the connection is near the end of the column (i.e., if the load is applied within a distance $d/2$ from the end), the strength given by AISC Equation K1-8 should be reduced by half.

In summary, to investigate the need for column stiffeners, the following limit states should be checked:

1. Local flange bending (AISC Equation K1-1)

2. Local web yielding (AISC Equation K1-2 or K1-3)

3. Web crippling *or* compression buckling of the web. (If the compressive load is applied to *one* flange only, check web crippling [AISC Equation K1-4 or K1-5]. If the compressive load is applied to *both* flanges, check compressive buckling of the web [AISC Equation K1-8].)

FIGURE 8.35 ≡

In the event stiffeners are required by AISC Equation K1-2 for local web yielding, the required cross-sectional area of the stiffeners can be found as follows: Assume the additional design strength can be obtained from an area of stiffener A_{st} that has yielded. Thus, from AISC Equation K1-2,

$$\phi R_n = \phi[(5k + N)F_{yw}t_w + A_{st}F_{yst}]$$

where F_{yst} is the yield stress of the stiffener. Equating the right side of this equation to the applied load, denoted by P_{bf}, and solving for A_{st},

$$A_{st} = \frac{P_{bf}/\phi - (5k + N)F_{yw}t_w}{F_{yst}}$$

$$= \frac{P_{bf} - (5k + t_b)F_{yw}t_w}{F_{yst}} \tag{8.6}$$

where $\phi = 1.0$ and t_b is the thickness of the beam flange or flange plate. Equation 8.6 can also be used to check the local web yielding strength of the column. Solve for A_{st}; if a negative result is obtained, no stiffeners are needed for this limit state.

If stiffeners are required because of any of these provisions, AISC K1.9 gives the following guidelines for their proportioning:

- The width of the stiffener plus half the column web thickness must be equal to at least one third of the width of the beam flange or plate delivering the force to the column (see Figure 8.36):

$$b + \frac{t_w}{2} \geq \frac{b_b}{3}$$

$$\therefore \quad b \geq \frac{b_b}{3} - \frac{t_w}{2}$$

- The stiffener thickness must be at least half the thickness of the beam flange or plate:

$$t_{st} \geq \frac{t_b}{2}$$

- The following width-thickness requirement should be met:

$$\frac{b}{t_{st}} \leq \frac{95}{\sqrt{F_y}}$$

Full-depth stiffeners are required for the compression buckling case, but half-depth stiffeners are permitted for the other limit states. This implies that full-depth stiffeners are required only when beams are connected to both sides of the column.

For any of the limit states, the decision on whether to weld the stiffener to the flange should be based on the following criteria:

- On the tension side, the stiffeners should be welded to both the web and flange.
- On the compression side, the stiffeners need only bear on the flange but may be welded.

FIGURE 8.36 ≡

Part 3 of the *Manual*, "Column Design," contains tabulated constants that expedite the evaluation of the need for stiffeners. Their use is illustrated in an example in the "General Notes" preceding the Column Load Tables and will not be covered here.

Shear in the Column Web

The transfer of a large moment to a column can produce large shearing stresses in the column web within the boundaries of the connection; for example, region ABCD in Figure 8.37. This region is sometimes called the *panel zone*. The *net* moment is of concern, so if beams are connected to both sides of the column, it is the algebraic sum of the moments that induces this web shear.

If the beam flange forces are assumed to act at a distance of $0.95d_b$ apart, where d_b is the beam depth, the flange forces can be taken as $(M_1 + M_2)/0.95d_b$. If the column shear adjacent to the panel is V_u and is directed as shown, the total shear force in the panel is

$$P = \frac{(M_1 + M_2)}{0.95d_b} - V_u \tag{8.7}$$

FIGURE 8.37 ≡

(See text for discussion of notation.)

$$P = \frac{M_1 + M_2}{0.95\,d_b} - V_u$$

The web shear strength is given in AISC K1.7 as ϕR_v, where $\phi = 0.90$ and R_v is a function of the axial load in the column:
When $P_u \leq 0.4P_y$,

$$R_v = 0.60F_y d_c t_w \qquad \text{(AISC Equation K1-9)}$$

When $P_u > 0.4P_y$,

$$R_v = 0.60F_y d_c t_w \left[1.4 - \left(\frac{P_u}{P_y} \right) \right] \qquad \text{(AISC Equation K1-10)}$$

where

$\quad P_u = $ factored axial load in column

$\quad P_y = $ axial yield strength of column $= AF_y$

$\quad A = $ cross-sectional area of column, including any reinforcement (i.e., doubler plates)

$\quad d_c = $ total column depth

$\quad t_w = $ column web thickness, including any reinforcing plates

$\quad F_y = $ yield stress of column web

If the column web has insufficient shear strength, it must be reinforced. A *doubler plate* with sufficient thickness to make up the deficiency can be welded to the web or a pair of diagonal stiffeners can be used. Stiffeners are usually more practical.

AISC K1.7 also provides equations to be used when frame stability, including plastic deformation of the panel zone, is considered in the analysis. They will not be covered here.

≡ EXAMPLE 8.10

Determine whether stiffeners or other column web reinforcement is required for the connection of Example 8.9. Assume that $V_u = 0$ and $P_u/P_y = 0.4$.

SOLUTION From Example 8.9, the flange force can be conservatively taken as

$$P_{bf} = H = 121.0 \text{ kips}$$

Check local flange bending with AISC Equation K1-1:

$$\phi R_n = \phi(6.25t_f^2 F_{yf})$$
$$= 0.90[6.25(0.780)^2(36)] = 123 \text{ kips} > 121 \text{ kips} \quad \text{(OK)}$$

Check local web yielding with AISC Equation K1-2:

$$\phi R_n = \phi[(5k + N)F_{yw} t_w]$$
$$= 1.0 \left[5(1.438) + \frac{5}{8} \right](36)(0.485) = 136 \text{ kips} > 121 \text{ kips} \quad \text{(OK)}$$

Check web crippling with AISC Equation K1-4:

$$\phi R_n = \phi 135 t_w^2 \left[1 + 3\left(\frac{N}{d}\right)\left(\frac{t_w}{t_f}\right)^{1.5} \right] \sqrt{\frac{F_{yw}\, t_f}{t_w}}$$

$$= 0.75(135)(0.485)^2 \left[1 + 3\left(\frac{5/8}{14.16}\right)\left(\frac{0.485}{0.780}\right)^{1.5} \right] \sqrt{\frac{36(0.780)}{0.485}}$$

$$= 193 \text{ kips} > 121 \text{ kips} \quad \text{(OK)}$$

Answer   Column stiffeners are not required.

Shear in column web:

From Equation 8.7, neglecting the thickness of shims in the computation of d_b, the factored load shear force in the column web panel zone is

$$P = \frac{(M_1 + M_2)}{0.95 d_b} - V_u$$

$$= \frac{210(12)}{0.95[20.83 + 2(5/8)]} - 0 = 120 \text{ kips}$$

Since $P_u = 0.4 P_y$, use AISC Equation K1-9:

$$R_v = 0.60 F_y\, d_c\, t_w = 0.60(36)(14.16)(0.485) = 148.3 \text{ kips}$$

The design strength is

$$\phi R_v = 0.90(148.3) = 134 \text{ kips} > 120 \text{ kips} \quad \text{(OK)}$$

Answer   Column web reinforcement is not required.

EXAMPLE 8.11

The beam-to-column connection shown in Figure 8.38 must transfer a factored moment of 142 ft-kips. This moment is caused by dead and live gravity loads. All steel is A36, and E70 electrodes are used. Investigate column stiffener and web panel–zone reinforcement requirements. Assume that $V_u = 0$ and $P_u < 0.4 P_y$.

SOLUTION   The flange force is

$$P_{bf} = \frac{M}{d_b - t_b} = \frac{142(12)}{17.90 - 0.525} = 98.07 \text{ kips}$$

Check local flange bending (AISC Equation K1-1):

$$\phi R_n = \phi(6.25 t_f^2 F_{yf})$$

$$= 0.90[6.25(0.560)^2(36)] = 63.50 \text{ kips} < 98.07 \text{ kips} \quad \text{(N.G.)}$$

∴   stiffeners are required to prevent local flange bending.

FIGURE 8.38 ≡

Check for local web yielding. Use Equation 8.6 in lieu of AISC Equation K1-2:

$$A_{st} = \frac{P_{bf} - (5k + t_b)F_{yw}\, t_w}{F_{yst}}$$

$$= \frac{98.07 - [5(1.062) + 0.525](36)(0.360)}{36} = 0.6236 \text{ in.}^2$$

Since A_{st} is positive, a pair of stiffeners with a combined cross-sectional area of at least 0.6236 in.^{2} will be required.

Check web crippling strength (AISC Equation K1-4):

$$\phi R_n = \phi 135 t_w^2 \left[1 + 3\left(\frac{N}{d}\right)\left(\frac{t_w}{t_f}\right)^{1.5} \right] \sqrt{\frac{F_{yw}\, t_f}{t_w}}$$

$$= 0.75(135)(0.360)^2 \left[1 + 3\left(\frac{0.525}{8.25}\right)\left(\frac{0.360}{0.560}\right)^{1.5} \right] \sqrt{\frac{36(0.560)}{0.360}}$$

$$= 107.9 \text{ kips} > 98.07 \text{ kips} \quad \text{(OK)}$$

The stiffener dimensions will be selected on the basis of the criteria given in AISC Section K1.9, and the area of the resulting cross section will then be checked.

Minimum width:

$$b \geq \frac{b_f}{3} - \frac{t_w}{2} = \frac{6.015}{3} - \frac{0.360}{2} = 1.825 \text{ in.}$$

If the stiffeners are not permitted to extend beyond the edges of the column flange, the maximum width is

$$b \leq \frac{8.07 - 0.360}{2} = 3.855 \text{ in.}$$

The minimum thickness is

$$\frac{t_b}{2} = \frac{0.525}{2} = 0.2625 \text{ in.}$$

Try a bar $3 \times \frac{5}{16}$:

$$A_{st} = 3\left(\frac{5}{16}\right) \times 2 \text{ stiffeners} = 1.875 \text{ in.}^2 > 0.6236 \text{ in.}^2 \quad \text{(OK)}$$

Check width-thickness ratio:

$$\frac{b}{t_{st}} = \frac{3}{5/16} = 9.6$$

$$\frac{95}{\sqrt{F_y}} = \frac{95}{\sqrt{36}} = 15.8 > 9.6 \quad \text{(OK)}$$

Since this is a one-sided connection, full-depth stiffeners are not required:

$$\frac{d}{2} = \frac{8.25}{2} = 4.125 \text{ in.} \qquad \text{Use } 4\frac{1}{2} \text{ in.}$$

Answer   Use 2 bars $3 \times \frac{5}{16} \times 0$ ft $4\frac{1}{2}$ in. (Clip the inside corners so as to avoid the fillets at the column flange-to-web intersection. Clip at a 45° angle for $\frac{5}{8}$ in.)

Welds — stiffener to column web:
   Minimum size $= \frac{3}{16}$ in. (AISC Table J2.4, based on web thickness).

Size required for strength:

$$s = \frac{\text{force resisted by stiffener}}{0.707L(\phi F_W)}$$

From Equation 8.6, the force to be resisted by the stiffener is

$$A_{st} F_{yst} = P_{bf} - (5k + t_b)F_{yw} t_w$$
$$= 98.07 - [5(1.062) + 0.525](36)(0.360) = 22.45 \text{ kips}$$

The length available for welding the stiffener to the web is

$$L = \left(4.5 - \frac{5}{8}\right) \times 2 \text{ sides} \times 2 \text{ stiffeners} = 15.5 \text{ in.} \qquad \text{(See Figure 8.39.)}$$

$$s = \frac{22.45}{0.707(15.5)(31.5)} = 0.0650 \text{ in.} < \frac{3}{16} \text{ in. minimum}$$

Shear strength of base metal:

$$\phi R_n = \phi F_{BM} t = 0.54 F_y t_{st} = 0.54(36)\left(\frac{5}{16}\right) = 6.075 \text{ kips/in.}$$

Required weld capacity (for one stiffener) $= 0.0650(0.707)(31.5)(2)$

$$= 2.90 \text{ kips/in.} < 6.075 \text{ kips/in.}$$

$$\text{(OK)}$$

Answer   Use a $\frac{3}{16}$-in. fillet weld.

Welds — stiffener to column flange:

Minimum size $= \frac{1}{4}$ in.   (AISC Table J2.4, based on flange thickness)

Capacity per inch $= 0.707\left(\frac{1}{4}\right)(31.5) = 5.568$ kips/in.

$$< 0.54 F_y t_{st} = 6.075 \text{ ksi} \quad \text{(OK)}$$

Length available for welding $= \left(3 - \frac{5}{8}\right)(2)(2) = 9.5$ in.

Size required for strength:

$$s = \frac{\text{force resisted by stiffener}}{0.707 L(\phi F_W)} = \frac{22.45}{0.707(9.5)(31.5)}$$

$$= 0.106 \text{ in.} < \frac{1}{4} \text{ in.}$$

FIGURE 8.39 ≡

Section

Answer   Use a $\frac{1}{4}$-in. fillet weld. (Since the applied moment is caused by gravity loads and is not reversible, the stiffeners opposite the compression flange of the beam can be fit to bear on the column flange and need not be welded, but this option will not be exercised here.)

Check the column web for shear:
From Equation 8.7,

$$P = \frac{(M_1 + M_2)}{0.95d_b} - V_u = \frac{142(12)}{0.95(17.90)} - 0 = 100.2 \text{ kips}$$

From AISC Equation K1-9,

$$R_v = 0.60F_y \, d_c \, t_w = 0.60(36)(8.25)(0.360) = 64.15 \text{ kips}$$

and the design strength is

$$\phi R_v = 0.90(64.15) = 57.74 \text{ kips} < 100.2 \text{ kips} \quad \text{(N.G.)}$$

Use AISC Equation K1-9 to find the required web thickness. Multiplying both sides by ϕ and solving for t_w,

$$t_w = \frac{\phi R_v}{\phi(0.60F_y \, d_c)} = \frac{100.2}{0.90(0.60)(36)(8.25)} = 0.625 \text{ in.}$$

Option 1: Web doubler plate:
Let

t_d = required doubler plate thickness
$\quad = 0.625 - 0.360 = 0.265 \text{ in.}$ Try $t_d = \frac{5}{16}$ in.

Provide a weld to match the shear strength of the *required* thickness of doubler plate: Let

$$\phi F_{BM} \, t_d = 0.707s(\phi F_W)$$

or

$$s = \frac{\phi F_{BM} \cdot t_d}{0.707(\phi F_W)} = \frac{0.54(36)(0.265)}{0.707(31.5)} = 0.231 \text{ in.}$$

To the nearest $\frac{1}{16}$ inch, the required weld size is $\frac{1}{4}$ in.
From AISC J2.2b, the maximum weld size is

$$t_d - \frac{1}{16} = \frac{5}{16} - \frac{1}{16} = \frac{1}{4} \text{ in. (OK)}$$

Answer   Use a $\frac{5}{16}$-in. doubler plate and a $\frac{1}{4}$-in. fillet weld.

Option 2: Diagonal stiffener:
Use full-depth horizontal stiffeners, as shown in Figure 8.40, with this alternative (this means that the corresponding weld size *could* be reduced).

FIGURE 8.40 ≡

The shear force that is to be resisted by the web reinforcement is $100.2 - 57.74 = 42.46$ kips.

If this is taken as the horizontal component of an axial force P in the stiffener,

$$P \cos \theta = 42.46 \text{ kips}$$

where

$$\theta = \tan^{-1}\left(\frac{d_b}{d_c}\right) = \tan^{-1}\left(\frac{17.90}{8.25}\right) = 65.26°$$

$$P = \frac{42.46}{\cos (65.26°)} = 101.5 \text{ kips}$$

Let

$$\phi R_n = \phi A_{st} F_y = 0.9 A_{st}(36) = 101.5 \text{ kips}$$

$$A_{st} = \frac{101.5}{0.9(36)} = 3.13 \text{ in.}^2$$

Use two stiffeners, $3 \times \frac{7}{16}$, one on each side of web.

$$A_{st} \text{ provided} = 2(3)\left(\frac{9}{16}\right) = 3.38 \text{ in.}^2 > 3.13 \text{ in.}^2 \text{ required} \quad \text{(OK)}$$

Check width-thickness ratio:

$$\frac{b}{t_{st}} = \frac{3}{9/16} = 5.3 < \frac{95}{\sqrt{36}} = 15.8 \quad \text{(OK)}$$

Welds:
The approximate length of each diagonal stiffener is

$$\frac{d_c}{\cos \theta} = \frac{8.25}{\cos (65.26°)} = 19.7 \text{ in.}$$

If welds are used on both sides of the stiffeners, the available length for welding is

$$L = 19.7(4) = 78.8 \text{ in.}$$

Weld size required for strength:

$$s = \frac{P}{0.707L(\phi F_w)} = \frac{101.5}{0.707(78.8)(31.5)} = 0.058 \text{ in.}$$

Use the minimum size of $\frac{1}{4}$ in. (AISC Table J2.4).

Because of the small size required for strength, investigate the possibility of using intermittent welds. From AISC J2.2b,

$$\text{Minimum length} = 4s = 4\left(\frac{1}{4}\right)$$

$$= 1.0 \text{ in. but not less than 1.5 in. (1.5 in. controls)}$$

Capacity of a group of four welds:

$$4(0.707)sL(\phi F_w) = 4(0.707)\left(\frac{1}{4}\right)(1.5)(31.5) = 33.41 \text{ kips}$$

$$\text{Required capacity per inch} = \frac{101.5}{19.7} = 5.152 \text{ kips/in.}$$

$$\text{Required spacing of welds} = \frac{33.41}{5.152} = 6.48 \text{ in.}$$

$$\text{Shear capacity of base metal} = 0.54F_y t_w = 0.54(36)(0.360) = 7.00 \text{ kips/in.}$$

$$\text{Required capacity of weld} = 0.707s(\phi F_w) = 0.707\left(\frac{1}{4}\right)(31.5)$$

$$= 5.57 \text{ kips/in.} < 7.00 \text{ kips/in.} \quad \text{(OK)}$$

Answer   Use $\frac{1}{4}$-in. × $1\frac{1}{2}$-in. intermittent fillet welds spaced at 6 in. on centers, on each side of each diagonal stiffener.

As previously mentioned, diagonal stiffeners are usually more practical than doubler plates, but the most economical alternative may be to simply use a larger column section. The labor costs associated with doubler plates and stiffeners (including those opposite the beam flanges) may outweigh the additional material cost of a larger column.

8.8 ≡ END PLATE CONNECTIONS

The end plate connection is a popular beam-to-column and beam-to-beam connection that has been in use since the mid-1950s. Figure 8.41 illustrates two

FIGURE 8.41 ≡

Simple — shear only

Rigid — moment resisting

types: the simple, or shear only, connection (Type PR construction) and the rigid connection (Type FR construction). The basis of both types is a plate that is shop welded to the end of a beam and field bolted to a column or another beam. This feature is one of the chief advantages of this connection, the other being that ordinarily fewer bolts are required than with other types of connections, thus making erection faster. On the other hand, there is little room for error in the beam length, and the end must be square. Cambering will make the fit even more critical. Some latitude can be provided in the beam length by fabricating it short and achieving final fit with shims.

In a simple connection, care must be taken to make the connection sufficiently flexible so that enough end rotation of the beam is possible. This requires that the plate be relatively short and thin as compared to the fully restrained version of this connection. The *Manual of Steel Construction*, in Part 9, "Simple Shear Connections," recommends that the thickness be between $\frac{1}{4}$ inch and $\frac{3}{8}$ inch to achieve this flexibility. This part of the *Manual* also presents other guidelines and a design example as an introduction to Table 9-5, which contains reaction capacities for various combinations of plates and bolts.

The design of moment-resisting end plate connections requires the determination of the plate thickness, the weld size or sizes, and the bolt details. The design of the welds and the bolts is a fairly straightforward matter, employing traditional analysis procedures. The determination of the plate thickness, however, relies on the results of experimental and statistical research.[51] The tension side of the connection is critical; the bolts on the compression side serve mainly to keep the connection in alignment. If the moment is reversible, the design for the tension side is used on both sides. The general approach is as follows:

1. Determine the force in the tension flange of the beam.

2. Select the bolts needed to resist this force and arrange them symmetrically about the tension flange. If the moment is reversible, use this same arrangement at the compression side. Otherwise, use a nominal number of bolts for alignment purposes. The total number of bolts must be adequate to resist the shear from the beam reaction.

3. Consider a portion of the beam flange and adjacent plate to act as a tee-shape subjected to a tensile load applied to its stem (Figure 8.42).

FIGURE 8.42 ≡

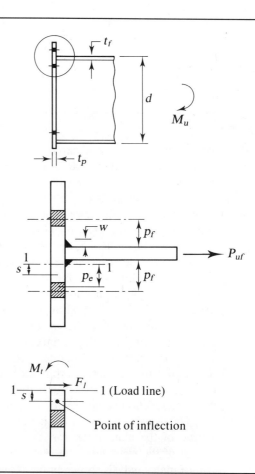

4. Select the width and thickness of the "flange" of this tee to satisfy flexural requirements, in much the same way as the design of a tee hanger is approached (Section 7.7).

5. Check the shear in the plate.

6. Design the welds.

The *Manual of Steel Construction, Volume II*, presents a detailed design procedure with examples in Part 10, "Fully Restrained (FR) Moment Connections." It is essentially the same as the procedure outlined above, sometimes called the *split-tee method*, with modifications.[51] The main difference is found in Step 4 in the computation of the bending moment in the plate. Traditional analysis accounts for prying forces in the manner discussed in Section 7.7. In the current approach, the selection of the bolts and the plate thickness is independent of any consideration of prying action. The moment computation is based on the statistical analysis of finite element studies that have been confirmed experimentally. The first step in the procedure is to compute the force in the beam tension flange:

$$P_{uf} = \frac{M_u}{d - t_f}$$

Next, bolts are selected to resist this tensile force and are arranged symmetrically in two rows about the beam tension flange. Additional bolts are placed at the compression side as required for the beam reaction, with a minimum of two.

The maximum moment in the split-tee will occur at the "load line," Section 1.1 of Figure 8.42, and is given by

$$M_t = F_1 s$$

where

$$F_1 = \text{shear force} = \frac{P_{uf}}{2}$$

$$s = \text{distance from load line to point of inflection} = \frac{p_e}{2}$$

$$p_e = p_f - 0.25 d_b - 0.707 w$$

From Figure 8.42, p_f is the distance from the bolt line to the beam flange, usually equal to the bolt diameter d_b plus $\frac{1}{2}$ inch, and w is the weld size. p_f is called the bolt distance, and p_e is the effective bolt distance or effective span. The moment M_t is modified by the factor α_m to obtain the effective moment M_{eu}:

$$M_{eu} = \alpha_m M_t$$

where

$$\alpha_m = C_a C_b (A_f/A_w)^{1/3} (p_e/d_b)^{1/4}$$

$C_a = $ a constant relating the material properties of the bolts and the plate

$$C_b = \sqrt{b_f/b_p}$$

b_f = width of beam flange

b_p = width of end plate   (Reference 51 recommends a maximum *effective* width of $b_f + 2w + t_p$, where t_p is the thickness of the end plate. The *Manual* recommends a maximum *actual* width of $b_f + 1$ in.)

A_f = area of beam flange

A_w = area of beam web between fillets

The constant C_a is a function of only the material properties and can be tabulated for the usual grades of structural steel and high-strength bolts. This has been done in Table 10-1 in Part 10 of the *Manual*. Table 10-2 gives values of A_f/A_w for commonly used beam shapes. Once the moment M_{eu} has been computed, it can be equated to the design strength, and the minimum required plate thickness $t_{p_{req}}$ can be found. For a rectangular cross section bent about its minor axis, the design strength is

$$\phi_b M_n = \phi_b M_p = \phi_b Z F_y = 0.90\left(\frac{b_p t_{p_{req}}^2}{4}\right)F_y$$

Equating this to the factored load moment and solving for the plate thickness,

$$0.90\left(\frac{b_p t_{p_{req}}^2}{4}\right)F_y = M_{eu}$$

$$t_{p_{req}} = \sqrt{\frac{4M_{eu}}{0.90 b_p F_y}}$$

The tension flange of the beam can be attached to the plate with either a full penetration groove weld or a fillet weld completely surrounding the flange. The full flange force must be developed on the tension side. The web should be welded on both sides with fillet welds capable of resisting the beam reaction. The following additional guidelines must be met in order to satisfy assumptions that underlie the method:

1. Both the plate and the beam must have the same yield stress F_y.

2. The bolt diameter should not exceed $1\frac{1}{2}$ inches.

3. The bolts must be tensioned according to AISC Table J3.1.

4. The vertical edge distance should be approximately $1\frac{3}{4}d_b$, but no less than $1\frac{1}{2}d_b$.

≡ EXAMPLE 8.12

Design an end plate connection for a W18 × 35 beam. This connection must be capable of transferring a factored load moment of 173 ft-kips. Use A36 steel, E70XX electrodes, and A325 bearing-type bolts.

SOLUTION   The flange force is

$$P_{uf} = \frac{M_u}{d - t_f} = \frac{173(12)}{17.7 - 0.425} = 120.2 \text{ kips}$$

Bolts:

Try two rows of two bolts each. The design strength in tension for one bolt is

$$\phi R_n = 0.75(90A_b)$$

and the required area of one bolt is

$$A_b = \frac{\text{Required } \phi R_n}{0.75(90)} = \frac{120.4/4}{0.75(90)} = 0.445 \text{ in.}^2$$

Answer   Use $\frac{7}{8}$-in. diameter A325 bolts ($A_b = 0.6013 \text{ in.}^2$).

The maximum reaction that can be supported by this connection can be determined from a consideration of the shear strength of the bolts. For one bolt,

$$\phi R_n = \phi(48A_b) = 0.75(48)(0.6013) = 21.65 \text{ kips}$$

Reaction capacity with six bolts = $6(21.65) = 129.9$ kips. (Since the thickness of the column flange is not given, and the end plate thickness is not yet known, bearing strength cannot be investigated at this time. Once all connected parts have been designed, however, it should be checked.)

Compare this reaction capacity with the shear strength of the web. From the Factored Uniform Load tables in Part 4 of the *Manual*,

$$\phi_v V_n = 103 \text{ kips} < 129.9 \text{ kips} \quad \text{(OK)}$$

The bolts are therefore capable of transferring any reaction that the beam can deliver.

Answer   Use six bolts, with four arranged symmetrically about the tension flange and two located at the compression flange.

Flange weld:

Available length $L = 2b_f + 2t_f - t_w = 2(6.00) + 2(0.425) - 0.300$
$$= 12.55 \text{ in.}$$

Required weld size:

$$s = \frac{P_{uf}}{0.707L(\phi F_W)} = \frac{120.2}{0.707(12.55)(31.5)} = 0.4301 \text{ in.}$$

Although the thickness of the end plate is not yet known, the minimum weld size from AISC Table J2.4 will never exceed $\frac{5}{16}$ in., so the 0.430 in. required for strength will control.

Answer   Use a $\frac{7}{16}$-in. fillet weld.

End plate:

Use $p_f = d_b + \frac{1}{2} = 0.875 + 0.500 = 1.375$ in.

$p_e = p_f - 0.25d_b - 0.707w = 1.375 - 0.25(0.875) - 0.3093 = 0.8470$ in.

For the plate width, use $b_p = b_f + 1 = 6.00 + 1 = 7.00$ in.

$$M_t = F_1 s = \left(\frac{P_{uf}}{2}\right)\left(\frac{p_e}{2}\right) = \left(\frac{120.2}{2}\right)\left(\frac{0.8470}{2}\right) = 25.45 \text{ in.-kips}$$

$C_a = 1.36$ (Table 10-1, Part 10 of *Manual*)

$$C_b = \sqrt{\frac{b_f}{b_p}} = \sqrt{\frac{6.00}{7.00}} = 0.9258$$

$\dfrac{A_f}{A_w} = 0.504$ (Table 10-2, Part 10 of *Manual*)

$$\alpha_m = C_a C_b \left(\frac{A_f}{A_w}\right)^{1/3}\left(\frac{p_e}{d_b}\right)^{1/4}$$

$$= 1.36(0.9258)(0.504)^{1/3}(0.8470/0.875)^{1/4} = 0.9939$$

$$M_{eu} = \alpha_m M_t = 0.9939(25.45) = 25.29 \text{ in.-kips}$$

$$t_{Preq} = \sqrt{\frac{4M_{eu}}{0.90b_p F_y}} = \sqrt{\frac{4(25.29)}{0.90(7.00)(36)}} = 0.668 \text{ in.}$$

Answer   Use a plate thickness of $\frac{3}{4}$ in.

From Ref. 51, the maximum effective plate width is

$$b_f + 2w + t_p = 6.00 + 2\left(\frac{7}{16}\right) + \frac{3}{4} = 7.62 \text{ in.} > 7.00 \quad \text{(OK)}$$

Check shear. The shear force in the plate is

$$F_1 = \frac{P_{uf}}{2} = \frac{120.2}{2} = 60.1 \text{ kips}$$

From AISC J5, the shear strength is

$$\phi R_n = 0.90(0.60A_g F_y) = 0.90(0.60)\left(7 \times \frac{3}{4}\right)(36)$$

$$= 102 \text{ kips} > 60.1 \text{ kips} \quad \text{(OK)}$$

FIGURE 8.43 ≡

Web weld: To match the shear strength of the web, the required weld strength (two welds, one each side of web) is

$$\frac{\phi_v V_n}{d} = \frac{103}{17.7} = 5.819 \text{ kips/in.}$$

The required weld size is

$$s = \frac{5.819/2}{0.707(31.5)} = 0.131 \text{ in.}$$

Determine the size required to resist bending in the web. When the plastic moment has been reached, the stress in the web is equal to the yield stress F_y, and the load per unit length of web is

$$\phi_b(F_y \times t_w \times 1) = 0.90(36)(0.300) = 9.720 \text{ kips/in.}$$

The load per weld is $\frac{9.720}{2} = 4.860$ kips/in.

The required weld size is

$$s = \frac{4.860}{0.707(31.5)} = 0.2182 \text{ in.} > 0.131 \text{ in.}$$

The minimum size is $\frac{1}{4}$ in. (AISC Table J2.4, based on plate thickness).

Answer   Use a $\frac{1}{4}$ in. fillet weld. (The design is summarized in Figure 8.43.)

FIGURE 8.44 ≡

(a)   (b)

Column Web Stiffener Considerations

A minor liberalization of AISC Equation K1-2, which prevents column web yielding in beam-to-column connections, can be made when end plates are used. This equation is based on limiting the stress on a cross section of the web formed by its thickness and a length equal to $t_b + 5k$ (Figure 8.44a). As can be seen in Figure 8.44b, a larger area will be available when the load is transmitted through the additional thickness of the end plate. If the beam flange-to-plate welds are taken into account and the load is assumed to disperse at a slope of $1:1$ through the plate, the loaded length of web will be $t_b + 2w + 2t_p + 5k$. Based on experimental studies,[52] the term $5k$ can be replaced by $6k$, resulting in the following equation for web yielding strength:

$$\phi R_n = \phi[(6k + t_b + 2w + 2t_p)F_{yw}\, t_w]$$

where w = weld size.

In addition, local flange bending and web stability (either web crippling or compression buckling) should be investigated. Part 5 of the *Manual* contains guidelines that may be used to account for local flange bending.

8.9 ≡ CONCLUDING REMARKS

This chapter has emphasized the design and analysis of bolts and welds rather than connection fittings such as framing angles and beam seats. In most cases, the provisions for bearing in bolted connections and base metal shear in welded connections will ensure adequate strength of these parts. Sometimes, however, additional shear investigation is needed. In other cases, direct tension or bending must be considered.

Flexibility of the connection is another important consideration. In a shear connection (simple framing), the connecting parts must be flexible enough to permit the connection to rotate under load. Type FR connections (rigid connections), however, should be stiff enough so that relative rotation of the connected members is kept to a minimum.

This chapter is intended to be introductory only, and it is by no means a complete guide to the design of building connections. Blodgett[53] is a useful source of detailed welded-connection information. Although somewhat dated, it contains a wealth of practical suggestions. Two AISC publications are also recommended;[54,55] the second is intended primarily for detailers, but it also contains information useful to the designer.

PROBLEMS

NOTES   **1.** Use elastic analysis for each connection subject to eccentric shear.

2. Unless otherwise indicated, bolt threads are in the plane of shear.

8.1   Find the force on the critical fastener in Figure P8.1. The 23 kip load is a factored load.

FIGURE P8.1

8.2   What size A325 bolts are required for the bracket connection shown in Figure P8.2? Slip is permitted. The given load has been factored. All structural steel is A36.

FIGURE P8.2

8.3 The fasteners in the connection in Figure P8.3 are placed as shown. What additional force is experienced by the critical fastener as a consequence of the fasteners not being on the centroidal axis of the member?

FIGURE P8.3

8.4 A double-angle shape, $2L6 \times 6 \times \frac{7}{8}$, is used as a tension member and is connected at each end to a gusset plate with a single row of bolts as shown in Figure P8.4. Use the usual gage distance given in the *Manual* and determine the moment in the member (in terms of the axial load P) caused by the resulting eccentricity.

FIGURE P8.4

8.5 The bolts in the bracket connection shown are 1-in.-diameter A325 bolts. The connection is slip-critical. Are the bolts adequate? The loads in Figure P8.5 are service loads, and A36 steel is used throughout.

FIGURE P8.5

8.6 A307 bolts are used in the connection shown in Figure P8.6. Determine the required size. All structural steel is A36, and the given load is factored.

FIGURE P8.6

In each of Problems 8.7, 8.8, and 8.9, determine the maximum fastener force caused by the given factored loads.

8.7

FIGURE P8.7

8.8

FIGURE P8.8

8.9

FIGURE P8.9

8.10   For the connection and factored load in Figure P8.10, compute the force in fastener A.

FIGURE P8.10

8.11   A325 bolts are used in the connection in Figure P8.11. Determine the required size if slip is permitted. The 10 kip load consists of 2.5 kips of service dead load and 7.5 kips of service live load. All structural steel is A36.

FIGURE P8.11

8.12 Compute the maximum fastener tensile force in Figure P8.12. The fasteners are A325 bolts. Assume that the entire surface of contact remains compressed.

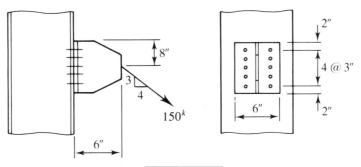

FIGURE P8.12

In each of Problems 8.13 through 8.17, check the adequacy of the fasteners. The given loads are service loads.

8.13

FIGURE P8.13

8.14

$D = 30^k$
$L = 90^k$

WT6 × 20

W14 × 61   6″

3″
3″
6″
3″

1½″   5″   1½″

1″ diameter A325 bolts
A36 steel
Slip is permitted.

FIGURE P8.14

8.15

6″ $D = 25^k$
$L = 75^k$

Ten ¾″ fasteners
A325 bolts
A36 steel
Slip is permitted.

2″

4 @ 3″

2″

WT15 × 49.5

W12 × 65

FIGURE P8.15

8.16

3″

1½″   8″

6″

3″

1½″

WT5 × 22.5

W18 × 50

W10 × 49

Reaction = 38^k
$(D = 10^k, L = 28^k)$

¾″ diameter A325 bolts
A36 steel
Slip is permitted.

FIGURE P8.16

8.17

$D = 18^k$
$L = 54^k$

3"

2"

10"

14"

2"

1" diameter A325 bolts
A36 steel
Slip is permitted.

W10 × 100   WT5 × 24.5

FIGURE P8.17

8.18 Design a simply supported beam and an all-bolted connection for the conditions shown in Figure P8.18. In addition to its own weight, the beam must support a service live load of 10 kips/ft. Use the following guidelines:

a. Design the connection without regard to eccentricity, then check for its effect.

b. Specify the exact beam length.

c. Furnish a detailed sketch of your recommended connection.

Use A36 steel and slip-critical A325 bolts. Assume continuous lateral support of the beam compression flange.

Col.
₵

Col.
₵

Beam ₵

12'-0"

W12 × 106

FIGURE P8.18

8.19   What size E70 fillet weld is required for the connection shown in Figure P8.19? The 30 kip load is a service load, consisting of 8 kips dead load and 22 kips live load. All structural steel is A36.

FIGURE P8.19

In Problems 8.20 and 8.21, check the adequacy of the welds. Assume that the shear in the base metal is acceptable. The given loads are factored.

8.20

FIGURE P8.20

8.21

FIGURE P8.21

In Problems 8.22 and 8.23, compute the load in the weld (in kips per inch of weld) caused by the *eccentricity alone.*

8.22

112^k

L3 × 3 × ½

4″

FIGURE P8.22

8.23

150^k

7″

L7 × 4 × ½

6″

FIGURE P8.23

8.24 An L6 × 6 × $\frac{3}{8}$, $F_y = 50$ ksi, is attached to a gusset plate with E70XX fillet welds. Design the welds to develop the design strength of the member. Use a placement of welds that will eliminate eccentricity. Assume that the gusset plate strength does not govern.

8.25 For the factored load in Figure P8.25, use the minimum weld size and design the connection for the following cases:

a. Balance the welds to eliminate eccentricity. Use side welds only.

b. Balance the welds and use both side and end welds.

c. Do not balance the welds. Use side and end welds, and account for the resulting eccentricity.

FIGURE P8.25

8.26   If the maximum permissible size E70 fillet weld is used, is the connection in Figure P8.26 satisfactory? All steel is A36 and the load is factored.

FIGURE P8.26

8.27   A factored load P_u in the plane of a two-sided weld acts 6 inches from the vertical leg as shown in Figure P8.27. The weld is a $\frac{5}{16}$-in. fillet weld made with an E70 electrode. What is the maximum permissible value of the load?

FIGURE P8.27

8.28   Determine the maximum load in the weld in Figure P8.28 (in kips per inch of length). The 15 kip load is factored.

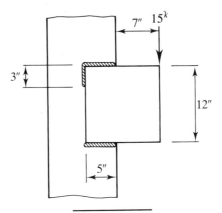

FIGURE P8.28

In Problems 8.29 through 8.31, determine the required fillet weld size. A36 steel and E70 electrodes are used in each case, and all loads are factored.

8.29

FIGURE P8.29

8.30

FIGURE P8.30

8.31

FIGURE P8.31

8.32   Use the maximum size E70 fillet weld and compute the maximum factored load reaction R (as limited by the strength of the weld) that can be supported by the connection of Figure P8.32. All steel is A36. Use minimum end returns at the top, rounded to the nearest $\frac{1}{4}$ in.

FIGURE P8.32

In Problems 8.33 and 8.34, determine the maximum load in the given weld (in kips per inch of length).

8.33

FIGURE P8.33

8.34

FIGURE P8.34

8.35   Investigate the adequacy of the weld in Figure P8.35. The 150 kip load is factored. Consider each of the following conditions:

   a. Use side welds with 1-in. end returns at the top.

   b. Ignore the end returns.

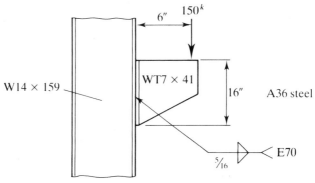

FIGURE P8.35

8.36   What is the moment strength of the connection in Figure P8.36? The fasteners are $\frac{5}{8}$-in.-diameter A325 bolts, and slip is permitted. Structural steel is A36.

FIGURE P8.36

In Problems 8.37 and 8.38, determine whether column stiffeners are required for the maximum force to be developed in the beam flange or flange plate. If they are, specify the required dimensions. Use A36 steel in each problem.

8.37

W16 × 31
W12 × 40

FIGURE P8.37

8.38

ℙ 9 × ¹¹⁄₁₆

W24 × 76
ℙ 9 × ¹¹⁄₁₆
W14 × 145

FIGURE P8.38

8.39 Design a connection of the type shown in Figure P8.39 for a W10 × 45 column and a W16 × 40 beam. The maximum factored load moment and shear, which are caused by gravity loads, are $M_u = 143$ ft-kips and $V_u = 60$ kips. Use A36 steel, E70XX electrodes, and A325 bolts. Assume that slip of the connection is permissible.

FIGURE P8.39

In Problems 8.40 and 8.41, investigate the adequacy of the bolts in the given end plate connection. The given loads are service loads, consisting of 25% dead load and 75% live load.

8.40

FIGURE P8.40

8.41

FIGURE P8.41

8.42 Design an end plate connection for a W12 × 30 beam and a W10 × 60 column. The shear consists of a 10 kip service dead load and a 25 kip service live load. The service dead load moment is 6 ft-kips, and the service live load moment is 14 ft-kips. Use A36 steel, E70XX electrodes, and A325 bolts. Slip of the connection is not permitted.

8.43 Design an end plate connection for a W18 × 40 beam to a W8 × 40 column. Design for the full moment and shear capacities of the beam. Use A36 steel and A325 bolts. Assume that slippage of the bolts is permissible.

8.44 Write a computer program to determine the shear force in the upper right-hand fastener in a connection of the type shown in Figure P8.44. There are two vertical lines of bolts at a uniform vertical spacing of s, and the two lines are separated by a distance g. The applied load P is vertical and acts at an eccentricity e. Use elastic analysis.

Input: The number of fasteners n; the vertical spacing s; the horizontal spacing g; and the eccentricity e.

Output: Same as input plus the resultant shear force on the critical fastener in terms of the load P.

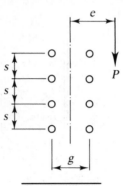

FIGURE P8.44

8.45   Write a computer program to determine the critical fastener force resulting from the eccentricity of the load in a connection of the type shown in Figure P8.45. The tension member is a double-angle shape and is connected with a single gage line of equally spaced fasteners. Use elastic theory and compute only the part of the force that is caused by the eccentricity.

Input: The length L of the angle leg; the distance y from the top edge to the centroid of the cross section; the number of bolts n; and the spacing s.

Output: Same as input plus the extra force in terms of P.

FIGURE P8.45

8.46   Write an *interactive* computer program to design welds for the connection of a single-angle tension member to a gusset plate (Figure P8.46). The welds should be placed along the end and both sides and balanced to eliminate eccentricity.

Input: The program should prompt the user for input data with messages in the following form:

LENGTH OF ANGLE LEG (IN.):

VERTICAL DISTANCE FROM TOP OF ANGLE TO CENTROID (IN.):

MAGNITUDE OF FACTORED LOAD (KIPS):

WELD ELECTRODE (ENTER 70 OR 80):

WELD SIZE (IN.):

Output: The values of L_1 and L_2 (see figure).

FIGURE P8.46

8.47   Write a computer program to produce a design aid for end plate moment connections. The program is to generate a table of values of required plate sizes corresponding to various bolt arrangements.

Input: M_u, d, b_f, t_f, t_w, k

where

M_u = moment to be transferred (ft-kips)

d = overall depth of the beam (in.)

b_f = beam flange width (in.)

t_f = beam flange thickness (in.)

t_w = beam web thickness (in.)

k = distance from outer face of flange to the web fillet (in.)

Output: Same as input plus the following table:

FLANGE WELD SIZE =   IN.

NO. BOLTS AT EACH FLANGE	REQUIRED BOLT DIAMETER (IN.)	PLATE WIDTH (IN.)	PLATE THICKNESS (IN.)
4	·	·	·
6	·	·	·
8	·	·	·

Additional instructions:

1. Restrict the program to A36 steel, E70XX electrodes, and A325 bearing-type bolts.

2. Select the flange weld size to resist the flange force. Do not check for maximum and minimum weld sizes.

3. Assume two rows of bolts at each flange.

4. For simplicity, do not round any of the results to practical values; leave them as calculated.

8.48   Write a computer program to compute the maximum bolt tension in a connection of the type shown in Figure P8.48. The load is not in the plane of shear; as a result, the fasteners are subjected to both shear and tension. Assume that the fasteners are properly tensioned high-strength bolts. Your program should work for any number of vertical lines of equally spaced bolts.

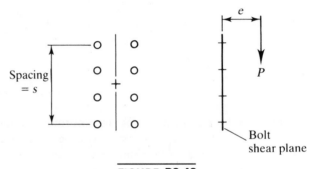

FIGURE P8.48

Input: The total number of bolts n; the number of vertical lines m; and the vertical bolt spacing s.

Output: Same as input plus the following: For values of the eccentricity e ranging from 3 inches to 12 inches in increments of 0.5 inches, compute and print the maximum bolt tension corresponding to a load P of 1 kip. Present your results in tabular form as follows:

15 BOLTS

3 VERTICAL LINES

VERTICAL SPACING = 3.0 IN.

ECCENTRICITY (IN.)	BOLT TENSION PER KIP OF LOAD (KIPS)
3.0	.
3.5	.
.	.
.	.
.	.
12.0	.

9

COMPOSITE CONSTRUCTION

INTRODUCTION

Composite construction employs structural members that are composed of two materials: structural steel and reinforced concrete. Strictly speaking, any structural member formed with two or more materials is composite. In buildings and bridges, however, we usually mean structural steel and reinforced concrete, and we usually mean composite beams or columns. Composite columns are being used in some structures today after a period of disuse; they will be covered later in this chapter. Our coverage of beams will be restricted to those that are part of a floor or roof system. Composite construction is covered in AISC Chapter I, "Composite Members."

Composite beams can take several forms. The earliest versions consisted of beams encased in concrete (Figure 9.1a). This was a practical alternative when the primary means of fireproofing structural steel was to encase it in concrete; the rationale was that if the concrete was there, we might as well account for its contribution to the strength of the beam. Currently, lighter and more economical methods of fireproofing are available, and encased composite beams are rarely used. Instead, composite behavior is achieved by connecting the steel beam to the reinforced concrete slab it supports, causing the two parts to act as a unit. In a floor or roof system, a portion of the slab acts with each steel beam to form a composite beam consisting of the rolled steel shape augmented by a concrete flange at the top (Figure 9.1b).

This unified behavior is possible only if horizontal slippage between the two components is prevented. This will be accomplished if the horizontal shear at the interface is resisted by connecting devices known as *shear connectors*. These devices — which can be headed studs, spiral reinforcing steel, or short lengths of small channel shapes — are welded to the top flange of the steel beam at prescribed intervals and provide the connection mechanically through anchorage in the hardened concrete (Figure 9.1c). Studs are the most commonly used type of shear connectors, and more than one can be used at each location if the flange is wide enough to accommodate them (this will depend on the allowable spacing, which will be considered in Section 9.4). One reason for the popularity of shear studs is their ease of installation. It is essentially a one-worker job made possible by an automatic tool that allows the operator to position the stud and weld it to the beam in one operation.

A certain number of shear connectors will be required to make a beam fully composite. Any fewer than this number will permit some slippage to occur

FIGURE 9.1 ≡

(a)

(b)

(c)

Section

between the steel and concrete; such a beam is said to be *partially composite*. Partially composite beams (which are actually more efficient than fully composite beams) will be covered in Section 9.7, "Composite Beams with Formed-Steel Deck."

Most composite construction in buildings utilizes formed steel deck, which serves as formwork for the concrete slab and is left in place after the concrete cures. This metal deck also contributes to the strength of the slab, the design of which will not be considered here. The deck can be used with its ribs oriented either transversely or parallel to the beams. In the usual floor system, the ribs will be perpendicular to the floor beams and parallel to the supporting girders. The shear studs are welded to the beams from above, through the deck. Since they can be placed only in the ribs, the spacing of the studs along the length of the beam is limited to multiples of the rib spacing. Figure 9.2 shows a slab with formed steel deck with the ribs perpendicular to the beam axis.

FIGURE 9.2 ≡

Almost all highway bridges that use steel beams are of composite construction, and composite beams are frequently the most economical alternative in buildings. Although smaller, lighter rolled steel beams can be used with composite construction, this advantage will sometimes be offset by the additional cost of the shear connectors. Even if this is so, other advantages may make composite construction attractive. Shallower beams can be used, and deflections will be smaller than with conventional noncomposite construction.

Elastic Stresses in Composite Beams

Although the design strength of composite beams is usually based on conditions at failure, an understanding of the behavior at service loads is important for several reasons. Deflections are always investigated at service loads, and in some cases, the design strength is based on the limit state of first yield.

Flexural and shearing stresses in beams of homogeneous materials can be computed from the formulas

$$f_b = \frac{Mc}{I} \quad \text{and} \quad f_v = \frac{VQ}{It}$$

A composite beam is not homogeneous, however, and these formulas are not valid. To be able to use them, an artifice known as the *transformed section* is employed to "convert" the concrete into an amount of steel that has the same effect as the concrete. This is done by requiring the strains in the fictitious steel to be the same as those in the concrete it replaces. Figure 9.3 shows a segment of a composite beam with stress and strain diagrams superimposed. If the slab is properly attached to the rolled steel shape, the strains must be as shown, consistent with small displacement theory, which states that cross sections that are plane before bending remain plane after bending. The linear stress distribution shown, however, is valid only if the beam is assumed to be homogeneous. We first require that the strain in the concrete at any point be equal to the strain in any replacement steel at that point:

$$\varepsilon_c = \varepsilon_s$$

FIGURE 9.3 ≡

Strain

Stress
(for transformed section)

or

$$\frac{f_c}{E_c} = \frac{f_s}{E_s}$$

and

$$f_s = \frac{E_s}{E_c} f_c = n f_c \qquad (9.1)$$

where

E_c = modulus of elasticity of concrete

$n = \dfrac{E_s}{E_c}$ = modular ratio

The ACI building code[22] gives the following formula for E_c:

$$E_c = w_c^{1.5}(33)\sqrt{f_c'}$$

where

w_c = unit weight of concrete (lb/ft³)

f_c' = 28-day compressive strength of the concrete (psi)

Normal-weight concrete weighs approximately 145 pounds per cubic foot.*

Equation 9.1 can be interpreted as follows: n square inches of concrete are required to resist the same force as one square inch of steel. To determine the area of steel that will resist the same force as the concrete, divide the concrete area by n. That is, replace A_c by A_c/n. The result is the *transformed area*.

---

* AISC I2.2 permits E_c to be computed from $E_c = w_c^{1.5}\sqrt{f_c'}$, where E_c and f_c' are in ksi.

FIGURE 9.4 ≡

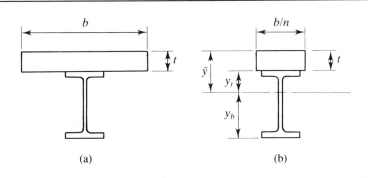

(a)   (b)

Consider the composite section shown in Figure 9.4a (determination of the effective flange width b when the beam is part of a floor system will be discussed presently). To transform the concrete area A_c, we must divide by n. The most convenient way to do this is to divide the width by n and leave the thickness unchanged. When this is done, the homogeneous steel section of Figure 9.4b is the result. To compute stresses, the neutral axis of this composite shape is located, and the corresponding moment of inertia is computed. Bending stresses can then be computed by the flexure formula. At the top of the steel,

$$f_{st} = \frac{My_t}{I_{tr}}$$

At the bottom of the steel,

$$f_{sb} = \frac{My_b}{I_{tr}}$$

where

M = applied bending moment

I_{tr} = moment of inertia about the neutral axis (same as the centroidal axis for this homogeneous section).

y_t = distance from the neutral axis to the top of the steel

y_b = distance from the neutral axis to the bottom of the steel

The stress in the concrete may be computed in the same way, but since the material under consideration is steel, the result must be divided by n (see Equation 9.1):

$$\text{Maximum } f_c = \frac{M\bar{y}}{nI_{tr}}$$

where $\bar{y}$ is the distance from the neutral axis to the top of the concrete.

This procedure is valid only for a positive bending moment, with compression at the top, since concrete has negligible tensile strength.

≡ **EXAMPLE 9.1**

A composite beam consists of a W16 × 36 of A36 steel with a 5-in.-thick × 87-in.-wide concrete slab at the top. The strength of the concrete is $f'_c = 4000$ psi. Determine the maximum stress in the steel resulting from a positive bending moment of 160 ft-kips.

SOLUTION

$$E_c = w_c^{1.5}(33)\sqrt{f'_c} = (145)^{1.5}(33)\sqrt{4000} = 3{,}644{,}000 \text{ psi}$$

$$n = \frac{E_s}{E_c} = \frac{29{,}000{,}000}{3{,}644{,}000} = 7.96 \quad \text{Use } n = 8.$$

Since the modulus of elasticity of concrete can only be approximated, the usual practice of rounding n to the nearest whole number is sufficiently accurate.

$$\frac{b}{n} = \frac{87}{8} = 10.88 \text{ in.}$$

The transformed section is shown in Figure 9.5.

The location of the neutral axis can be found by applying the principle of moments with the axis of moments at the top of the slab. The computations are summarized in Table 9.1, and the distance from the top of the slab to the centroid is

$$\bar{y} = \frac{\sum Ay}{\sum A} = \frac{273.1}{65.00} = 4.202 \text{ in.}$$

Applying the parallel axis theorem and tabulating the computations in Table 9.2, we obtain the moment of inertia of the transformed section as

$$I_{tr} = 1526 \text{ in.}^4$$

FIGURE 9.5 ≡

TABLE 9.1 ≡

Component	A	y	Ay
Concrete	54.40	2.50	136.0
W16 × 36	10.6	12.93	137.1
	65.00		273.1

TABLE 9.2 ≡

Component	A	y	Ī	d	Ī + Ad²
Concrete	54.40	2.50	113.3	1.702	270.9
W16 × 36	10.6	12.93	448	8.728	1255
					1525.9

Stress at top of steel:

$$y_t = \bar{y} - t = 4.202 - 5.000 = -0.7980 \text{ in.}$$

where t is the thickness of the slab.

$$f_{st} = \frac{My_t}{I_{tr}} = \frac{(160 \times 12)(0.7980)}{1526} = 1.00 \text{ ksi (tension)}$$

(Since the top of the steel is below the neutral axis, f_{st} is a *tensile* stress.)

Stress at bottom of steel:

$$y_b = t + d - \bar{y} = 5 + 15.86 - 4.202 = 16.66 \text{ in.}$$
$$f_{sb} = \frac{My_b}{I_{tr}} = \frac{(160 \times 12)(16.66)}{1526} = 21.0 \text{ ksi (tension)}$$

If the concrete is assumed to have no tensile strength, the concrete below the neutral axis should be discounted. The geometry of the transformed section will then be different from what was originally assumed; to obtain an accurate result, the location of the neutral axis should be recomputed on the basis of this new geometry. Referring to Figure 9.6 and Table 9.3, we can compute the new location of the neutral axis as follows:

$$\bar{y} = \frac{\sum Ay}{\sum A} = \frac{5.44\bar{y}^2 + 137.1}{10.88\bar{y} + 10.6}$$
$$\bar{y}(10.88\bar{y} + 10.6) = 5.44\bar{y}^2 + 137.1$$
$$5.44\bar{y}^2 + 10.6\bar{y} - 137.1 = 0$$
$$\bar{y} = 4.140 \text{ in.}$$

FIGURE 9.6 ≡

TABLE 9.3 ≡

Component	A	y	Ay
Concrete	$10.88\bar{y}$	$\dfrac{\bar{y}}{2}$	$5.44\bar{y}^2$
W16 × 36	10.6	12.93	137.1

The moment of inertia of this revised composite area is

$$I_{tr} = \frac{1}{3}(10.88)(4.140)^3 + 448 + 10.6(12.93 - 4.140)^2 = 1524 \text{ in.}^4$$

and the stresses are

$$f_{st} = \frac{(160 \times 12)(5 - 4.140)}{1524} = 1.08 \text{ ksi (tension)}$$

$$f_{sb} = \frac{(160 \times 12)(5 + 15.86 - 4.140)}{1524} = 21.1 \text{ ksi (tension)}$$

The difference between the two analyses is negligible, and the refinement in locating the neutral axis is not necessary.

Answer   The maximum stress in the steel is 21.1 ksi tension.

Flexural Strength

In most cases, the nominal flexural strength will be reached when the entire steel cross section yields and the concrete crushes in compression. The corresponding stress distribution on the composite section is called a *plastic stress distribution*. The AISC Specification gives the design strength for *positive* bending as $\phi_b M_n$, which is defined as follows:

For shapes with compact webs — that is, $h/t_w \leq 640/\sqrt{F_y}$ — the resistance factor ϕ_b is 0.85, and M_n is obtained from the plastic stress distribution.

For shapes with noncompact webs ($h/t_w > 640/\sqrt{F_y}$), ϕ_b is 0.9, and M_n is obtained from the elastic stress distribution corresponding to first yielding of the steel.

Since all shapes tabulated in the *Manual* have compact webs, the first condition will govern for all composite beams except for those with built-up steel shapes. We will consider only compact shapes in this chapter.

When a composite beam has reached the plastic limit state, the stresses will be distributed in one of the three ways shown in Figure 9.7. In this figure, the concrete stress is shown as a uniform compressive stress of $0.85 f_c'$, extending from the top of the slab to a depth that may be equal to or less than the total

FIGURE 9.7 ≡

(a)

(b)

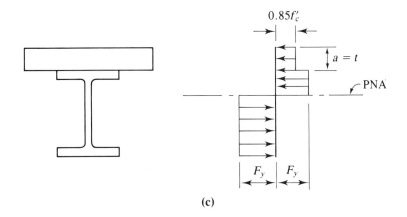

(c)

slab thickness. This is the *Whitney equivalent stress distribution*, which has a resultant that matches that of the actual stress distribution.[22] Figure 9.7a shows the distribution corresponding to full tensile yielding of the steel and partial compression of the concrete, with the plastic neutral axis (PNA) in the slab. The tensile strength of concrete is small and is discounted, so no stress is shown where tension is applied to the concrete. This condition will usually prevail when there are enough shear connectors provided to completely prevent slip — that is, to ensure full composite behavior. In Figure 9.7b, the concrete stress block extends the full depth of the slab, and the PNA is in the flange of the steel shape. Part of the flange will therefore be in compression to augment the compressive force in the slab. The third possibility, the PNA in the web, is shown in Figure 9.7c. Note that the concrete stress block need not extend the full depth of the slab for any of these three cases.

In each case shown in Figure 9.7, the nominal moment capacity is found by computing the moment of the couple formed by the compressive and tensile resultants. This can be accomplished by summing the moments of the resultants about any convenient point. Because of the connection of the steel shape to the concrete slab, lateral-torsional buckling is no problem once the concrete has cured and composite action has been achieved.

To determine which of the three cases governs, compute the compressive resultant as the smallest of

1. $A_s F_y$
2. $0.85 f'_c A_c$
3. $\sum Q_n$

where

A_s = cross-sectional area of steel shape

A_c = area of concrete

   = tb (see Figure 9.7)

$\sum Q_n$ = total shear strength of the shear connectors

Each possibility represents a horizontal shear force at the interface between the steel and the concrete. When the first possibility controls, the steel is being fully utilized, and the stress distribution of Figure 9.7a applies. The second possibility corresponds to the concrete controlling, and the PNA will be in the steel (Figure 9.7b or c). The third case governs only when there are fewer shear connectors than required for full composite behavior, resulting in partial composite behavior. Although partial composite action can exist with either solid slabs or slabs with formed steel deck, it will be covered in Section 9.7, "Composite Beams with Formed Steel Deck."

≡ **EXAMPLE 9.2**

Compute the design strength of the composite beam of Example 9.1. Assume that sufficient shear connectors are provided for full composite behavior.

FIGURE 9.8 ≡

 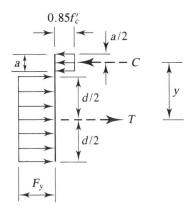

SOLUTION
Determine the compressive force C in the concrete (horizontal shear force at the interface between the concrete and steel). Since there will be full composite action, this force will be the smaller of $A_s F_y$ and $0.85 f'_c A_c$.

$$A_s F_y = 10.6(36) = 381.6 \text{ kips}$$
$$0.85 f'_c A_c = 0.85(4)(5 \times 87) = 1479 \text{ kips}$$

The steel controls; $C = 381.6$ kips. This means that the full depth of the slab is not needed to develop the required compression force. The stress distribution of Figure 9.8 will result.

The resultant compressive force can also be expressed as

$$C = 0.85 f'_c ab$$

From this we obtain

$$a = \frac{C}{0.85 f'_c b} = \frac{381.6}{0.85(4)(87)} = 1.290 \text{ in.}$$

The force C will be located at the centroid of the compressive area at a depth of $a/2$ from the top of the slab. The resultant tensile force T (equal to C) will be located at the centroid of the steel area. The moment arm of the couple formed by C and T is

$$y = \frac{d}{2} + t - \frac{a}{2} = \frac{15.86}{2} + 5 - \frac{1.290}{2} = 12.28 \text{ in.}$$

The nominal strength is the moment of the couple:

$$M_n = Cy = Ty = 381.6(12.28) = 4686 \text{ in.-kips} = 390.5 \text{ ft-kips}$$

and the design strength is

$$\phi_b M_n = 0.85(390.5) = 332 \text{ ft-kips}$$

Answer
Design strength = 332 ft-kips.

When full composite behavior exists, the conditions of the previous example will be the norm. Analysis for the case of the plastic neutral axis located within the steel section will be deferred until partial composite action has been covered.

9.2 ≡ SHORED VERSUS UNSHORED CONSTRUCTION

Until the concrete has cured and attained its design strength (at least 75% of its 28-day compressive strength f'_c), there can be no composite behavior, and the weight of the slab must be supported by some other means. Once the concrete has cured, composite action is possible, and all subsequently applied loads will be resisted by the composite beam. If the steel shape is supported at a sufficient number of points along its length before the slab is placed, the weight of the wet concrete will be supported by these temporary *shores* rather than by the steel. Once the concrete has cured, the temporary shoring can be removed, and the weight of the slab, as well as any additional loads, will be carried by the composite beam. If shoring is not used, however, the rolled steel shape must support not only its own weight, but also the weight of the slab and its formwork during the curing period. Once composite behavior is attained, additional loads, both dead and live, will be supported by the composite beam. These different conditions will now be considered in more detail.

Unshored: Before Concrete Cures

AISC I3.4 requires that when temporary shoring is not provided, the steel shape alone have sufficient strength to resist all loads applied before the concrete attains 75% of its strength f'_c. The flexural strength is computed in the usual way, based on Chapter F of the Specification (Chapter 5 of this book). Depending on its design, the formwork for the concrete slab may or may not provide lateral support for the steel beam. If not, the unbraced length L_b must be taken into account, and lateral-torsional buckling may control the flexural strength. If temporary shoring is not used, the steel beam may also be called on to resist incidental construction loads. To account for these loads, an additional 20 pounds per square foot is recommended.[18]

Although the weight of the beam and slab are known with some degree of precision, there are some live load characteristics associated with these loads. If a construction load is considered, it is certainly a live load. Doe and Smith[18] recommend a load factor of 1.2 for all construction loads, including forms and wet concrete, applied to an unshored beam before the attainment of composite behavior. In keeping with this recommendation, in this book we will use a load factor of 1.2 for all loads applied to an unshored beam before the concrete cures.

Unshored: After Concrete Cures

After composite behavior is achieved, all loads subsequently applied will be supported by the composite beam. At failure, however, all loads will be resisted by the internal couple corresponding to the stress distribution at failure. Thus the composite section must have adequate strength to resist all loads, including those applied to the steel beam before the concrete cures.

Shored Construction

In shored construction, only the composite beam need be considered, since the steel shape will not be required to support anything other than its own weight.

Shear Strength

AISC I3.6 conservatively requires that all shear be resisted by the web of the steel shape, as provided for in Chapter F of the Specification.

EXAMPLE 9.3

A W12 × 50 acts compositely with a 4-in.-thick concrete slab. The effective width of the slab is 72 inches. Shoring is not used, and the applied bending moments are as follows: from the beam weight, $M_{\text{beam}} = 13$ ft-kips; from the slab weight, $M_{\text{slab}} = 77$ ft-kips; and from the live load, $M_L = 38$ ft-kips. (Do not consider any additional construction loads.) Steel is A36, and $f'_c = 4000$ psi. Determine whether the flexural strength of this beam is adequate. Assume full composite action and assume that the formwork provides lateral support of the steel section before curing of the concrete.

SOLUTION

Before the concrete cures, the factored load moment is

$$M_u = 1.2(13 + 77) = 108 \text{ ft-kips}$$

From the Beam Design Charts in Part 4 of the *Manual*, for A36 steel,

$$\phi_b M_n = 195 \text{ ft-kips} > 108 \text{ ft-kips} \quad (\text{OK})$$

After the concrete cures, the factored load moment that must be resisted by the composite beam is

$$M_u = 108 + 1.6(38) = 168.8 \text{ ft-kips}$$

The compressive force C is the smaller of

$$A_s F_y = 14.7(36) = 529.2 \text{ kips}$$

or

$$0.85 f'_c A_c = 0.85(4)(4 \times 72) = 979.2 \text{ kips}$$

The PNA is in the concrete, and $C = 529.2$ kips. From Figure 9.8, the depth of the compressive stress block is

$$a = \frac{C}{0.85 f'_c b} = \frac{529.2}{0.85(4)(72)} = 2.162 \text{ in.}$$

The moment arm is

$$y = \frac{d}{2} + t - \frac{a}{2} = \frac{12.19}{2} + 4 - \frac{2.162}{2} = 9.014 \text{ in.}$$

The design moment is

$$\phi_b M_n = \phi_b C y = 0.85(529.2)(9.014)$$

$$= 4055 \text{ in.-kips} = 338 \text{ ft-kips} > 168.8 \text{ ft-kips} \quad \text{(OK)}$$

Answer   The beam has sufficient flexural strength.

Obviously, shored construction is more efficient than unshored construction, because the steel section is not called on to support anything other than its own weight. In some situations, the use of shoring will enable a smaller steel shape to be used. Most composite construction is unshored, however, because the additional cost of the shores, especially labor cost, outweighs the small savings in steel weight that may result. Consequently, the remainder of this chapter will be devoted to unshored composite construction.

9.3 ≡ EFFECTIVE FLANGE WIDTH

The portion of the floor slab that acts compositely with the steel beam is a function of several factors, including the span length and beam spacing. AISC I3.1 requires that the effective width of floor slab on *each side* of the beam centerline be taken as the smallest of the following:

(a) one eighth of the span length

(b) one half of the beam center-to-center spacing

(c) the distance from the beam centerline to the edge of the slab

The third criterion will apply to edge beams only, so for *interior* beams, the full effective width will be the smaller of one fourth the span length or the center-to-center spacing of the beams (assuming the beams are evenly spaced).

≡ **EXAMPLE 9.4**

A composite floor system consists of W21 × 44 steel beams spaced at 9 feet and supporting a 4.5-in.-thick reinforced concrete slab. The span length is 30 feet. In addition to the weight of the slab, there is a 20 psf partition load and a live load of 125 psf (light manufacturing). The steel is A36, and the concrete strength is $f'_c = 4000$ psi. Investigate a typical interior beam for compliance with the AISC Specification if no temporary shores are used. Assume full lateral support during construction and an additional construction load of 20 psf. Sufficient shear connectors are provided for full composite action.

SOLUTION

Loads applied before the concrete cures:

$$\text{slab:} \quad \frac{4.5}{12}(150) = 56.25 \text{ psf}$$

For a beam spacing of 9 ft,

56.25(9)	= 506 lb/ft
Beam weight:	44 lb/ft
Construction loads:	
20(9)	= 180 lb/ft
Total:	730 lb/ft

The factored load and moment are

$$w_u = 1.2(730) = 876 \text{ lb/ft}$$

$$M_u = \frac{1}{8}(0.876)(30)^2 = 98.6 \text{ ft-kips}$$

From the Load Factor Design Selection Table,

$$\phi_b M_n = \phi_b M_p = 258 \text{ ft-kips} > 98.6 \text{ ft-kips} \quad \text{(OK)}$$

After the concrete cures:

Treating the partition load as a live load,

$$w_L = (20 + 125)(9) = 1305 \text{ lb/ft}$$

Omitting the construction load of 20 psf, the dead load is

$$w_D = 506 + 44 = 550 \text{ lb/ft}$$

The factored load is

$$w_u = 1.2(550) + 1.6(1305) = 2748 \text{ lb/ft}$$

and

$$M_u = \frac{1}{8}(2.748)(30)^2 = 309 \text{ ft-kips}$$

FIGURE 9.9 ≡

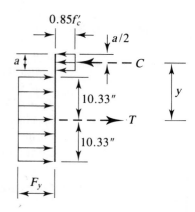

Effective flange width:

1. $\dfrac{\text{Span}}{4} = \dfrac{30(12)}{4} = 90$ in.

2. Beam spacing $= 9(12) = 108$ in.

Since this is an interior beam, the third criterion is not applicable. Use $b = 90$ inches. Referring to Figure 9.9, the compressive force is the smaller of

$$A_s F_y = 13(36) = 468 \text{ kips}$$

or

$$0.85 f'_c A_c = 0.85(4)(4.5)(90) = 1377 \text{ kips}$$

Use $C = 468$ kips. From Figure 9.9,

$$a = \dfrac{C}{0.85 f'_c b} = \dfrac{468}{0.85(4)(90)} = 1.529 \text{ in.}$$

$$y = \dfrac{d}{2} + t - \dfrac{a}{2} = 10.33 + 4.5 - \dfrac{1.529}{2} = 14.07 \text{ in.}$$

$$\phi_b M_n = \phi_b C y = 0.85(468)(14.07)$$
$$= 5595 \text{ in.-kips} = 466 \text{ ft-kips} > 309 \text{ ft-kips} \quad \text{(OK)}$$

Check shear:

$$V_u = \dfrac{w_u L}{2} = \dfrac{2.748(30)}{2} = 41.2 \text{ kips}$$

From the Factored Uniform Load tables,

$$\phi_v V_n = 141 \text{ kips} > 41.2 \text{ kips} \quad \text{(OK)}$$

Answer   The beam is in compliance with the AISC Specification.

9.4 ≡ SHEAR CONNECTORS

As we have seen, the horizontal shear force to be transferred between the concrete and the steel is equal to the compressive force in the concrete, C. We will denote this horizontal shear force as V_h. Thus, V_h is given by the smallest of $A_s F_y$, $0.85 f'_c A_c$, or $\sum Q_n$. If $A_s F_y$ or $0.85 f'_c A_c$ control, full composite action will exist and the number of shear connectors required between the points of zero moment and maximum moment is

$$N_1 = \frac{V_h}{Q_n} \tag{9.2}$$

where Q_n is the nominal shear strength of one connector. The N_1 connectors should be uniformly spaced within the length where they are required. The AISC Specification gives equations for the strength of both stud and channel shear connectors. As indicated at the beginning of this chapter, stud connectors are the most common. We will consider only these. For one stud shear connector,

$$Q_n = 0.5 \, A_{sc} \sqrt{f'_c E_c} \leq A_{sc} F_u \qquad \text{(AISC Equation I5-1)}$$

where

A_{sc} = cross-sectional area of stud (in.²)

f'_c = 28-day compressive strength of the concrete (ksi)

E_c = modulus of elasticity of the concrete (ksi)

F_u = minimum tensile strength of stud (ksi)

For studs used as shear connectors in composite beams, the tensile strength F_u is 60 ksi.[47] Values given by AISC Equation I5-1 are based on experimental studies.[56] There is no resistance factor applied to Q_n; the overall flexural resistance factor ϕ_b accounts for all strength uncertainties.

Equation 9.2 gives the number of shear connectors required between the point of zero moment and the point of maximum moment. Consequently, for a simply supported, uniformly loaded beam, $2N_1$ connectors will be required, and they should be equally spaced. When concentrated loads are present, AISC I5.6 requires that enough of the N_1 connectors be placed between the concentrated load and the adjacent point of zero moment to develop the moment required at the load. Denoting this portion as N_2, this requirement is illustrated in Figure 9.10. Note that the total number of shear connectors is not affected by this requirement.

Miscellaneous Requirements (AISC I5.6)

For headed studs (all spacings are center-to-center):

- Maximum diameter = 2.5 × flange thickness of steel shape

FIGURE 9.10 ≡

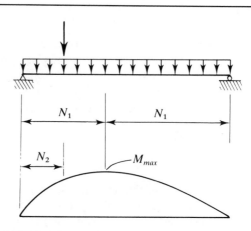

- Minimum longitudinal spacing = 6 × stud diameter
- Maximum longitudinal spacing = 8 × slab thickness
- Minimum transverse spacing = 4 × stud diameter

For all types of connectors:

- Minimum lateral concrete cover = 1 inch
  (There is no minimum *vertical* cover.)

≡ **EXAMPLE 9.5**

Design shear connectors for Example 9.4.

SOLUTION   Summary of data from Example 9.4:

W21 × 44, A36 steel

$f'_c = 4000$ psi

Slab thickness $t = 4.5$ in.

Span length = 30 ft

From Example 9.4, the horizontal shear force V_h corresponding to full composite action is

$$V_h = C = 468 \text{ kips}$$

Try $\frac{1}{2}$ in. × 2 in. studs. The maximum permissible diameter is

$$2.5t_f = 2.5(0.450) = 1.125 \text{ in.} > 0.5 \text{ in.} \quad \text{(OK)}$$

The cross-sectional area of one shear connector is

$$A_{sc} = \frac{\pi(0.5)^2}{4} = 0.1963 \text{ in.}^2$$

FIGURE 9.11 ☰

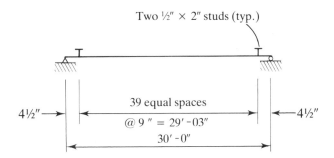

Assuming normal weight concrete, the modulus of elasticity of the concrete is

$$E_c = w_c^{1.5}(33)\sqrt{f_c'} = (145)^{1.5}(33)\sqrt{4000} = 3,644,000 \text{ psi} = 3644 \text{ ksi}$$

From AISC Equation I5-1, the shear strength of one connector is

$$Q_n = 0.5 \, A_{sc}\sqrt{f_c' \, E_c} \le A_{sc} \, F_u$$
$$= 0.5(0.1963)\sqrt{4(3644)}$$
$$= 11.85 \text{ kips}$$

$A_{sc} \, F_u = 0.1963(60) = 11.78 \text{ kips} < 11.85 \text{ kips} \quad \therefore \quad \text{use } Q_n = 11.78 \text{ kips}$

Minimum longitudinal spacing $= 6d = 6(0.5) = 3$ in.

Minimum transverse spacing $= 4d = 4(0.5) = 2$ in.

Maximum longitudinal spacing $= 8t = 8(4.5) = 36$ in.

The number of studs required between the end of the beam and midspan is

$$N_1 = \frac{V_h}{Q_n} = \frac{468}{11.78} = 39.7 \qquad \text{Use 40 for half the beam, 80 total.}$$

If one stud is used at each section, the approximate spacing will be

$$s = \frac{30(12)}{80} = 4.5 \text{ in.}$$

For two studs per section,

$$s = \frac{30(12)}{80/2} = 9.0 \text{ in.}$$

Either arrangement is satisfactory; either spacing will be between the lower and upper limits. The layout shown in Figure 9.11 will be recommended.

Answer   Use 80 studs, $\frac{1}{2}$ in. × 2 in., placed as shown in Figure 9.11.

9.5 ☰ DESIGN

The first step in the design of a floor system is to select the thickness of the floor slab, whether it is solid or ribbed (formed with steel deck). The thickness will be a function of the beam spacing, and several combinations of slab thickness and beam spacing may need to be investigated so that the most economical system can be found. The design of the slab is beyond the scope of this text, however, and we will assume that the slab thickness and beam spacing are known. Having made this stipulation, the following steps can be taken to complete the design of an unshored floor system:

1. Compute the factored load moments acting before and after the concrete cures.

2. Select a steel shape for trial.

3. Compute the design strength of the steel shape and compare it with the factored moment acting before the concrete cures. Account for the unbraced length if the formwork does not provide adequate lateral support. If this shape is not satisfactory, try a larger one.

4. Compute the design strength of the composite section and compare with the *total* factored load moment. If the composite section is inadequate, select another steel shape for trial.

5. Check the shear strength of the steel shape.

6. Design the shear connectors:

   a. Compute V_h, the horizontal shear force at the interface between the concrete and the steel.

   b. Divide this force by Q_n, the shear capacity of a single connector, to obtain the total number of shear connectors required. This number of connectors will provide full composite action. If partial composite behavior is desired, the number of connectors can be reduced (this will be covered in Section 9.7).

7. Check deflections (this will be covered in Section 9.6).

The major task in the trial and error procedure just outlined is the selection of a trial steel shape. A formula that will give the required area (or, alternately, the required weight per foot of length) can be developed if a beam depth is assumed. Assuming full composite action and the PNA in the slab (i.e., steel controlling, the most common case), we can write the design strength as (see Figure 9.12)

$$\phi_b M_n = \phi_b(Ty) = \phi_b(A_s F_y\, y)$$

Equating the design strength to the factored load moment and solving for A_s, we obtain

$$\phi_b A_s F_y\, y = M_u \qquad A_s = \frac{M_u}{\phi_b F_y\, y}$$

or

$$A_s = \frac{M_u}{\phi_b F_y(d/2 + t - a/2)} \tag{9.3}$$

FIGURE 9.12 ≡

Equation 9.3 can also be written in terms of weight rather than area. Since one foot of length has a volume of $A_s/144$ cubic feet and structural steel weighs 490 pounds per cubic foot,

$$w = \frac{A_s}{144}(490) = 3.4A_s \text{ lb/ft (for } A_s \text{ in square inches)}$$

From Equation 9.3, the estimated weight per foot is therefore

$$w = \frac{3.4\,M_u}{\phi_b\,F_y(d/2 + t - a/2)} \text{ lb/ft} \tag{9.4}$$

where M_u is in in.-kips; F_y is in ksi; and d, t, and a are in inches. Either Equation 9.3 or Equation 9.4 can be used to select a trial shape. Both require an assumed depth and an estimate of $a/2$. The stress block depth will generally be very small; consequently, an error in the estimate of $a/2$ will have only a slight effect on the estimated value of A_s. An assumed value of $a/2 = 1.0$ is suggested.

If Equation 9.4 is used, and a nominal depth is assumed for d, the selection of a trial shape can be quickly made. The use of this equation also provides an estimate of the beam weight directly.

≡ **EXAMPLE 9.6**

The span length of a certain floor system is 30 ft, and the beam spacing is 10 ft center-to-center. Select a rolled steel shape and the shear connectors needed to achieve full composite behavior with a 3.5-in.-thick reinforced concrete floor slab. Superimposed loading consists of a 10 psf partition load and a 55 psf live load. Concrete strength is $f'_c = 4000$ psi, and A36 steel is to be used. Assume that the beam has full lateral support during construction.

SOLUTION   Loads applied before the concrete cures:

$$\text{Slab}: \frac{3.5}{12}(150) = 43.75 \text{ psf}$$

Weight per linear foot = 43.75(10) = 437.5 lb/ft

(The beam weight will be accounted for later.)

Loads applied after the concrete cures (treat the partition load as a live load):

$$w_L = (10 + 55)(10) = 650 \text{ lb/ft}$$

$$w_u = 1.2(0.4375) + 1.6(0.650) = 1.565 \text{ kip/ft}$$

$$M_u = \frac{1}{8}(1.565)(30)^2 = 176.1 \text{ ft-kips}$$

Try a nominal depth of $d = 16$ in. From Equation 9.4, the estimated beam weight is

$$w = \frac{3.4M_u}{\phi_b F_y(d/2 + t - a/2)} = \frac{3.4(176.1 \times 12)}{0.85(36)(16/2 + 3.5 - 1)} = 22.36 \text{ lb/ft}$$

Try a W16 × 26.

Check the unshored steel shape for loads applied before the concrete cures (the weight of the slab and the steel beam).

$$w_u = 1.2(0.4375 + 0.026) = 0.5562 \text{ kips/ft}$$

$$M_u = \frac{1}{8}(0.5562)(30)^2 = 62.6 \text{ ft-kips}$$

From the Load Factor Design Selection Table,

$$\phi_b M_n = \phi_b M_p = 119 \text{ ft-kips} > 62.6 \text{ ft-kips}\quad \text{(OK)}$$

After the concrete cures and composite behavior has been achieved,

$$w_u = 1.2w_D + 1.6w_L = 0.5562 + 1.6(0.650) = 1.596 \text{ kips/ft}$$

$$M_u = \frac{1}{8}(1.596)(30)^2 = 179.6 \text{ kips/ft}$$

Before computing the design strength of the composite section, we must first determine the effective slab width. For an interior beam, the effective width is the smaller of

$$\frac{\text{span}}{4} = \frac{30(12)}{4} = 90 \text{ in.}$$

or

$$\text{beam spacing} = 10(12) = 120 \text{ in.}$$

Use $b = 90$ in. For full composite behavior, the compressive force in the concrete at ultimate (equal to the horizontal shear at the interface between the concrete and steel) will be the smaller of

$$A_s F_y = 7.68(36) = 276.5 \text{ kips}$$

or

$$0.85f'_c A_c = 0.85(4)(90)(3.5) = 1071 \text{ kips}$$

Use $C = V_h = 276.5$ kips. The depth of the compressive stress block in the slab is

$$a = \frac{C}{0.85f'_c b} = \frac{276.5}{0.85(4)(90)} = 0.9036 \text{ in.}$$

and the moment arm of the internal resisting couple is

$$y = \frac{d}{2} + t - \frac{a}{2} = \frac{15.69}{2} + 3.5 - \frac{0.9036}{2} = 10.89 \text{ in.}$$

The design flexural strength is

$$\phi_b M_n = \phi_b(Cy) = 0.85(276.5)(10.89)$$
$$= 2550 \text{ in.-kips} = 213 \text{ ft-kips} > 179.6 \text{ ft-kips} \quad \text{(OK)}$$

Check shear:

$$V_u = \frac{w_u L}{2} = \frac{1.596(30)}{2} = 23.9 \text{ kips}$$

From the Factored Uniform Load tables,

$$\phi_v V_n = 76.3 \text{ kips} > 23.9 \text{ kips} \quad \text{(OK)}$$

Answer   Use a W16 × 26.

The modulus of elasticity of the concrete will be needed for the shear connector design. From Example 9.5, $E_c = 3644$ ksi for normal weight concrete with $f'_c = 4000$ psi.

Try $\frac{1}{2}$-in. × 2-in. studs ($A_{sc} = 0.1963$ in.^{2}):

$$\text{Maximum diameter} = 2.5t_f = 2.5(0.345) = 0.8625 \text{ in.} > 0.5 \text{ in.} \quad \text{(OK)}$$

From AISC Equation I5-1, the shear strength of one connector is

$$Q_n = 0.5A_{sc}\sqrt{f'_c E_c} \leq A_{sc} F_u$$
$$= 0.5(0.1963)\sqrt{4(3644)}$$
$$= 11.85 \text{ kips}$$
$$A_{sc} F_u = 0.1963(60) = 11.78 \text{ kips} < 11.85 \text{ kips} \quad \therefore \quad \text{use } Q_n = 11.78 \text{ kips}$$

FIGURE 9.13 ≡

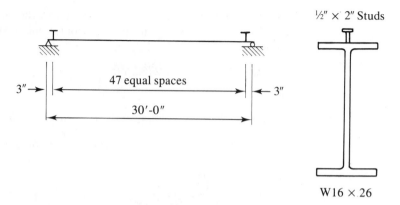

½″ × 2″ Studs

W 16 × 26

The number of studs required between the end of the beam and midspan is

$$N_1 = \frac{V_h}{Q_n} = \frac{276.5}{11.78} = 23.5 \qquad \text{Use 24 for half the beam, 48 total.}$$

Minimum longitudinal spacing $= 6d = 6(0.5) = 3$ in.
Minimum transverse spacing $= 4d = 4(0.5) = 2$ in.
Maximum longitudinal spacing $= 8t = 8(3.5) = 28$ in.

If one stud is used at each section, the approximate spacing will be

$$s = \frac{30(12)}{48} = 7.5 \text{ in.}$$

This is between the upper and lower limits and is therefore satisfactory.

Answer   Use the design shown in Figure 9.13.

9.6 ≡ DEFLECTIONS

Because of the large moment of inertia of the transformed section, deflections in composite beams are smaller than in noncomposite beams. This larger moment of inertia is available only after the concrete slab has cured, however. Deflections caused by loads applied before the concrete cures must be computed with the moment of inertia of the steel shape. An additional complication arises if the beam is subject to sustained loading, such as the weight of partitions, after the concrete cures. In positive moment regions, the concrete will be in compression continuously and is subject to a phenomenon known as *creep*. Creep is a deformation that takes place under sustained compressive load. After the initial deformation, additional deformation will take place at a very slow rate, but over a long period of time. The effect on a composite beam is to increase the curvature and hence the vertical deflection. The long-term deflection can only be estimated; the usual technique is to use a reduced area of concrete in the transformed section so as to obtain a smaller moment of inertia and a larger computed deflection. The reduced area is computed by using $2n$ or $3n$ instead of the actual modular ratio n. A value of $2n$ will be used in this book. The increased deflection resulting from creep is not covered by the AISC Specification.

For unshored construction, as many as three different moments of inertia can be required for computing the total long-term deflection:

1. Use I_s, the moment of inertia of the rolled steel shape, for deflection caused by loads applied before the concrete cures.
2. Use I_{tr}, the moment of inertia of the transformed section, computed with b/n for deflection caused by *live* loads and for the initial deflection caused by dead loads applied after the concrete cures.
3. Use I_{tr} computed with $b/2n$ for long-term deflections caused by dead loads applied after the concrete cures.

EXAMPLE 9.7

Compute the immediate and long-term deflections for the beam of Example 9.4.

SOLUTION Summary of data from Example 9.4:

W21 × 44, A36 steel.

Slab thickness $t = 4.5$ in. Effective width $b = 90$ in.

$f'_c = 4000$ psi

Dead load applied before concrete cures $= w_D = 550$ lb/ft (slab plus beam)

Construction load $= w_{const} = 180$ lb/ft

Live load $= w_L = 125(9) = 1125$ lb/ft

Partition load $= w_{part} = 20(9) = 180$ lb/ft

Immediate deflection (beam plus slab plus construction load):

$$w = w_D + w_{const} = 550 + 180 = 730 \text{ lb/ft}$$

$$\Delta_1 = \frac{5wL^4}{384EI_s} = \frac{5(0.730/12)(30 \times 12)^4}{384(29,000)(843)} = 0.5442 \text{ in.}$$

For the remaining deflections, the moments of inertia of two transformed sections will be needed: I_{tr} with a transformed slab width of b/n and I_{tr} with a transformed slab width of $b/2n$. For a normal weight concrete and $f'_c = 4000$ psi, $E_c = 3644$ ksi, and the modular ratio is

$$n = \frac{E_s}{E_c} = \frac{29,000}{3644} = 7.95 \qquad \text{Use } n = 8.$$

FIGURE 9.14 ≡

W21 × 44

For deflections of the composite section that do not involve creep, the effective width is

$$\frac{b}{n} = \frac{90}{8} = 11.25 \text{ in.}$$

Figure 9.14 shows the corresponding transformed section, and the computations for the neutral axis location and the moment of inertia are summarized in Table 9.4.

Deflection caused by partition weight (not long-term):

$$\Delta_2 = \frac{5w_{part}L^4}{384EI_{tr}} = \frac{5(0.180/12)(30 \times 12)^4}{384(29,000)(2566)} = 0.0441 \text{ in.}$$

Deflection caused by live load:

$$\Delta_3 = \frac{5w_L L^4}{384EI_{tr}} = \frac{5(1.125/12)(30 \times 12)^4}{384(29,000)(2566)} = 0.2755 \text{ in.}$$

Long-term deflection caused by creep:
Use a transformed slab width of

$$\frac{b}{2n} = \frac{90}{2(8)} = 5.625 \text{ in.}$$

TABLE 9.4 ≡

Component	A	y	Ay	$\bar{I}$	d	$\bar{I} + Ad^2$
Concrete	50.62	2.25	113.9	85.43	2.571	420
W21 × 44	13.00	14.83	192.8	843	10.01	2146
	63.62		306.7			2566 in.⁴

$$\bar{y} = \frac{\sum Ay}{\sum A} = \frac{306.62}{63.62} = 4.821 \text{ in.}$$

FIGURE 9.15 ≣

The transformed section is shown in Figure 9.15. The calculations for the centroid and moment of inertia are summarized in Table 9.5. Calling this transformed moment of inertia I'_{tr}, we can compute the long-term deflection caused by creep as

$$\Delta_4 = \frac{5w_{part} L^4}{384EI'_{tr}} = \frac{5(0.180/12)(30 \times 12)^4}{384(29,000)(2245)} = 0.0504 \text{ in.}$$

Answer The following is a summary of the deflections:

Immediate deflection, no superimposed loads:

$$\Delta_1 = 0.544 \text{ in.}$$

Short-term deflection with partitions (but no live load):

$$\Delta_1 + \Delta_2 = 0.5442 + 0.0441 = 0.588 \text{ in.}$$

Short-term deflection, live load added:

$$\Delta_1 + \Delta_2 + \Delta_3 = 0.5442 + 0.0441 + 0.2755 = 0.864 \text{ in.}$$

Long-term deflection, without live load:

$$\Delta_1 + \Delta_4 = 0.5442 + 0.0504 = 0.595 \text{ in.}$$

TABLE 9.5 ≣

Component	A	y	Ay	$\bar{I}$	d	$\bar{I} + Ad^2$
Concrete	25.31	2.25	56.95	42.71	4.269	504
W21 × 44	13.00	14.83	192.8	843	8.311	1741
	38.31		249.8			2245 in.⁴

$$\bar{y} = \frac{\sum Ay}{\sum A} = \frac{249.8}{38.31} = 6.519 \text{ in.}$$

Long-term deflection, with live load:

$$\Delta_1 + \Delta_3 + \Delta_4 = 0.5442 + 0.2755 + 0.0504 = 0.870 \text{ in.}$$

Because of the small dead load applied after the concrete cures, the deflection resulting from creep is insignificant in this example.

9.7 ≡ COMPOSITE BEAMS WITH FORMED STEEL DECK

The floor slab in many steel-framed buildings is formed on ribbed steel deck, which is left in place to become an integral part of the structure. Although there are exceptions, the ribs of the deck are usually oriented perpendicular to floor beams and parallel to supporting girders. In Figure 9.16 the ribs are shown perpendicular to the beam. The installation of shear studs is done in the same way as without the deck; the studs are welded to the beam flange directly through the deck. The attachment of the deck to the beam can be considered to provide lateral support for the beam before the curing of the concrete. The design or analysis of composite beams with formed steel deck is essentially the same as with slabs of uniform thickness, with the following exceptions:

1. The concrete in the ribs — that is, below the top of the deck — is neglected when the ribs are perpendicular to the beam (AISC I3.5b). When the ribs are parallel to the beam, it may be included in determining section properties, and it *must* be included in computing A_c (AISC I3.5c).

2. The capacity of the shear connectors will possibly be reduced.

3. Full composite behavior will not usually be possible. This is because the spacing of the shear connectors is limited by the spacing of the ribs, and the exact number of required connectors cannot always be used.

FIGURE 9.16 ≡

Although partial composite design can be used without formed steel deck, it is covered here since it is almost a necessity with formed steel deck. This is not a disadvantage; in fact, it will be the most economical alternative.

Most composite beams with formed steel deck are floor beams with the deck ribs oriented perpendicular to the beam, and we will limit our coverage to this case. Special requirements that apply when the ribs are parallel to the beam are given in AISC I3.5c.

Reduced Capacity of Shear Connectors

Based on tests, AISC I3.5b requires that Q_n, the shear strength of the shear connectors, be multiplied by the following reduction factor when the ribs are perpendicular to the beam:

$$\frac{0.85}{\sqrt{N_r}}\left(\frac{w_r}{h_r}\right)\left[\left(\frac{H_s}{h_r}\right) - 1.0\right] \le 1.0 \qquad \text{(AISC Equation I3-1)}$$

where

N_r = number of studs per rib at a beam intersection (limited to three in the computations)

w_r = average width of rib in inches

h_r = height of rib in inches

H_s = length of stud in inches, not to exceed $(h_r + 3)$ in the computations.

The above dimensions are illustrated in Figure 9.17.

Partial Composite Action

Partial composite action exists when there are not enough shear connectors to completely prevent slip between the concrete and steel. Neither the full strength of the concrete nor that of the steel can be developed, and the compressive force is limited to the maximum force that can be transferred across the interface between the steel and the concrete — that is, the strength of the shear connectors, $\sum Q_n$. Recall that C is the smallest of $A_s F_y$, $0.85 f'_c A_c$, and $\sum Q_n$.

With partial composite action, the plastic neutral axis (PNA) will usually fall within the steel cross section. This will make the strength analysis somewhat more difficult than what has been considered so far, but the basic principles are the same.

When an elastic analysis must be made, as when computing deflections, an estimate of the moment of inertia of the partially composite section must be made. A parabolic transition from I_s (for the steel shape alone) to I_{tr} (for the

FIGURE 9.17 ≡

fully composite section) works well.[18] This approximation results in the following equation for an effective moment of inertia, given in the Commentary to the AISC Specification:

$$I_{eff} = I_s + \sqrt{\sum Q_n / C_f} \, (I_{tr} - I_s) \qquad \text{(AISC Equation C-I3-6)}$$

where C_f is the compressive force in the concrete for the fully composite condition — the smaller of $A_s F_y$ and $0.85 f'_c A_c$. Since $\sum Q_n$ is the actual compressive force for the partially composite case, the ratio $\sum Q_n / C_f$ is the fraction of "compositeness" that exists. If the ratio is less than 0.25, AISC Equation C-I3-6 should not be used.[18]

Since the steel strength will not be fully developed in a partially composite beam, a larger shape will be required than with full composite behavior. On the other hand, fewer shear connectors will be required, and the costs of both the steel and the shear connectors (including the cost of installation) must be taken into account in any economic analysis. Whenever a fully composite beam has excess capacity, and this is almost always the case, the design can be "fine-tuned" by eliminating some of the shear connectors, thereby creating a partially composite beam.

Miscellaneous Requirements

The following requirements are from AISC Sections I3.5a and b. Only those provisions not already discussed are listed:

- Maximum rib height $h_r = 3$ inches.
- Minimum average width of rib $w_r = 2$ inches, but the value of w_r used in the calculations shall not exceed the clear width at the top of the deck.

- Minimum slab thickness above the top of the deck = 2 inches.
- Maximum stud diameter = $\frac{3}{4}$ inches. This requirement for formed steel deck is in addition to the usual maximum diameter of $2.5t_f$.
- Minimum height of stud above the top of the deck = 1.5 inches.
- Maximum longitudinal spacing of shear studs = 36 inches.
- The deck must be attached to the beam flange at intervals of no more than 18 inches, either by the studs or spot welds. This is for the purpose of resisting uplift.

EXAMPLE 9.8

Floor beams are to be used with the formed steel deck shown in Figure 9.18 and a reinforced concrete slab whose total thickness is 4.75 inches. The deck ribs are perpendicular to the beams. The span length is 30 feet, and the beams are spaced at 10 feet center-to-center. The structural steel is A36, and the concrete strength is $f'_c = 3000$ psi. The total load applied before the concrete cures is 0.356 kips/ft, excluding the beam weight. The dead load applied after the concrete cures is 0.130 kips/ft. The live load is 0.520 kips/ft. No shoring is used.

   a. Select a W-shape.

   b. Design the shear connectors.

   c. Check deflections. The maximum permissible total long-term deflection is $\frac{1}{240}$ of the span length.

SOLUTION

a. Beam design:

Select a trial shape based on full composite behavior.

$$w_u = 1.2w_D + 1.6w_L = 1.2(0.356 + 0.130) + 1.6(0.520)$$
$$= 1.415 \text{ kips/ft}$$

$$M_u = \frac{1}{8}(1.415)(30)^2 = 159.2 \text{ ft-kips}$$

Assume $d = 16$ in., $a/2 = 1.0$ in., and obtain an estimated beam weight from Equation 9.4:

$$w = \frac{3.4M_u}{\phi_b F_y(d/2 + t - a/2)} = \frac{3.4(159.2 \times 12)}{0.85(36)(16/2 + 4.75 - 1.0)} = 18.07 \text{ lb/ft}$$

Try a W16 × 26.

Check the flexural strength before the concrete cures.

$$w_u = 1.2(0.356 + 0.026) = 0.4584 \text{ kips/ft}$$

$$M_u = \frac{1}{8}(0.4584)(30)^2 = 51.6 \text{ ft-kips}$$

FIGURE 9.18 ≡

A W16 × 26 is compact for A36 steel, and since the steel deck will provide adequate lateral support, the nominal strength M_n is equal to the plastic moment strength M_p. From the Load Factor Design Selection Table,

$$\phi_b M_p = 119 \text{ ft-kips} > 51.6 \text{ ft-kips} \quad \text{(OK)}$$

After the concrete cures, the total factored load to be resisted by the composite beam, adjusted for the weight of the steel shape, is

$$w_u = 1.2(0.356 + 0.026 + 0.130) + 1.6(0.520)$$
$$= 1.446 \text{ kips/ft}$$

and the factored moment is

$$M_u = \frac{1}{8}(1.446)(30)^2 = 162.7 \text{ ft-kips}$$

The effective slab width of the composite section, assuming full composite action, will be the smaller of

$$\frac{\text{span}}{4} = \frac{30(12)}{4} = 90 \text{ in.}$$

or

$$\text{beam spacing} = 10(12) = 120 \text{ in.}$$

Use $b = 90$ in. For full composite action, the compressive force C in the concrete is the smaller of

$$A_s F_y = 7.68(36) = 276.5 \text{ kips}$$

or

$$0.85f'_c A_c = 0.85(3)[90(4.75 - 1.5)] = 745.9 \text{ kips}$$

where only the concrete above the top of the deck has been accounted for in the second equation (see Figure 9.19). With $C = 276.5$ kips, the depth of the compressive stress distribution in the concrete is

$$a = \frac{C}{0.85f'_c b} = \frac{276.5}{0.85(3)(90)} = 1.205 \text{ in.}$$

The moment arm of the internal resisting couple is

$$y = \frac{d}{2} + t - \frac{a}{2} = \frac{15.69}{2} + 4.75 - \frac{1.209}{2} = 11.99 \text{ in.}$$

FIGURE 9.19 ≡

and the design strength is

$$\phi_b M_n = 0.85(276.5)(11.99)/12$$
$$= 235 \text{ ft-kips} > 162.7 \text{ ft-kips} \quad \text{(OK)}$$

Check shear:

$$V_u = \frac{w_u L}{2} = \frac{1.446(30)}{2} = 21.7 \text{ kips}$$

From the Factored Uniform Load tables,

$$\phi_v V_n = 76.3 \text{ kips} > 21.7 \text{ kips} \quad \text{(OK)}$$

Answer   **a.** Use a W16 × 26.

b. Shear connectors:

Because this beam has a substantial excess of moment strength, it will benefit from partial composite behavior. We must first find the shear connector requirements for full composite behavior and then reduce the number of connectors. For the fully composite beam, $C = V_h = 276.5$ kips.
   Try $\frac{3}{4}$-in. × 3-in. studs ($A_{sc} = 0.4418 \text{ in.}^2$), one at each section:

$$\text{Maximum diameter} = 2.5t_f = 2.5(0.345) = 0.8625 \text{ in.}$$
$$\text{or } \tfrac{3}{4} \text{ in.} \quad \text{(controls)}$$
$$\text{Actual diameter} = \tfrac{3}{4} \text{ in.} \quad \text{(OK)}$$

Compute the stud strength-reduction factor:

$$N_r = 1$$

Height of stud above top of deck $= 3 - 1.5 = 1.5$ in. $=$ permissible value
(OK)

From AISC Equation I3-1,

$$\text{Reduction factor} = \frac{0.85}{\sqrt{N_r}} \left(\frac{w_r}{h_r}\right)\left[\left(\frac{H_s}{h_r}\right) - 1.0\right] \leq 1.0$$
$$= \frac{0.85}{1.0}\left(\frac{2.25}{1.5}\right)\left(\frac{3}{1.5} - 1.0\right) = 1.275 > 1.0$$

No reduction of stud strength is required. For $f'_c = 3000$ psi, the modulus of elasticity of the concrete is

$$E_c = w_c^{1.5}(33)\sqrt{f'_c} = 145^{1.5}(33)\sqrt{3000} = 3,156,000 \text{ psi} = 3156 \text{ ksi}$$

From AISC Equation I5-1, the shear strength of one connector is

$$\begin{aligned}
Q_n &= 0.5A_{sc}\sqrt{f'_c E_c} \leq A_{sc} F_u \\
&= 0.5(0.4418)\sqrt{3(3156)} \\
&= 21.49 \text{ kips}
\end{aligned}$$

$$A_{sc} F_u = 0.4418(60) = 26.51 \text{ kips} > 21.49 \text{ kips} \quad \therefore \quad \text{use } Q_n = 21.49 \text{ kips}$$

The number of studs required between the end of the beam and midspan is

$$N_1 = \frac{V_h}{Q_n} = \frac{276.5}{21.49} = 12.9$$

Use 13 for half the beam, 26 total.

With one stud in each rib, the spacing is 6 inches, and the maximum number that can be accommodated is

$$\frac{30(12)}{6} = 60 > 26 \text{ required}$$

With one stud in every other rib, 30 will be furnished; this is still too many. With one stud in every third rib, the spacing will be $3(6) = 18$ inches, and the number of studs will be

$$\frac{30(12)}{18} = 20$$

This is fewer than are needed for full composite action, but there is an excess of flexural strength, so partial composite action may be adequate.

Try 20 studs per beam, N_1 provided $= \frac{20}{2} = 10$.

$$\begin{aligned}
\sum Q_n &= 10(21.49) \\
&= 214.9 \text{ kips} < 276.5 \text{ kips} \quad \therefore \quad C = V_h = 214.9 \text{ kips}
\end{aligned}$$

Since C is smaller than $A_s F_y$, part of the steel section must be in compression, and the plastic neutral axis is in the steel section.

To analyze this case, we must first determine whether the PNA is in the top flange or in the web. This can be done as follows. If the PNA were at the bottom of the flange, the entire flange would be in compression, and the resultant compressive force would be (see Figure 9.20)

$$P_{yf} = b_f t_f F_y = 5.500(0.345)(36) = 68.31 \text{ kips}$$

The net force to be transferred at the interface between the steel and the concrete would be

$$T - C_s = T - P_{yf} = (A_s F_y - P_{yf}) - P_{yf} = 276.5 - 2(68.31)$$
$$= 139.9 \text{ kips}$$

This is less than the actual net tension force of 214.9 kips, so the top flange does not need to be in compression for its full thickness. This means that the PNA is in the flange. Referring to Figure 9.21, the horizontal shear force to be transferred is

$$T - C_s = (A_s F_y - b_f t' F_y) - b_f t' F_y = V_h$$
$$276.5 - 2[5.500t'(36)] = 214.9$$

Solving for the depth of compression in the flange,

$$t' = 0.1556 \text{ in.}$$

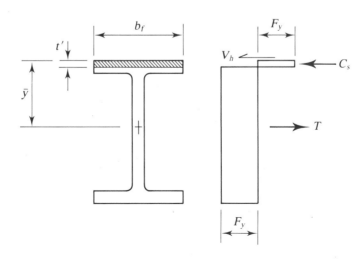

The tensile resultant force will act at the centroid of the area below the PNA. Before the moment strength can be computed, the location of this centroid must be determined. The calculations for $\bar{y}$, the distance from the top of the steel shape, are summarized in Table 9.6.

The depth of the compressive stress block in the concrete is

$$a = \frac{C}{0.85f'_c b} = \frac{214.9}{0.85(3)(90)} = 0.9364 \text{ in.}$$

Moment arm for concrete compressive force

$$= \bar{y} + t - \frac{a}{2} = 8.819 + 4.75 - \frac{0.9364}{2} = 13.10 \text{ in.}$$

Moment arm for compressive force in the steel

$$= \bar{y} - \frac{t'}{2} = 8.819 - \frac{0.1556}{2} = 8.741 \text{ in.}$$

Taking moments about the tensile force and using the notation of Figure 9.20, we obtain the nominal strength as

$$
\begin{aligned}
M_n &= C(13.10) + C_s(8.741) \\
&= 214.9(13.10) + 0.1556(5.500)(36)(8.741) \\
&= 3084 \text{ in.-kips} = 257.0 \text{ ft-kips}
\end{aligned}
$$

The design strength is

$$\phi_b M_n = 0.85(257.0) = 218.5 \text{ ft-kips} > 162.7 \text{ ft-kips} \quad \text{(OK)}$$

Since the deck will be attached to the beam flange at intervals of 18 inches, no spot welds will be needed to resist uplift.

TABLE 9.6 ≡

Component	A	y	Ay
W16 × 26	7.68	15.69/2 = 7.845	60.25
Flange segment	−0.1556(5.500) = −0.856	0.1556/2 = 0.0778	−0.07
	6.824		60.18

$$\bar{y} = \frac{\sum Ay}{\sum A} = \frac{60.18}{6.824} = 8.819 \text{ in.}$$

FIGURE 9.22 ≡

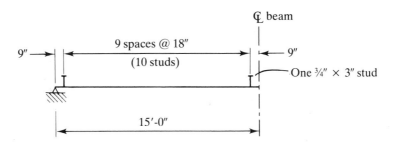

Answer   **b.** Use the shear connectors shown in Figure 9.22.

c. Deflections:

Before the concrete cures:

$$w_D = 0.356 + 0.026 = 0.382 \text{ kips/ft}$$

$$\Delta_1 = \frac{5w_D L^4}{384EI_s} = \frac{5(0.382/12)(30 \times 12)^4}{384(29,000)(301)} = 0.7976 \text{ in.}$$

For deflections that occur after the concrete cures, the moments of inertia of two transformed sections will be needed: I_{tr} with a transformed slab width of b/n and I_{tr} with a transformed slab width of $b/2n$. The modular ratio is

$$n = \frac{E_s}{E_c} = \frac{29,000}{3156} = 9.19 \qquad \text{Use } n = 9.$$

For deflections of the composite section that do not involve creep, the effective width is

$$\frac{b}{n} = \frac{90}{9} = 10 \text{ in.}$$

Figure 9.23 shows the corresponding transformed section, and the computations for the neutral axis location and the moment of inertia are summarized in Table 9.7.

FIGURE 9.23 ≡

TABLE 9.7 ≣

Component	A	y	Ay	$\bar{I}$	d	$\bar{I} + Ad^2$
Concrete	32.50	1.625	52.81	28.61	2.098	172
W16 × 26	7.68	12.60	96.77	301	8.877	906
	40.18		149.58			1078 in.⁴

$$\bar{y} = \frac{\sum Ay}{\sum A} = \frac{149.6}{40.18} = 3.723 \text{ in.}$$

Since partial composite action is being used, a reduced transformed moment of inertia must be used. This effective moment of inertia is

$$I_{eff} = I_s + \sqrt{\sum Q_n/C_f}(I_{tr} - I_s)$$
$$= 301 + \sqrt{214.9/276.5}(1078 - 301) = 986.0 \text{ in.}^4$$

(AISC Equation C-I3-6)

The deflection caused by live load is

$$\Delta_2 = \frac{5w_L L^4}{384EI_{eff}} = \frac{5(0.520/12)(30 \times 12)^4}{384(29,000)(986.0)} = 0.3314 \text{ in.}$$

Deflection caused by dead load applied after the concrete cures should be based on a transformed moment of inertia obtained with 2n rather than n. Use a transformed slab width of

$$\frac{b}{2n} = \frac{90}{2(9)} = 5 \text{ in.}$$

From Figure 9.24 and Table 9.8, the transformed moment of inertia is

$$I'_{tr} = 943.5 \text{ in.}^4$$

The effective moment of inertia, which we will call I'_{eff}, is

$$I'_{eff} = I_s + \sqrt{\sum Q_n/C_f}(I'_{tr} - I_s)$$
$$= 301 + \sqrt{214.9/276.5}(943.5 - 301) = 867.4 \text{ in.}^4$$

FIGURE 9.24 ≣

TABLE 9.8 ≡

Component	A	y	Ay	$\bar{I}$	d	$\bar{I} + Ad^2$
Concrete	16.25	1.625	26.41	14.30	3.522	215.9
W16 × 26	7.68	12.60	96.77	301	7.453	727.6
	23.93		123.2			943.5 in.⁴

$$\bar{y} = \frac{\sum Ay}{\sum A} = \frac{123.2}{23.93} = 5.147 \text{ in.}$$

The deflection caused by the dead load applied after the concrete cures is

$$\Delta_3 = \frac{5(0.130/12)(30 \times 12)^4}{384(29,000)(867.4)} = 0.0942 \text{ in.}$$

Total deflection $= \Delta_1 + \Delta_2 + \Delta_3 = 0.7976 + 0.3314 + 0.0942 = 1.22$ in.

$$\frac{L}{240} = \frac{30(12)}{240} = 1.50 \text{ in.} > 1.22 \text{ in.}   \text{(OK)}$$

Answer   c. Deflection is satisfactory.

9.8 ≡ TABLES FOR COMPOSITE BEAM ANALYSIS AND DESIGN

When the plastic neutral axis is within the steel section, computation of the flexural strength can be laborious. Formulas to expedite this computation have been developed,[18] but the tables presented in Part 5 of the *Manual* are more convenient. Two sets of tables are presented: design strengths of various combinations of shapes and slabs, for $F_y = 36$ ksi and $F_y = 50$ ksi; and tables of "lower bound" moments of inertia for these same combinations.

The design strength table, called "Composite Beam Selection Table," is limited to shapes with compact webs and a total shear connector strength $\sum Q_n \geq 0.25 A_s F_y$ (the recommended lower limit for partially composite beams).

The design strength $\phi_b M_n$ is given for seven specific locations of the plastic neutral axis (see Figure 9.25): top of the top flange, bottom of the top flange, three equally spaced levels within the top flange, and two locations in the web.*

---

* The table uses the notation ϕM_n for the design strength of composite shapes and $\phi_b M_p$ for the design strength of the steel shape alone. The resistance factor is denoted in two different ways because two different numerical values are used.

FIGURE 9.25 ≡

To use the tables for analysis of a composite beam, first find the portion of the table corresponding to the steel shape and proceed as follows:

1. Select $\sum Q_n$. This is the *Manual*'s notation for the compressive force C, which is the smallest of $A_s F_y$, $0.85f'_c A_c$, and the total shear connector strength (which we have been calling $\sum Q_n$).

2. Select $Y2$, the distance from the top of the steel shape to the resultant compressive force in the concrete, computed as

$$Y2 = t - \frac{a}{2}$$

This is illustrated in Figure 9.26.

3. Read ϕM_n, interpolating if necessary.

For design, the tables can be entered with the required ϕM_n, and a combination of steel shape and $\sum Q_n$ can be selected. A value of $Y2$ will be needed,

FIGURE 9.26 ≡

so the depth of the concrete compressive stress distribution will need to be assumed and then revised after an iteration. The discussion in the *Manual* preceding the tables gives the equivalent of Equation 9.4 for estimating the beam weight, but if the tables are used, it is not really needed.

The tables also give values of $\phi_b M_p$, which may be needed when checking unshored beams during the curing of the concrete; and $Y1$, the distance from the top of the steel to the plastic neutral axis.

The tables for lower-bound moment of inertia, denoted by I_{LB}, give a conservative estimate of the moment of inertia of the transformed section for the same beams that are in the design strength tables. If computed deflections based on a lower bound moment of inertia are acceptable, an elastic analysis can be avoided.

EXAMPLE 9.9

Compute the design strength of the composite beam of Examples 9.1 and 9.2. Use the tables in Part 5 of the *Manual*.

SOLUTION

From Example 9.1, the composite beam consists of a W16 × 36 of A36 steel, a slab with a thickness $t = 5$ in., and an effective width $b = 87$ in. The 28-day compressive strength of the concrete $f'_c = 4000$ psi.

The compressive force in the concrete is the smaller of

$$A_s F_y = 10.6(36) = 381.6 \text{ kips}$$

or

$$0.85 f'_c A_c = 0.85(4)(5 \times 87) = 1487 \text{ kips}$$

Use $C = 381.6$ kips. The depth of the compressive stress block is

$$a = \frac{C}{0.85 f'_c b} = \frac{381.6}{0.85(4)(87)} = 1.290 \text{ in.}$$

The distance from the top of the steel to the compressive force C is

$$Y2 = t - \frac{a}{2} = 5 - \frac{1.290}{2} = 4.36 \text{ in.}$$

Enter the tables with $\sum Q_n = 382$ kips and $Y2 = 4.36$. By interpolation,

$$\phi M_n = 332 \text{ ft-kips}$$

which checks with the results of Example 9.2 but involves about the same amount of effort. The value of the tables becomes obvious when the plastic neutral axis is within the steel shape.

Answer   Design strength = 332 ft-kips.

EXAMPLE 9.10

Rework Example 9.8 with the aid of the tables in Part 5 of the *Manual*.

SOLUTION   **a.** Beam design:

From Example 9.8, $M_u = 159.2$ ft-kips (excluding the beam weight).
Assuming $a = 2$ in.,

$$Y2 = t - \frac{a}{2} = 4.75 - \frac{2}{2} = 3.75 \text{ in.}$$

From the Composite Beam Selection Table, any combination of steel shape, $\sum Q_n$, and $Y2$ that furnishes a design strength of more than 159.2 ft-kips will be an acceptable trial beam. Two possibilities are summarized in Table 9.9.

The W14 × 22 is the lighter shape, but because $\sum Q_n$ is larger, it will require more shear connectors — almost twice as many. For this reason, try the W16 × 26.

Refine the value for $Y2$:

$$a = \frac{C}{0.85 f'_c b} = \frac{\sum Q_n}{0.85 f'_c b} = \frac{104}{0.85(3)(90)} = 0.4532 \text{ in.}$$

$$Y2 = t - \frac{a}{2} = 4.75 - \frac{0.4532}{2} = 4.52 \text{ in.}$$

(where $b = 90$ in. is from Example 9.8).

This will result in a design strength of $\phi M_n = 182.2$ ft-kips (by interpolation). This is far greater than what is needed, so the value of $\sum Q_n$ can be reduced. Try PNA location 7, with $\sum Q_n = 69.1$ kips.

$$a = \frac{69.1}{0.85(3)(90)} = 0.3011 \text{ in.}$$

$$Y2 = 4.75 - \frac{0.3011}{2} = 4.60 \text{ in.}$$

$$\phi M_n = 164.4 \text{ ft-kips}$$

From Example 9.8, $M_u = 162.7$ ft-kips with the beam weight included. This is less than the design strength of 164.4 ft-kips, so this selection is acceptable. Also from Example 9.8, both the flexural strength during the construction phase and the shear strength are adequate for a W16 × 26.

TABLE 9.9 ≡

Shape	PNA location	$\sum Q_n$	ϕM_n (by interpolation)
W16 × 26	6	104 kips	177 ft-kips
W14 × 22	2	203 kips	167.5 ft-kips

Answer **a.** Use a W16 × 26.

b. Shear connectors:

Try $\frac{3}{4}$-in. × 3-in. studs.

$$N_1 = \frac{\sum Q_n}{Q_n} = \frac{69.1}{21.49} = 3.2$$

This requires four studs for half the beam, or eight total. This corresponds to a spacing of

$$\frac{30(12)}{8} = 45 \text{ in.}$$

This is greater than the maximum permissible spacing of 36 inches, so more studs must be used. If a stud is placed in every sixth rib, the spacing will be equal to 36 inches, and the total number of studs will be

$$\frac{30(12)}{36} = 10 \qquad N_1 = 5$$

The corresponding shear force to be transferred is

$$\sum Q_n = 5(21.49) = 107.5 \text{ kips}$$

For convenience in using the tables, conservatively approximate this as $\sum Q_n = 104$ kips.

$$a = \frac{104}{0.85(3)(90)} = 0.4532 \text{ in.}$$

$$Y2 = 4.75 - \frac{0.4532}{2} = 4.523 \text{ in.}$$

From the Composite Beam Selection Table, the design strength is

$$\phi M_n = 182 \text{ ft-kips} > 162.7 \text{ ft-kips} \quad \text{(OK)}$$

Answer **b.** Use ten $\frac{3}{4}$-in. × 3-in. studs, equally spaced. To resist uplift, use spot welds at 18 in., midway between the studs.

c. Deflections:

From Example 9.8, the deflection of the steel beam before composite behavior has been achieved is

$$\Delta_1 = \frac{5 w_D L^4}{384 E I_s} = \frac{5(0.382/12)(30 \times 12)^4}{384(29,000)(301)} = 0.7976 \text{ in.}$$

For deflections that occur after the concrete has cured, the lower bound moment of inertia from the tables can be used. For a W16 × 26 with $\sum Q_n = 104$ kips (PNA location 6) and $Y2 = 4.523$ inches,

$$I_{LB} = 623 \text{ in.}^4$$

No separate moment of inertia is available for calculation of the additional deflection caused by creep. From Example 9.8, the load applied after the concrete cures is

$$w_T = w_D + w_L = 0.130 + 0.520 = 0.650 \text{ kips/ft}$$

and the corresponding deflection is

$$\Delta_2 = \frac{5w_T L^4}{384EI_{LB}} = \frac{5(0.650/12)(30 \times 12)^4}{384(29,000)(623)} = 0.6557 \text{ in.}$$

Total deflection $= \Delta_1 + \Delta_2 = 0.7976 + 0.6557 = 1.45$ in.

The maximum permissible deflection is

$$\frac{L}{240} = \frac{30(12)}{240} = 1.5 \text{ in.} > 1.45 \text{ in.} \quad \text{(OK)}$$

Answer   c. Deflection is satisfactory.

The preceding example illustrates the usefulness of the tables. In particular, the Composite Beam Selection Table greatly simplifies the design of a partially composite beam, in which the PNA lies within the steel shape.

9.9 ≡ CONTINUOUS BEAMS

In a simply supported beam, the point of zero moment is at the support. The number of connectors required between each support and the point of maximum positive moment will be half the total number required. In a continuous beam, the points of inflection are also points of zero moment and, in general, $2N_1$ connectors will be required for each span. Figure 9.27a shows a typical continuous beam and the regions in which shear connectors would be required. In the negative moment zones, the concrete slab will be in tension and therefore ineffective. In these regions there will be no composite behavior in the sense that we have considered thus far. The only type of composite behavior possible is that between the structural steel beam and the longitudinal reinforcing steel in the slab. The corresponding composite cross section is shown in Figure 9.27b. If this concept is used, a sufficient number of shear connectors must be provided to achieve a degree of continuity between the steel shape and the reinforcing.

The AISC Specification in Section I3.2 offers two alternatives for negative moment:

    1. Rely on the strength of the steel shape only.

    2. Include the reinforcing steel in the composite section, subject to the following conditions:

FIGURE 9.27 ≡

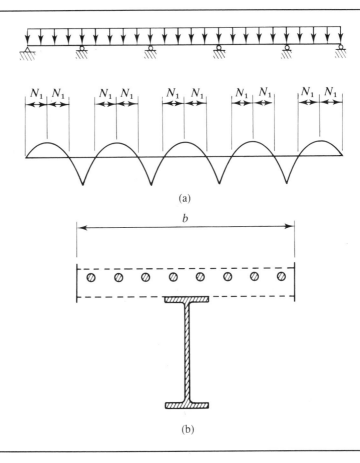

(a)

(b)

a. The steel shape must be compact with adequate lateral support.

b. Shear connectors must be present in the negative moment region (between the point of zero moment and the point of maximum negative moment).

c. The reinforcement within the effective width must be adequately developed (anchored).

The strength of the composite section should be based on a plastic stress distribution with $\phi_b = 0.85$.

If composite behavior is accounted for, AISC I5.2 requires that the horizontal shear force to be transferred between the point of maximum negative moment and the point of zero moment be taken as the smaller of $A_r F_{yr}$ and $\sum Q_n$, where

A_r = area of reinforcing steel within the effective width of the slab

F_{yr} = yield stress of the reinforcing steel

The additional strength gained from including the reinforcing steel is relatively small, however, and cover plates are sometimes used in the negative moment regions.

9.10 ≡ COMPOSITE COLUMNS

Composite columns can take one of two forms: a pipe or tube filled with plain concrete or a rolled steel shape encased in concrete with vertical reinforcement and lateral ties, just as in a reinforced concrete column. Figure 9.28 illustrates these two types.

The analysis of composite columns is done in the same way as for ordinary structural steel compression members, using the same equations from AISC Chapter E, but with values of F_y, E, and r that have been modified to give results that match experimental and theoretical results. Before considering the AISC equations for these values, we will examine the basis of the equations. If stability is ensured, the strength of a composite compression member can be expressed as the sum of the axial strengths of the steel shape, the reinforcing bars, and the concrete. This is called the "squash load" and is given by

$$P_n = A_s F_y + A_r F_{yr} + 0.85 f'_c A_c \qquad (9.5)$$

where

A_s = cross-sectional area of the rolled steel shape

A_r = total cross-sectional area of the vertical reinforcing steel

F_{yr} = yield stress of the reinforcing steel

A_c = cross-sectional area of the concrete

Modern reinforcing steel bars are *deformed* — that is, the surface of the bars has protrusions that help the steel grip or bond with the concrete. The cross-sectional area A_r used in computations is a nominal area equal to the area of a plain bar that has the same weight per foot of length as the deformed bar. Table 9.10 gives nominal diameters and areas for standard bar sizes as defined by ASTM[11] and ACI.[22]

FIGURE 9.28 ≡

TABLE 9.10 ≡

Bar Number	Diameter (in.)	Area (in.²)
3	0.375	0.11
4	0.500	0.20
5	0.625	0.31
6	0.750	0.44
7	0.875	0.60
8	1.000	0.79
9	1.128	1.00
10	1.270	1.27
11	1.410	1.56
14	1.693	2.25
18	2.257	4.00

To obtain an overall stress, divide the load from Equation 9.5 by the area of the structural steel shape:

$$\frac{P_n}{A_s} = F_{my} = F_y + A_r\frac{F_{yr}}{A_s} + 0.85f'_c\frac{A_c}{A_s} \tag{9.6}$$

The value of F_{my} obtained from Equation 9.6, when used in place of F_y in the compression member equations, gives good results for concrete-filled pipes and tubes, where the concrete is contained within the steel shape. (Longitudinal reinforcing bars are not normally used with pipes and tubes; therefore, A_r would be zero for this type of composite column.)

For encased shapes, the benefit of confinement is absent, and the Structural Stability Research Council[57] recommends that the ACI code[22] capacity reduction factor of 0.7 be applied to the reinforcing steel and concrete terms of Equation 9.6 as follows:

$$\begin{aligned}
F_{my} &= F_y + 0.7A_r\frac{F_{yr}}{A_s} + 0.7(0.85)f'_c\frac{A_c}{A_s} \\
&= F_y + 0.7A_r\frac{F_{yr}}{A_s} + 0.595f'_c\frac{A_c}{A_s}
\end{aligned} \tag{9.7}$$

To account for slenderness effects, the member flexural stiffness, which is proportional to the quantity EI/L, must be adjusted. This is done by modifying the value of E as follows:

$$E_m = E + \text{constant} \times E_c\frac{A_c}{A_s} \tag{9.8}$$

where

E = modulus of elasticity of the structural steel shape

E_c = modulus of elasticity of concrete

Although the stiffness is proportional to the moment of inertia, a ratio of areas gives better results for a composite column than a ratio of moments of inertia.[57] The constant in Equation 9.8 is 0.4 for a concrete-filled pipe or tube, representing an allowance of 40% of the concrete stiffness, and 0.2 for an encased shape.

The radius of gyration of a composite shape is larger than that of both the steel shape and the concrete area. It is conservative to use the larger of the radius of gyration of the steel shape or the radius of gyration of the concrete outline, which can be taken as 0.3 times the dimension in the plane of buckling. Denoting the radius of gyration of the composite shape by r_m,

$$r_m = r \geq 0.3b \tag{9.9}$$

where

r = radius of gyration of the steel shape in the plane of buckling

b = dimension of the concrete outline in the plane of buckling

Specification Requirements

The AISC provisions for composite columns are essentially the same as outlined above. Equations E2-1 and E2-3 from Chapter E of the Specification are used to determine the design strength, but the values of F_y, E, and r are modified. The modified values are based on Equations 9.6 through 9.9 given in the preceding discussion. From AISC Section I2.2, the modified value of F_y is

$$F_{my} = F_y + c_1 F_{yr}(A_r/A_s) + c_2 f'_c(A_c/A_s) \qquad \text{(AISC Equation I2-1)}$$

where the constants c_1 and c_2 account for the differences between encased sections and concrete-filled pipes and tubes:

$c_1 = 1.0$ and $c_2 = 0.85$ for pipes and tubes

$c_1 = 0.7$ and $c_2 = 0.6$ for encased shapes

The AISC modified value of E is the same as that given by Equation 9.8:

$$E_m = E + c_3 E_c(A_c/A_s) \qquad \text{(AISC Equation I2-2)}$$

where

$c_3 = 0.4$ for pipes and tubes

    $= 0.2$ for encased shapes

AISC I2.2 specifies the value of r_m to be that given by Equation 9.9—that is,

$$r_m = r \geq 0.3b$$

To qualify as a composite column, the following limitations, given in AISC I2.1, must be observed:

1. The structural steel must make up at least 4% of the total cross-sectional area, or the compression member will act more like a concrete column than a composite column.

2. Encased sections must conform to the following details:

    a. Both longitudinal and lateral reinforcement must be used. The spacing of the lateral ties must be no greater than two thirds of the least dimension of the concrete. The cross-sectional area of both longitudinal and lateral reinforcement must be at least 0.007 square inches per inch of bar spacing.

    b. There must be at least $1\frac{1}{2}$ inches of clear concrete cover for both the lateral and the transverse reinforcement.

    c. *Load-carrying* longitudinal reinforcement must be continuous at framed levels. Longitudinal bars whose purpose is to help restrain the concrete may be spliced at framed levels.

3. The concrete strength f'_c must be between 3 ksi and 8 ksi for normal-weight concrete (no test data are available for f'_c greater than 8 ksi) and at least 4 ksi for lightweight concrete.

4. The yield stress of both the structural steel and the longitudinal reinforcing bars may not exceed 55 ksi in the computations. This limit can be derived from a consideration of local stability. As long as the steel is confined by the concrete, there will be no local buckling. The concrete will be able to adequately confine the steel if it does not spall (a disintegration at the concrete surface). If it is assumed that the spalling occurs at a concrete strain of 0.0018, the corresponding stress in the steel is

$$F_{max} = \varepsilon_{max} E = 0.0018(29{,}000) = 52.2 \text{ ksi}$$

This is rounded to the limiting value of 55 ksi.

5. To prevent local buckling in pipes or tubes filled with concrete, the wall thickness should be at least

$$t = b\sqrt{F_y/3E}$$

for each rectangular section face of width b or

$$t = D\sqrt{F_y/8E}$$

for circular sections of outer diameter D.

≡ EXAMPLE 9.11

A composite compression member consists of a W12 × 136 encased in a 20-in. × 22-in. concrete column as shown in Figure 9.29. Four #10 bars are used for longitudinal reinforcement, and #3 ties spaced 13 in. center-to-center provide the lateral reinforcement. The steel yield stress is $F_y = 50$ ksi, and Grade 60 reinforcing bars are used. The concrete strength is $f'_c = 5000$ psi. Compute the design strength for an effective length of 16 ft with respect to both axes.

FIGURE 9.29 ≡

22"

20"

W12 × 136
4 #10
#3 ties @ 13"

SOLUTION   The modified values F_{my} and E_m will be determined from AISC Equations I2-1 and I2-2. Values needed for these equations are as follows:

$F_{yr} = 55$ ksi, the limiting value imposed by AISC I2.1

$A_r = 4(1.27) = 5.08$ in.^{2}

$A_c =$ net area of concrete $= 20(22) - A_s - A_r = 440 - 39.9 - 5.08$

$= 395.0$ in.^{2}

For $f'_c = 5000$ psi,

$$E_c = w_c^{1.5}(33)\sqrt{f'_c} = (145)^{1.5}(33)\sqrt{5000}$$
$$= 4,074,000 \text{ psi} = 4074 \text{ ksi}$$

From AISC Equation I2-1, the modified yield stress is

$$F_{my} = F_y + c_1 F_{yr}\left(\frac{A_r}{A_s}\right) + c_2 f'_c\left(\frac{A_c}{A_s}\right)$$
$$= 50 + 0.7(55)\left(\frac{5.08}{39.9}\right) + 0.6(5)\left(\frac{395}{39.9}\right) = 84.60 \text{ ksi}$$

From AISC Equation I2-2, the modified modulus of elasticity is

$$E_m = E + c_3 E_c\left(\frac{A_c}{A_s}\right) = 29,000 + 0.2(4074)\left(\frac{395}{39.9}\right) = 37,070 \text{ ksi}$$

The radius of gyration r_m to be used in the compression member equations of AISC Chapter E will be either r for the steel shape or $0.3b$, whichever is larger. In this example, buckling will be about the y-axis of the member, so r for the shape $= r_y = 3.16$ inches. In the plane of buckling,

$0.3b = 0.3(20) = 6$ in.   (controls)

Therefore, $r_m = 6$ in. The design strength can now be computed as for a noncomposite compression member by using the modified values F_m, E_m, and r_m in place of F_y, E, and r.

$$\lambda_e = \frac{KL}{r_m \pi}\sqrt{\frac{F_{my}}{E_m}} = \frac{16(12)}{6\pi}\sqrt{\frac{84.60}{37,070}} = 0.4866 < 1.5$$

$$F_{cr} = (0.658)^{\lambda_e^2}F_{my} = (0.658)^{(0.4866)^2}(84.60) = 76.62 \text{ ksi}$$

The nominal strength is

$$P_n = A_s F_{cr} = 39.9(76.62) = 3057 \text{ kips}$$

and the design strength is

$$\phi_c P_n = 0.85(3057) = 2600 \text{ kips}$$

Answer The design compressive strength is 2600 kips.

Tables for Analysis and Design

Part 5 of the *Manual* contains tables that greatly simplify the analysis and design of composite columns. These tables are similar to the column strength tables in Part 3 of the *Manual*. Axial compressive design strengths are given as a function of effective length for concrete-filled pipes and tubes as well as for encased W-shapes. For the encased columns, longitudinal bars and transverse ties that satisfy AISC requirements are included. Values of r_{mx}/r_{my} are given for those cases where $K_x L \neq K_y L$.

≡ EXAMPLE 9.12

An 18-ft compression member must support a total service load of 1000 kips, consisting of equal parts of dead and live load. The member is pinned at both ends with additional support at midheight in the weak direction. Use the tables in Part 5 of the *Manual* to select a square encased W-shape, using the smallest area of concrete possible. Use A36 steel, Grade 60 reinforcing bars, and $f'_c = 3.5$ ksi.

SOLUTION The factored axial load is

$$P_u = 1.2(500) + 1.6(500) = 1400 \text{ kips}$$

An examination of the tables shows that for $f'_c = 3.5$ ksi, the value of r_{mx}/r_{my} varies from 1.0 to 1.22, with most values equal to 1.0. Since

$$\frac{K_x L}{K_y L} = 2 > 1.22$$

$K_x L$ will control. Assume $r_{mx}/r_{my} = 1.0$ and enter the tables with

$$KL = \frac{K_x L}{r_{mx}/r_{my}} = \frac{18}{1.0} = 18 \text{ ft}$$

The following tabulation shows the possible choices.

Concrete Dimensions	Shape	r_{mx}/r_{my}	$\phi_c P_n$
18 × 18	W10 × 112	1.0	1450 kips
20 × 20	W12 × 87	1.0	1420 kips

Although the 20 × 20 column requires the lighter structural steel shape, the criterion is minimum concrete dimensions, so the 18 × 18 will be selected.

Answer   Use an 18-in. × 18-in. column with a W10 × 112, four #8 bars, and #3 ties at 12 in. center-to-center.

≡ PROBLEMS ≡

NOTES   Unless otherwise indicated, the following conditions apply to each composite beam problem:

1. No shoring is used.

2. The slab formwork provides continuous lateral support for the beam during the construction phase.

3. Other than the slab and beam weight, there are no construction loads.

4. Normal-weight concrete is used.

9.1   W12 × 50 floor beams are used compositely with a 4.5-in.-thick reinforced concrete floor slab. The effective slab width b is 90 inches. Locate the neutral axis of the transformed section. $f'_c = 3000$ psi.

9.2   Compute the moment of inertia of the transformed section for the composite beam shown in Figure P9.2. The given width is the effective width, not the transformed width.

$b = 80''$

4″

W24 × 84
$f'_c = 4000$ psi

FIGURE P9.2

9.3 For the composite beam and service live load in Figure P9.3, compute the maximum service live load stress in the steel.

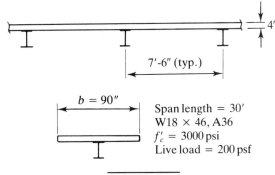

Span length = 30'
W18 × 46, A36
f'_c = 3000 psi
Live load = 200 psf

FIGURE P9.3

9.4 A composite floor system uses W30 × 99 beams and a 5.5-in.-thick floor slab with an effective width of 8.33 feet. A36 steel is used, and f'_c = 4000 psi. If the service load moment resisted by composite action is 700 ft-kips, find

    a. the maximum service-load compressive stress in the steel

    b. the maximum service-load tensile stress in the steel

9.5 Compute the flexural design strength for the composite beam of Problem 9.1 if A36 steel is used.

9.6 Compute the flexural design strength for the composite beam of Problem 9.2 if A36 steel is used.

9.7 Compute the flexural design strength for the composite beam of Problem 9.3.

9.8 Compute the flexural design strength for the composite beam of Problem 9.4.

9.9 A W16 × 57 is used compositely with a 5-in.-thick floor slab. The effective slab width is 88.5 inches. Compute the design strength. A36 steel is used and f'_c = 3000 psi.

9.10 Determine whether the given beam has adequate strength. In addition to the weight of the beam and slab, assume an additional construction load of 20 psf. Use an effective slab width of 60 inches.

W18 × 50, A36
f'_c = 3000 psi
Live load = 100 psf
Span length = 25'

FIGURE P9.10

9.11   A composite floor system consists of W24 × 131 beams and a 5-in.-thick reinforced concrete floor slab. The span length is 30 feet, and the beams are spaced at 7 feet. In addition to the weight of the slab, there is a uniform live load of 800 psf. Review this beam for compliance with the AISC Specification. Assume that there is no lateral support during the construction phase. A36 steel is used, and $f'_c = 4000$ psi.

9.12   Evaluate the adequacy of the beam of Problem 9.3.

9.13   Evaluate the adequacy of floor beam AB in Figure P9.13. A36 steel is used, and $f'_c = 4000$ psi.

FIGURE P9.13

9.14   A composite floor system consists of W16 × 31 beams spaced at 9 feet with a 4-in.-thick floor slab. The span length is 30 feet. A36 steel and 3000 psi concrete are used.

   a. Investigate the adequacy of this floor system if the service live load is 100 psf.

   b. How many $\frac{1}{2}$-in. × 2-in. shear studs are required for each beam for full composite action?

9.15   A cross section of a floor system is shown in Figure P9.15.

   a. Check the floor beams for compliance with the AISC Specification if the service live load is 250 psf.

   b. Select shear connectors for full composite action and specify the total number required for each beam.

FIGURE P9.15

9.16   If the beam in Problem 9.9 is simply supported and uniformly loaded, what is the total number of $\frac{3}{4}$-in.-diameter $\times$ 3-in.-long shear studs required for full composite action?

9.17   Design the shear connectors for the beam of Problem 9.11 for full composite action.

9.18   Select a W-shape and design the shear connectors for the following composite floor system: The slab is 4.5-in. thick with beams spaced at 6.5 feet. The beams are simply supported with a span length of 36 feet and are subjected to a service live load of 175 psf. A36 steel and 4000 psi concrete are to be used. Design for full composite behavior.

9.19   **a.** Select a W-shape to be used with a 4-in.-thick slab to resist a service live load of 125 psf and a partition load of 15 psf. The span length is 25 feet, and the beams will be spaced at 5 feet. Use A36 steel and 3000 psi concrete, and design for full composite behavior.
         b. Design the shear connectors.

9.20   A floor system consists of rolled steel beams spaced at 8 feet and supporting a 4.5-in.-thick reinforced concrete slab. The span length is 35 feet, and the service live load is 160 psf.
         a. Select a W-shape. Use A36 steel and 3000 psi concrete.
         b. Design the shear connectors.

In Problems 9.21 through 9.26, assume full composite behavior.

9.21   A portion of a composite floor system is shown in Figure P9.21. The superimposed loading consists of an 80 psf live load, 10 psf for partitions, and 5 psf for a suspended ceiling. Compute the following deflections for a typical floor beam: (a) the maximum immediate deflection, and (b) the maximum long-term deflection.

4″ slab
$f'_c = 3000\,\text{psi}$
W14 × 22
A36 steel

4 @ 7′-6″ = 30′-0″

30′-0″

FIGURE P9.21

9.22 Compute the live load deflection for the composite beam of Problem 9.3.

9.23 Compute the dead load, live load, and total deflections for the beam of Problem 9.11.

9.24 Compute the dead load, live load, and total deflections for the composite beam of Problem 9.13

9.25 Compute the dead load, live load, and total deflections for the composite beam designed in Problem 9.19. If the maximum total deflection is not to exceed $\frac{1}{240}$ of the span length, is the design satisfactory?

9.26 Design the shear connectors for full composite behavior for the floor system in figure P9.26. The span length is 40 feet, and the beams are uniformly loaded. Use A36 steel and 4000 psi concrete.

6″

W36 × 135

6′-0″   6′-0″

Steel deck   1½″

6″   2¼″

FIGURE P9.26

9.27 A composite floor system uses formed steel deck of the type shown in Figure P9.27. The beams are W18 × 50, and the slab has a total thickness of 4.5 in. from top of slab to bottom of deck. The effective slab width b is 85 in., and the span length is 30 feet. The structural steel is A36, and the concrete strength is $f'_c = 3000$ psi.

    a. Compute the flexural strength with two $\frac{7}{8}$-in. diam. × $3\frac{1}{2}$-in. studs per rib.

    b. Compute the flexural strength with one $\frac{7}{8}$-in. diam. × $3\frac{1}{2}$-in. stud in every other rib.

2″

6″   2½″

FIGURE P9.27

9.28 Cross sections of a composite floor system, including the formed steel deck, are shown in Figure P9.28. The span length is 40 feet, and the floor must support a live load of 175 psf. The structural steel is A36, and lightweight concrete weighing 110 pounds per cubic foot and having a strength of $f'_c = 3000$ psi is used.

    a. Check this floor system for compliance with the AISC Specification.

    b. Design the shear connectors.

4″

W24 × 55

8′-0″

1½″

6″   2¼″

FIGURE P9.28

9.29 The composite floor system in Figure P9.29 consists of W21 × 57 floor beams, the formed steel deck shown, and a floor slab whose total thickness is 5.5 inches. The live load is 100 psf, and there is a partition load of 10 psf. A36 steel and 3000 psi concrete are used.

    a. Check all applicable provisions of the AISC Specification.

    b. Design the shear connectors.

    c. Compute the maximum total long-term deflection.

FIGURE P9.29

9.30 Select a W-shape of A36 steel, and design shear connectors for the composite floor system shown in Figure P9.30, subject to the following conditions:

Total slab thickness = 4 in.

$f'_c = 3000$ psi, lightweight concrete ($w_c = 110$ pcf)

FIGURE P9.30

Live load = 60 psf

Ceiling weight = 5 psf

Partition load = 10 psf

Maximum total deflection = $\frac{1}{240}$ of span length

9.31 Select a W-shape of A36 steel, and design shear connectors for a floor system consisting of beams spaced at 8 ft with a span length of 36 ft. The formed steel deck shown in Figure P9.31 is used with a 4-in.-thick slab (2 in. of concrete above the top of the deck) having a 28-day compressive strength of $f'_c = 3000$ psi. In addition to the weight of the beam and slab, the service loads consist of a partition load of 20 psf, a suspended ceiling weighing 8 psf, and a live load of 100 psf. The maximum permissible live load deflection is $\frac{1}{360}$ of the span length.

2″

2½″   6″

FIGURE P9.31

9.32 Select a W-shape and design the shear connectors for the following conditions (the tables in Part 5 of the *Manual* may be used):

Span length $L = 30$ ft

Beam spacing = 7.5 ft

Solid slab thickness = 4 in.

Loads: live load = 80 psf
partition load = 10 psf
suspended ceiling weight = 5 psf

Maximum permissible live load deflection = $\frac{L}{360}$

Use A36 steel and $f'_c = 4000$ psi.

9.33 Select a W-shape and design the shear connectors for the following conditions (the tables in Part 5 of the *Manual* may be used):

Span length $L = 40$ ft

Beam spacing = 6 ft

Total slab thickness = 6 in.

Formed steel deck with rib spacing = 6 in., $h_r = 1.5$ in., $w_r = 2.25$ in.

Live load = 180 psf

Maximum permissible live load deflection = $\frac{L}{360}$

Use A36 steel and $f'_c = 4000$ psi.

9.34 Select a W-shape and design the shear connectors for the following conditions (the tables in Part 5 of the *Manual* may be used):

Span length $L = 40$ ft

Beam spacing $= 8$ ft

Total slab thickness $= 4$ in.

Formed steel deck with rib spacing $= 6$ in., $h_r = 1.5$ in., $w_r = 2.25$ in.

Live load $= 250$ psf

Maximum permissible total deflection $= \frac{L}{240}$

Use steel with a yield stress $F_y = 50$ ksi and lightweight concrete weighing 110 pcf with $f'_c = 4000$ psi.

9.35 Select a W-shape and design the shear connectors for the following conditions (the tables in Part 5 of the *Manual* may be used):

Span length $L = 35$ ft

Beam spacing $= 12$ ft

Total slab thickness $= 5.5$ in.

Formed steel deck with rib spacing $= 6$ in., $h_r = 2$ in., $w_r = 2.25$ in.

Live load $= 100$ psf, partition load $= 15$ psf

Maximum permissible live load deflection $= \frac{L}{360}$

Use steel with a yield stress $F_y = 50$ ksi and lightweight concrete weighing 110 pcf with $f'_c = 4000$ psi.

9.36 An $8 \times 6 \times \frac{1}{2}$ structural tube filled with concrete is used as a composite column as shown in Figure P9.36. The steel has a yield stress of $F_y = 46$ ksi, and the concrete has a compressive strength of $f'_c = 3500$ psi. Compute the design strength of the column, and check your answer with the tables in Part 5 of the *Manual*.

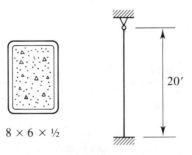

$8 \times 6 \times \frac{1}{2}$

20'

FIGURE P9.36

9.37 The composite compression member shown in Figure P9.37 has lateral support at the one-third points in the weak direction only. Compute the design strength and check your answer with the tables in Part 5 of the *Manual*. The steel yield stress is 50 ksi, and $f'_c = 5000$ psi.

W10 × 49
4 #8
#3 ties @ 12″

FIGURE P9.37

9.38   Use the tables in Part 5 of the *Manual* and select a concrete-filled pipe for the conditions shown in Figure P9.38. Select the smallest diameter pipe that will work. Use A36 steel and $f'_c = 3500$ psi.

$D = 90^k$
$L = 200^k$

15′-4″

FIGURE P9.38

9.39   Design a square composite column with an encased W-shape to resist a factored axial load of 2000 kips. The effective lengths with respect to the two principal axes are $K_x L = 27$ ft and $K_y L = 15$ ft. Select the smallest concrete dimensions possible. Use $F_y = 50$ ksi and $f'_c = 8000$ psi. The tables in Part 5 of the *Manual* may be used.

9.40   Write a computer program to determine the location of the neutral axis of the transformed section of composite beams with solid slabs.

Input data:

b = effective width of the concrete slab (in.)

t = thickness of the concrete slab (in.)

f'_c = 28-day compressive strength of the concrete (ksi)

A_s = area of steel shape (in.^{2})

d = depth of steel shape (in.)

I = moment of inertia of steel shape about its major principal axis (in.^{4})

F_y = yield stress of steel shape (ksi)

Output: Same as input plus $\bar{y}$, the distance from the top of the slab to the neutral axis of the transformed section.

9.41   Modify the computer program of Problem 9.38 to include computation of the moment of inertia about the neutral axis of the transformed section.

9.42   Modify the computer program of either Problem 9.40 or Problem 9.41 (as directed by your instructor) to include computation of the flexural design strength. Specialize this program for the case of the PNA in the slab. Inspection of the output can verify whether this assumption is valid.

Additional input:

F_y = yield stress of the steel shape (ksi)

Additional output:

Compressive force for a fully compressed slab ($0.85f'_c A_c$)

Tensile force for the steel shape fully yielded in tension ($A_s F_y$)

Design strength $\phi_b M_n$

9.43   Modify the program of Problem 9.40, 9.41, or 9.42, as directed by your instructor, to accommodate composite beams with formed steel deck. The following modifications in the input data are required:

1. The value of t is the thickness of slab above the top of the deck, rather than the total thickness.

2. An additional item of input, h_r (the height of the deck rib), is required.

This version of the program can easily handle solid slabs by using appropriate input data. If formed steel deck is not used, the value of t is the total thickness, and h_r is zero.

9.44   Write a computer program to assist in the selection of shear connectors.

Input data

F_y = yield stress of the steel (ksi)

f'_c = 28-day compressive strength of the concrete (ksi)

b = effective width of slab (in.)

t = thickness of slab above the top of the deck (in.). If formed steel deck is not used, t is the total thickness of the slab.

A_s = area of the steel shape (in.^{2})

I_s = moment of inertia of the steel shape about the x-axis (in.^{4})

I_{tr} = moment of inertia of the transformed section (in.^{4})

Output: Same as input plus a table of the following form:

			EFFECTIVE I FOR SUMQ/C =		
DIAM	Q	N1	0.75	0.50	0.25
1/2	9.55	28.9	973.9	850.4	689.5
5/8	.	.	.	.	.
3/4	.	.	.	.	.
7/8	.	.	.	.	.

Note: Stud reduction factor not used.

PLATE GIRDERS

INTRODUCTION

A plate girder is usually thought of as a flexural member whose cross section is composed of plate elements. (As discussed in Chapter 5 of this book and to be discussed again presently, AISC distinguishes beams from plate girders on the basis of the web slenderness.) A plate girder is basically a large beam, both in span and cross section, the large cross section usually being a consequence of the long span. If the largest available hot-rolled steel shape is inadequate for a given span and loading, the first alternative would normally be a rolled shape with cover plates added to one or both flanges. If this could not provide enough moment strength, a cross section with exactly the properties needed could be fabricated from plate elements. If the span is long enough, however, the depth and weight of a built-up girder may be excessive, and other alternatives, such as a truss, may be necessary.

A plate girder cross section can take several forms. Figure 10.1 shows some of the possibilities. The usual configuration is a single web with two equal flanges, with all parts connected by welding. The box section, which has two webs as well as two flanges, is a torsionally superior shape and can be used when large unbraced lengths are necessary. Hybrid girders, in which the steel in the flanges is of a higher strength than that in the web or webs, are sometimes used.

Before the widespread use of welding, connecting the components of the cross section was a major consideration in the design of plate girders. Since all of the connections were made by riveting, there was no way to attach the flange directly to the web, and additional cross-sectional elements were introduced for the specific purpose of transmitting the load from one component to the other. The usual technique was to use a pair of angles, placed back-to-back, to attach the flange to the web; one pair of legs was attached to the web, and the other pair attached to the flange (Figure 10.1b). If web stiffeners were needed, pairs of angles were used for that purpose also. To avoid a conflict between the stiffener angles and the flange angles, filler plates were added to the web so that the stiffeners could clear the flange angles (Figure 10.1c). If a variable cross section was desired, one or more cover plates of different lengths were riveted to the flanges. (Although cover plates can also be used with welded plate girders, a simpler approach is to use different thicknesses of flange plate, welded end-to-end, at different locations along the length of the girder.) It should be evident that the welded plate girder is far superior to the riveted or bolted girder in

FIGURE 10.1 ≡

(a) Welded (b) Riveted without Stiffeners

Flange plate

Flange angle

Stiffener angle

Web plate

Filler plate

(rivets not shown)

(c) Riveted with Stiffeners

terms of simplicity and efficiency. We will consider only I-shaped welded plate girders in this chapter.

Before considering specific AISC Specification requirements for plate girders, we will examine, in a very general way, the peculiarities of plate girders as opposed to ordinary rolled beams. Although flexural members have been covered in Chapter 5, "Beams," there are special flexural and shear strength requirements for plate girders.

10.2 ≡ GENERAL CONSIDERATIONS

Structural steel design is largely a matter of providing for stability, either locally or in an overall sense. Many standard hot-rolled structural shapes are proportioned so that local stability problems have been eliminated or minimized. When a plate girder is used, however, the designer must account for factors that in many cases would not be a problem with a rolled shape. Deep, thin webs account for many of the special problems associated with plate girders, including local instability. A thorough understanding of the basis of the AISC provisions for plate girders requires a background in stability theory, particularly plate stability. Such a treatment is beyond the scope of this book, however, and the emphasis in this chapter will be on the qualitative basis of the Specification requirements and their application. For those interested in delving deeper, the *Guide to Stability Design Criteria for Metal Structures*[33] is a good starting point, and *Buckling Strength of Metal Structures*[31] and *Theory of Elastic Stability*[36] will provide the fundamentals of stability theory.

Plate girders rely on the strength available after the web has buckled, so most of the flexural strength will come from the flanges. The limit states considered are yielding of the tension flange and buckling of the compression flange. The compression flange buckling can take the form of vertical buckling into the web or flange local buckling (FLB), or it can be caused by lateral-torsional buckling (LTB).

At a location of high shear in a girder web, usually near the support and at or near the neutral axis, the principal planes will be inclined with respect to the longitudinal axis of the member, and the principal stresses will be diagonal tension and diagonal compression. The diagonal tension poses no particular problem, but the diagonal compression can cause buckling of the web. This problem can be addressed in one of three ways: (1) the depth-to-thickness ratio of the web can be made small enough so that the problem is eliminated, (2) web stiffeners can be used to form panels with increased shear strength, or (3) web stiffeners can be used to form panels that resist the diagonal compression through *tension-field action*. Figure 10.2 illustrates the concept of tension-field action. At the point of impending buckling, the web loses its ability to support the diagonal compression, and this stress is shifted to the transverse stiffeners and the flanges. The stiffeners resist the vertical component of the diagonal compression, and the flanges resist the horizontal component. The web will need to resist only the diagonal tension, hence the term *tension-field action*. This behavior can be likened to that of a Pratt truss, in which the vertical web members carry compression and the diagonals carry tension (Figure 10.2b). Since the tension field does not actually exist until the web begins to buckle, its contribution to the web shear strength will not exist until the web buckles. The total strength will consist of the strength prior to buckling plus the post-buckling strength deriving from tension-field action.

If an unstiffened web is incapable of resisting the applied shear, appropriately spaced stiffeners are used to develop tension-field action. Cross-section requirements for these stiffeners, called *intermediate stiffeners*, are minimal, since their primary purpose is to provide stiffness rather than resist directly applied loads.

Additional stiffeners may be required at points of concentrated loads for the

FIGURE 10.2 ≡

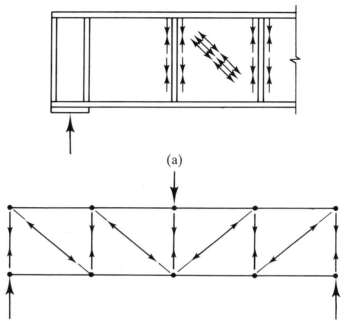

(a)

Note: The arrows indicate forces acting on members, rather than
on joints

(b)

purpose of protecting the web from the direct compressive load. These are
called *bearing stiffeners*, and they must be proportioned to resist the applied
loads. They can also simultaneously serve as intermediate stiffeners. Figure 10.3
shows a bearing stiffener consisting of two rectangular plates, one on each side
of the girder web. The plates are notched at the inside top and bottom corners
so as to avoid the flange-to-web welds. If the stiffeners are conservatively assumed to resist the total applied load P (this neglects any contribution by the
web), the bearing stress on the contact surfaces may be written as

$$f_p = \frac{P}{A_{pb}}$$

where

A_{pb} = projected bearing area
     = $2at$ (see Figure 10.3)

or, expressing the bearing load in terms of the stress,

$$P = f_p A_{pb} \tag{10.1}$$

FIGURE 10.3 ≡

Section

In addition, the pair of stiffeners, together with a short length of web, is treated as a column with an effective length less than the web depth and is investigated for compliance with the same Specification provisions as any other compression member. This cross section is illustrated in Figure 10.4. The compressive strength should always be based on the radius of gyration about an axis in the plane of the web, since instability about the other principal axis is prevented by the web itself.

Other limit states resulting from the application of concentrated loads to the top flange are web yielding, web crippling (buckling), and sidesway web

FIGURE 10.4 ≡

Section

buckling. Sidesway web buckling occurs when the compression in the web causes the *tension* flange to buckle laterally. This phenomenon can occur if the flanges are not adequately restrained against movement relative to one another by stiffeners or lateral bracing.

The welds for connecting the components of a plate girder are designed in much the same way as for other welded connections. The flange-to-web welds must resist the horizontal shear at the interface between the two components. This applied shear, called the *shear flow*, is usually expressed as a force per unit length of girder to be resisted by the weld. From Chapter 5 of this book, the shear flow, based on elastic behavior, is given by

$$f = \frac{VQ}{I_x}$$

where Q is the moment, about the neutral axis, of the area between the horizontal shear plane and the outside face of the section. This is Equation 5.6 for shearing stress multiplied by the width of the shear plane. Since the applied shear force V will usually be variable, the spacing of intermittent welds, if used, can also vary.

10.3 ≡ AISC REQUIREMENTS

Requirements for plate girders are addressed in Chapter G of the AISC Specification. Chapter G does little more than refer the reader to Appendix G, so for all practical purposes, plate girders are covered in Appendix G.

Whether a flexural member is classified as a beam or a plate girder is a function of the web slenderness h/t_w, where h is the depth of the web from inside face of flange to inside face of flange and t_w is the web thickness. If h/t_w is less than $970/\sqrt{F_{yf}}$, the member is classified as a beam, and the provisions of AISC Chapter F apply, regardless of whether the member is built up from plates or is a hot-rolled shape. If h/t_w is greater than $970/\sqrt{F_{yf}}$, the member is treated as a plate girder, and the provisions of Chapter G apply. This means that flexural members with slender webs, where slenderness is defined in AISC Chapter B, are treated as plate girders. Note that when hybrid girders are used, the limiting value of h/t_w is based on F_{yf}, the yield stress of the flange. This is because the stability of the web against flexural buckling is dependent on the strain in the flange.[58]

To prevent vertical buckling of the flange into the web, AISC Appendix G2 imposes an upper limit on the web width-thickness ratio h/t_w. The limiting value is a function of the *aspect ratio* a/h of the girder panels, which is the ratio of intermediate stiffener spacing to web depth (see Figure 10.5):
For $a/h \le 1.5$,

$$\frac{h}{t_w} \le \frac{2000}{\sqrt{F_{yf}}} \qquad \text{(AISC Equation A-G1-1)}$$

FIGURE 10.5 ≡

For $a/h > 1.5$,

$$\frac{h}{t_w} \leq \frac{14{,}000}{\sqrt{F_{yf}(F_{yf} + 16.5)}}$$

(AISC Equation A-G1-2)

where a is the clear distance between stiffeners.

10.4 ≡ FLEXURAL STRENGTH

The design flexural strength of plate girders is $\phi_b M_n$, where $\phi_b = 0.90$. The nominal flexural strength M_n is based on either tension flange yielding or compression flange buckling. The compression flange buckling strength will be determined by either flange local buckling (FLB) or lateral-torsional buckling (LTB). Vertical buckling of the compression flange into the web has been eliminated from consideration by the limitations imposed by AISC Equations A-G1-1 and A-G1-2.[17]

Tension Flange Yielding

From Chapter 5 of this book, the maximum bending stress in a flexural member bent about its strong axis is given by

$$f_b = \frac{M}{S_x}$$

where S_x is the elastic section modulus about the strong axis. Expressing the bending moment as a function of the section modulus and stress,

$$M = S_x f_b$$

(10.2)

AISC Appendix G2 gives the nominal flexural strength based on tension flange yielding as

$$M_n = S_{xt} R_e F_{yt}$$ (AISC Equation A-G2-1)

where

S_{xt} = elastic section modulus referred to the tension side

R_e = hybrid girder factor

F_{yt} = yield stress of tension flange

The hybrid girder factor R_e is equal to 1.0 for nonhybrid girders. For hybrid girders,

$$R_e = \frac{12 + \dfrac{A_w}{A_f}(3m - m^3)}{12 + 2\dfrac{A_w}{A_f}} \le 1.0$$

where

A_w = area of the web $\le 10A_f$

A_f = area of the compression flange

$m = F_{yw}/F_{yf}$ (this definition is for the limit state of tension flange yielding; for compression flange buckling, the hybrid girder factor will also be employed, but m will be defined differently.)

Compression Flange Buckling

The nominal flexural strength corresponding to compression flange buckling is also based on Equation 10.2, since an *elastic* buckling limit state is involved. From AISC Appendix G2, this strength is given by

$$M_n = S_{xc} R_{PG} R_e F_{cr}$$ (AISC Equation A-G2-2)

where

S_{xc} = elastic section modulus referred to the compression side

R_{PG} = strength reduction factor to account for elastic web buckling

F_{cr} = critical compression flange stress, based on either LTB or FLB

R_e = hybrid girder factor (computed by the same equation as for tension flange yielding, but with a value of $m = F_{yw}/F_{cr}$).

The hybrid girder factor is again equal to 1.0 for nonhybrid girders. The plate girder strength reduction factor R_{PG} is given by

$$R_{PG} = 1 - \frac{a_r}{1200 + 300a_r}\left(\frac{h_c}{t_w} - \frac{970}{\sqrt{F_{cr}}}\right) \le 1.0$$ (AISC Equation A-G2-3)

where

$$a_r = A_w/A_f \leq 10$$

h_c = twice the distance from the centroid to the inside face of the compression flange ($h_c = h$ for girders with equal flanges)

The critical stress F_{cr} is based on either lateral-torsional buckling or flange local buckling. The AISC Specification uses the generic notation λ, λ_p, and λ_r to deal with the slenderness parameters for both of these limit states, and a common set of equations is used for F_{cr}. The equations will be presented here in a somewhat expanded form in an attempt to clarify the provisions. For lateral torsional buckling, the slenderness of a portion of the compression region of the girder is used:

$$\lambda = \frac{L_b}{r_T}$$

$$\lambda_p = \frac{300}{\sqrt{F_{yf}}} \qquad \text{(AISC Equations A-G2-7, 8, and 9)}$$

$$\lambda_r = \frac{756}{\sqrt{F_{yf}}}$$

where L_b is the unbraced length and r_T is the radius of gyration about the weak axis for a portion of the cross section consisting of the compression flange and one third of the compressed part of the web. For a doubly symmetric girder, this will be one sixth of the web depth (see Figure 10.6).

FIGURE 10.6 ≡

If $\lambda \le \lambda_p$, failure will be by yielding, and

$$F_{cr} = F_{yf} \qquad \text{(AISC Equation A-G2-4)}$$

If $\lambda_p < \lambda \le \lambda_r$, failure will be by inelastic LTB, and

$$F_{cr} = C_b F_{yf}\left[1 - \frac{1}{2}\left(\frac{\lambda - \lambda_p}{\lambda_r - \lambda_p}\right)\right] \le F_{yf} \qquad \text{(AISC Equation A-G2-5)}$$

If $\lambda > \lambda_r$, failure will be by elastic LTB, and

$$F_{cr} = \frac{C_{PG}}{\lambda^2} \qquad \text{(AISC Equation A-G2-6)}$$

where

$$C_{PG} = 286{,}000 C_b \qquad \text{(AISC Equation A-G2-10)}$$

For compression flange buckling based on flange local buckling, the relevant width-thickness ratio and its limits are

$$\lambda = \frac{b_f}{2t_f}$$

$$\lambda_p = \frac{65}{\sqrt{F_{yf}}} \qquad \text{(AISC Equations A-G2-11, 12, and 13)}$$

$$\lambda_r = \frac{230}{\sqrt{F_{yf}/k_c}}$$

where

$$k_c = 4/\sqrt{h/t_w} \qquad (0.35 \le k_c \le 0.763)$$

If $\lambda \le \lambda_p$, failure will be by yielding, and

$$F_{cr} = F_{yf} \qquad \text{(AISC Equation A-G2-4)}$$

If $\lambda_p < \lambda \le \lambda_r$, failure will be by inelastic FLB, and

$$F_{cr} = C_b F_{yf}\left[1 - \frac{1}{2}\left(\frac{\lambda - \lambda_p}{\lambda_r - \lambda_p}\right)\right] \le F_{yf} \qquad \text{(AISC Equation A-G2-5)}$$

where $C_b = 1.0$.

If $\lambda > \lambda_r$, failure will be by elastic FLB, and

$$F_{cr} = \frac{C_{PG}}{\lambda^2} \qquad \text{(AISC Equation A-G2-6)}$$

where

$$C_{PG} = 26{,}200 k_c \qquad \text{(AISC Equation A-G2-14)}$$

10.5 ≡ SHEAR STRENGTH

The design shear strength of a plate girder is $\phi_v V_n$, where $\phi = 0.9$. The shear strength is a function of the depth-to-thickness ratio of the web and the spacing of any intermediate stiffeners that may be present. The shear capacity has two components: the strength before buckling and the post-buckling strength. The post-buckling strength relies on tension field action, which is made possible by the presence of intermediate stiffeners. If stiffeners are not present or are spaced too far apart, tension-field action will not be possible, and the shear capacity will consist only of the strength before buckling. The nominal shear strength with tension-field action is given in AISC Appendix G3 as follows:

For $\dfrac{h}{t_w} \le 187 \sqrt{\dfrac{k_v}{F_{yw}}}$,

$$V_n = 0.6 A_w F_{yw} \qquad\qquad \text{(AISC Equation A-G3-1)}$$

For $\dfrac{h}{t_w} > 187 \sqrt{\dfrac{k_v}{F_{yw}}}$,

$$V_n = 0.6 A_w F_{yw}\left(C_v + \frac{1 - C_v}{1.15\sqrt{1 + (a/h)^2}}\right) \qquad \text{(AISC Equation A-G3-2)}$$

where

$$k_v = 5 + \frac{5}{(a/h)^2}$$

$$= 5 \ \text{if} \ \frac{a}{h} > 3 \qquad\qquad \text{(AISC Equation A-G3-4)}$$

$$= 5 \ \text{if} \ \frac{a}{h} > \left[\frac{260}{(h/t_w)}\right]^2$$

The first of the above equations, AISC Equation A-G3-1, gives the shear strength when tension-field action is not present and web failure is by yielding. The second equation, AISC Equation A-G3-2, accounts for tension-field action. The first term in this equation gives the web shear buckling strength and the second term gives the post-buckling strength. The factor C_v is the ratio of the critical web buckling stress to the web shear yield stress and is defined as follows:

For $187 \sqrt{\dfrac{k_v}{F_{yw}}} \le \dfrac{h}{t_w} \le 234 \sqrt{\dfrac{k_v}{F_{yw}}}$,

$$C_v = \frac{187\sqrt{k_v/F_{yw}}}{h/t_w} \qquad\qquad \text{(AISC Equation A-G3-5)}$$

For $\dfrac{h}{t_w} > 234\sqrt{\dfrac{k_v}{F_{yw}}}$,

$$C_v = \frac{44{,}000 k_v}{(h/t_w)^2 F_{yw}}$$ (AISC Equation A-G3-6)

Solution of AISC Equations A-G3-1 and A-G3-2 is facilitated by tables given in the Numerical Values section of the Specification. Tables 9-36 and 10-36 relate the parameters of these two equations for A36 steel, and Tables 9–50 and 10-50 do the same for steels with a yield stress of 50 kips per square inch. Details of the use of these tables will be discussed in a forthcoming example.

A tension field cannot ordinarily be fully developed in an end panel. This can be understood by considering the horizontal components of the tension fields shown in Figure 10.7. (The vertical components are resisted by the stiffeners.) The tension field in panel *CD* is balanced on the left side in part by the tension field in panel *BC*. Thus, interior panels are anchored by adjacent panels. Panel *AB*, however, has no such anchorage on the left side. Although anchorage could be provided by an end stiffener specially designed to resist the bending induced by a tension field, this is not usually done. (Since the tension field does not cover the full depth of the web, the interior stiffeners are also subjected to a certain amount of bending caused by the offset of the tension fields in the adjacent panels, but this bending is not significant.) Hence, the anchorage for panel *BC* must be provided on the left side by a *beam-shear panel* rather than the tension-field panel shown in the figure. This means that the nominal shear strength is the same as for a flexural member without tension-field action — that is,

$$V_n = 0.6 A_w F_{yw} C_v$$ (AISC Equation A-G3-3)

Note that this is the first term of AISC Equation A-G3-2. Tension-field action is also not permitted for hybrid girders or when $a/h > 3$ or when $a/h > [260/(h/t_w)]^2$. (Each of these last two cases corresponds to $k_v = 5$.) Therefore, AISC Equation A-G3-3 applies in these situations.

FIGURE 10.7

AISC G4 states that intermediate stiffeners are not required when $h/t_w \leq 418/\sqrt{F_{yw}}$. This can be shown as follows. When there are no intermediate stiffeners, a/h is greater than 3, and $k_v = 5$. This results in

$$C_v = \frac{187\sqrt{\dfrac{5}{F_{yw}}}}{h/t_w} = \frac{418/\sqrt{F_{yw}}}{h/t_w}$$

and a nominal shear strength of

$$V_n = \frac{0.6A_w F_{yw}\left[\dfrac{418}{\sqrt{F_{yw}}}\right]}{h/t_w}$$

This means that when $h/t_w \leq 418/\sqrt{F_{yw}}$,

$$V_n \geq 0.6A_w F_{yw}$$

But this corresponds to the upper limit (shear yielding), so the greater-than symbol is dropped, and

$$V_n = 0.6A_w F_{yw} \qquad \text{(AISC Equation A-G3-1)}$$

which is the nominal shear strength when $h/t_w \leq 187\sqrt{k_v/F_{yw}}$.

If tension-field action is not used, the shear strength is obtained from AISC Appendix F2 as follows:

For $\dfrac{h}{t_w} \leq 187\sqrt{\dfrac{k_v}{F_{yw}}}$,

$$V_n = 0.6F_{yw} A_w \qquad \text{(AISC Equation A-F2-1)}$$

For $187\sqrt{\dfrac{k_v}{F_{yw}}} < \dfrac{h}{t_w} \leq 234\sqrt{\dfrac{k_v}{F_{yw}}}$,

$$V_n = 0.6F_{yw} A_w \frac{187\sqrt{k_v/F_{yw}}}{h/t_w} \qquad \text{(AISC Equation A-F2-2)}$$

For $\dfrac{h}{t_w} > 234\sqrt{\dfrac{k_v}{F_{yw}}}$,

$$V_n = A_w \frac{26{,}400k_v}{(h/t_w)^2} \qquad \text{(AISC Equation A-F2-3)}$$

To summarize, the nominal shear strength is determined as follows:

1. Compute the aspect ratio a/h.
2. Compute k_v and C_v.
3. For hybrid girders, $V_n = 0.6A_w F_{yw} C_v$

For end panels of nonhybrid girders, $V_n = 0.6A_w F_{yw} C_v$. For other panels of nonhybrid girders with tension-field action:

$$\text{If } \frac{h}{t_w} \leq 187\sqrt{\frac{k_v}{F_{yw}}}, \quad V_n = 0.6A_w F_{yw}$$

$$\text{If } \frac{h}{t_w} > 187\sqrt{\frac{k_v}{F_{yw}}}, \quad V_n = 0.6A_w F_{yw}\left(C_v + \frac{1 - C_v}{1.15\sqrt{1 + (a/h)^2}}\right)$$

If tension-field action is not used, the provisions of Appendix F2 will apply:

$$\text{If } \frac{h}{t_w} \leq 187\sqrt{\frac{k_v}{F_{yw}}}, \quad V_n = 0.6F_{yw} A_w$$

$$\text{If } 187\sqrt{\frac{k_v}{F_{yw}}} < \frac{h}{t_w} \leq 234\sqrt{\frac{k_v}{F_{yw}}}, \quad V_n = 0.6F_{yw} A_w \frac{187\sqrt{k_v/F_{yw}}}{h/t_w}$$

$$\text{If } \frac{h}{t_w} > 234\sqrt{\frac{k_v}{F_{yw}}}, \quad V_n = A_w \frac{26{,}400k_v}{(h/t_w)^2}$$

Intermediate Stiffeners

If intermediate stiffeners are necessary to obtain enough shear strength when tension-field action is being used, the minimum cross-sectional area of either a single stiffener or a pair is given by AISC Appendix G4 as

$$A_{st} = \frac{F_{yw}}{F_{yst}}\left[0.15Dht_w(1 - C_v)\frac{V_u}{\phi_v V_n} - 18t_w^2\right] \geq 0 \quad \text{(AISC Equation A-G4-1)}$$

where

A_{st} = total cross-sectional area of the stiffener required when tension-field action is used

F_{yst} = yield stress of the stiffener

D = a function of the configuration of the stiffener

= 1.0 for stiffeners in pairs (angles or plates)

= 1.8 for single-angle stiffeners

= 2.4 for single-plate stiffeners

The area specified by AISC Equation A-G4-1 is needed to resist the vertical component of diagonal compression in the panel. Tables 10-36 and 10-50 in the Numerical Values section of the Specification, which give the shear strength based on tension-field action, also give required stiffener areas expressed as a percentage of the web area for various values of a/h and h/t_w.

The minimum moment of inertia of the stiffener, taken about an axis in the plane of the web (or for a single stiffener, about the face of the stiffener in contact with the web), is given in AISC Appendix F2.3 as

$$I_{st} = at_w^3 j$$

where

$$j = \frac{2.5}{(a/h)^2} - 2 \geq 0.5 \qquad\qquad \text{(AISC Equation A-F2-4)}$$

Although intermediate stiffeners are not designed as compression members, a width-thickness ratio limitation for avoiding local buckling can be used as a guide in proportioning the stiffener cross section. Table B5.1 in AISC B5 does not contain a recommendation for plate girder stiffeners. The width-thickness ratio limit for outstanding legs of pairs of angles in continuous contact will be used here — that is,

$$\frac{b}{t} = \frac{95}{\sqrt{F_y}}$$

Unless they also serve as bearing stiffeners, intermediate stiffeners are not required to bear against the tension flange, so their length can be somewhat less than the girder depth h, and fabrication problems associated with close fit can be avoided. According to Appendix F2.3 of the Specification, the length should be within limits established by the distance between the weld connecting the stiffener to the web and the weld connecting the web to the tension flange. This distance, labeled c in Figure 10.8, should be between four and six times the web thickness.

Proportioning the intermediate stiffeners by the AISC rules does not require the computation of any forces, but a force must be transmitted from the stiffener to the web, and the connection should be designed for this force. Basler[59] recommends the use of a shear flow of

$$f = 0.045h \sqrt{\frac{F_y^3}{E}} \text{ kips/in.} \qquad\qquad (10.3)$$

FIGURE 10.8 ≡

The minimum intermittent fillet weld will likely be adequate.[60] The only guidance from AISC is in Appendix F2.3, which requires the clear distance between intermittent fillet welds be no more than $16t_w$ or 10 inches.

10.6 ≡ INTERACTION OF FLEXURE AND SHEAR

In general, the stress at a point in the web of a plate girder will be a combination of shear and bending, with a major principal stress larger than either component. This will usually not be a problem if there is no tension-field action, and the interaction of shear and bending may be ignored in that case. If there is a tension field, the diagonal tensile stress will reach much higher levels, and the combination of stresses must be considered. AISC requires that the interaction be accounted for when there is a tension field with shear and bending moment levels of

$$0.6\phi V_n \leq V_u \leq \phi V_n$$

and

$$0.75\phi M_n \leq M_u \leq \phi M_n$$

where

$$\phi = 0.90$$

The following interaction equation must then be satisfied:

$$\frac{M_u}{\phi M_n} + 0.625 \frac{V_u}{\phi V_n} \leq 1.375 \qquad \text{(AISC Equation A-G5-1)}$$

10.7 ≡ BEARING STIFFENERS

Bearing stiffeners are required when the web has insufficient strength for any of the limit states of web yielding, web crippling, or sidesway web buckling. These limit states are covered in Chapter K of the Specification ("Concentrated Forces, Ponding, and Fatigue"). For web yielding, the design strength of the web is ϕR_n, where $\phi = 1.0$, and when the load is at least a distance equal to the girder depth from the end,

$$R_n = (5k + N)F_{yw}t_w \qquad \text{(AISC Equation K1-2)}$$

When the load is less than this distance from the end,

$$R_n = (2.5k + N)F_{yw}t_w \qquad \text{(AISC Equation K1-3)}$$

where

k = distance from the outer face of the flange to the toe of the fillet in the web (for rolled beams) or to the toe of the weld (for welded girders)

N = length of bearing of the concentrated load, measured in the direction of the girder longitudinal axis (not less than k for an end reaction)

(This limit state has been covered in Chapter 5 of this book.)

For web crippling, the resistance factor $\phi = 0.75$, and when the load is at least *half* the girder depth from the end,

$$R_n = 135t_w^2\left[1 + 3\left(\frac{N}{d}\right)\left(\frac{t_w}{t_f}\right)^{1.5}\right]\sqrt{\frac{F_{yw}t_f}{t_w}} \qquad \text{(AISC Equation K1-4)}$$

When the load is less than this distance from the end of the girder,

$$R_n = 68t_w^2\left[1 + 3\left(\frac{N}{d}\right)\left(\frac{t_w}{t_f}\right)^{1.5}\right]\sqrt{\frac{F_{yw}t_f}{t_w}} \quad \text{for } \frac{N}{d} \leq 0.2$$

$$\text{(AISC Equation K1-5a)}$$

and

$$R_n = 68t_w^2\left[1 + \left(4\frac{N}{d} - 0.2\right)\left(\frac{t_w}{t_f}\right)^{1.5}\right]\sqrt{\frac{F_{yw}t_f}{t_w}} \quad \text{for } \frac{N}{d} > 0.2$$

$$\text{(AISC Equation K1-5b)}$$

where

d = overall depth of girder

t_f = thickness of girder flange

(This limit state has also been covered in Chapter 5 of this book.)

Bearing stiffeners are required to prevent sidesway web buckling only under a limited number of circumstances. Sidesway web buckling should be checked when the compression flange is not restrained against movement relative to the tension flange. The design strength is ϕR_n, where

$$\phi = 0.85$$

If the flange is restrained against rotation,

$$R_n = \frac{C_r t_w^3 t_f}{h^2}\left[1 + 0.4\left(\frac{h/t_w}{\ell/b_f}\right)^3\right] \qquad \text{(AISC Equation K1-6)}$$

[This equation need not be checked if $(h/t_w)/(\ell/b_f) > 2.3$.]

If the flange is not restrained against rotation,

$$R_n = \frac{C_r t_w^3 t_f}{h^2}\left[0.4\left(\frac{h/t_w}{\ell/b_f}\right)^3\right] \qquad \text{(AISC Equation K1-7)}$$

[This equation need not be checked if $(h/t_w)/(\ell/b_f) > 1.7$.]

where

C_r = 960,000 when $M_u < M_y$ at the location of the force
= 480,000 otherwise
ℓ = largest unbraced length of the flange

Although the web can be proportioned to resist directly any applied concentrated loads, bearing stiffeners are usually provided. If stiffeners are used at each concentrated load, the above limit states of web yielding, web crippling, and sidesway web buckling do not need to be checked.

The bearing strength of a stiffener is given in AISC J8 as ϕR_n, where

$$\phi = 0.75$$

and

$$R_n = 1.8 F_y A_{pb} \qquad \text{(AISC Equation J8-1)}$$

This is the same as Equation 10.1 with a bearing stress of $f_p = 1.8 F_y$.

AISC K1.9 requires that full-depth bearing stiffeners be used in pairs and analyzed as axially loaded columns subject to the following guidelines:

1. The cross section of the axially loaded member consists of the stiffener plates and a symmetrically placed length of the web (see Figure 10.4). This length can be no greater than 12 times the web thickness for an end stiffener or 25 times the web thickness for an interior stiffener.

2. The effective length should be taken as 0.75 times the actual length — that is, $KL = 0.75h$.

The provisions of AISC Chapter E apply.

AISC K1.9 also gives the following additional criteria for bearing stiffeners:

• The width-thickness ratio should satisfy the following requirement:

$$\frac{b}{t} \le \frac{95}{\sqrt{F_y}}$$

• The weld connecting the stiffener to the web should have the capacity to transfer the unbalanced shear force. Conservatively, the weld can be designed to carry the entire concentrated load.

≡ **EXAMPLE 10.1**

The plate girder shown in Figure 10.9 must be investigated for compliance with the AISC Specification. The given loads are service loads with a live-load–to–dead-load ratio of 3.0. The uniform load of 4 kips/ft includes the weight of the girder. The compression flange has lateral support at the ends and at the points of application of the concentrated loads. The compression flange is restrained

FIGURE 10.9 ≡

Detail at *A* & *E*   Detail at *B, C, D*   Typical section

against rotation at these same points. Bearing stiffeners are provided as shown at the ends and at the concentrated loads. They are clipped one inch at the inside edge, both top and bottom, to clear the flange-to-web welds. There are no intermediate stiffeners, and A36 steel is used throughout. Assume that all welds are adequate and check the following:

a. flexural strength

b. shear strength

c. flexure-shear interaction

d. bearing stiffeners

SOLUTION   Figure 10.10 shows the loading, shear, and bending moment diagrams based on *factored* loads. Verification of the values shown is left as an exercise for the reader.

The first step in the analysis is to determine whether this member satisfies the AISC definition of a plate girder:

$$\frac{h}{t_w} = \frac{63}{3/8} = 168$$

$$\frac{970}{\sqrt{F_{yf}}} = \frac{970}{\sqrt{36}} = 161.7$$

FIGURE 10.10 ≡

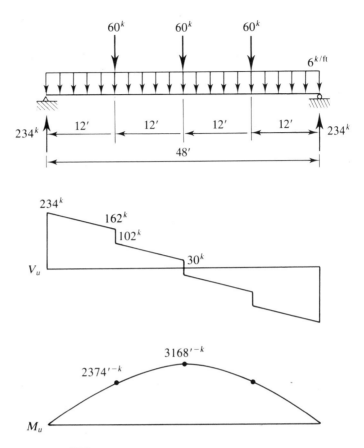

Since $h/t_w > 970/\sqrt{F_{yf}}$, this flexural member is a plate girder, and the provisions of AISC Appendix G apply.

The web must satisfy the slenderness limitation of AISC G1. The limiting value of h/t_w will depend upon the aspect ratio a/h. For this plate girder, the bearing stiffeners will serve as intermediate stiffeners, and

$$\frac{a}{h} \approx \frac{12(12)}{63} = 2.286$$

This is greater than 1.5, so AISC Equation A-G1-2 governs:

$$\frac{14{,}000}{\sqrt{F_{yf}(F_{yf} + 16.5)}} = \frac{14{,}000}{\sqrt{36(36 + 16.5)}} = 322 > \frac{h}{t_w} = 168 \quad \text{(OK)}$$

a. Flexural strength:

The flexural strength will be limited by the strength of either the tension flange or the compression flange. In either case, the elastic section modulus will be needed. Because of symmetry,

$$S_{xt} = S_{xc} = S_x$$

TABLE 10.1 ≡

Component	A	$\bar{I}$	d	$\bar{I} + Ad^2$
Web	—	7814	—	7,814
Flange	16	—	32	16,380
Flange	16	—	32	16,380
				40,574

The computations for I_x, the moment of inertia about the strong axis, are summarized in Table 10.1. The section modulus is

$$S_x = \frac{I_x}{c} = \frac{40{,}570}{32.5} = 1248 \text{ in.}^3$$

The hybrid girder factor R_e will be needed for both the tension flange strength and the compression flange strength. Since this girder is nonhybrid,

$$R_e = 1.0$$

From AISC Equation A-G2-1, the tension flange strength, based on yielding, is

$$\begin{aligned} M_n = S_{xt}\, R_e\, F_{yt} &= 1248(1.0)(36) = 44{,}930 \text{ in.-kips} \\ &= 3744 \text{ ft-kips} \end{aligned}$$

The compression flange buckling strength is given by AISC Equation A-G2-2:

$$M_n = S_{xc}\, R_{PG}\, R_e\, F_{cr}$$

where the critical buckling stress F_{cr} is based on either lateral-torsional buckling or flange local buckling. To check lateral-torsional buckling, the radius of gyration r_T is needed. Referring to Figure 10.11,

$$I_y = \frac{1}{12}(1)(16)^3 + \frac{1}{12}(10.5)(3/8)^3 = 341.4 \text{ in.}^4$$

$$A = 16(1.0) + 10.5(3/8) = 19.94 \text{ in.}^2$$

$$r_T = \sqrt{\frac{I_y}{A}} = \sqrt{\frac{341.4}{19.94}} = 4.138 \text{ in.}$$

The unbraced length of the compression flange is 12 feet, and the slenderness parameters for lateral-torsional buckling are

$$\lambda = \frac{L_b}{r_T} = \frac{12(12)}{4.138} = 34.80$$

$$\lambda_p = \frac{300}{\sqrt{F_{yf}}} = \frac{300}{\sqrt{36}} = 50$$

FIGURE 10.11 ≡

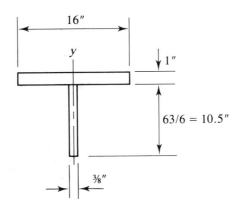

Since $\lambda < \lambda_p$, $F_{cr} = F_{yf} = 36$ ksi.

The critical buckling stress F_{cr} will now be computed for flange local buckling. The relevant slenderness parameters are

$$\lambda = \frac{b_f}{2t_f} = \frac{16}{2(1.0)} = 8$$

$$\lambda_p = \frac{65}{\sqrt{F_{yf}}} = \frac{65}{\sqrt{36}} = 10.83$$

Once again, $\lambda < \lambda_p$ and $F_{cr} = F_{yf} = 36$ ksi.

To compute the plate girder strength reduction factor R_{PG}, the value of a_r will be needed:

$$a_r = \frac{A_w}{A_f} = \frac{63(3/8)}{16(1.0)} = \frac{23.62}{16} = 1.477 < 10$$

From AISC Equation A-G2-3,

$$R_{PG} = 1 - \frac{a_r}{1200 + 300a_r}\left(\frac{h_c}{t_w} - \frac{970}{\sqrt{F_{cr}}}\right) \le 1.0$$

$$= 1 - \frac{1.477}{1200 + 300(1.477)}\left(168 - \frac{970}{\sqrt{36}}\right) = 0.9943$$

This is almost 1.0 because this flexural member barely qualifies as a plate girder and is very close to being classified as a beam. From AISC Equation A-G2-2, the nominal flexural strength for the compression flange is

$$M_n = S_{xc} R_{PG} R_e F_{cr} = 1248(0.9943)(1.0)(36)$$
$$= 44,670 \text{ in.-kips} = 3723 \text{ ft-kips}$$

This is slightly smaller than the nominal strength corresponding to the tension flange and therefore controls. The design strength is

$$\phi_b M_n = 0.90(3723) = 3350 \text{ ft-kips}$$

From Figure 10.10, the maximum factored load moment is

$$M_u = 3168 \text{ ft-kips} < 3350 \text{ ft-kips} \quad (OK)$$

Answer **a.** The flexural strength is adequate.

b. Shear strength:

The shear strength is a function of the web slenderness ratio h/t_w and the aspect ratio a/h. We will first determine whether tension-field action can be used in regions other than the end panels. This will be the case when a/h is less than 3.0 and less than

$$\left[\frac{260}{(h/t_w)}\right]^2 = \left[\frac{260}{(168)}\right]^2 = 2.395$$

The approximate value of a/h is 2.286; therefore, tension-field action can be used. Determine k_v and C_v. From AISC Equation A-G3-4,

$$k_v = 5 + \frac{5}{(a/h)^2} = 5 + \frac{5}{(2.286)^2} = 5.957$$

Establish the range of h/t_w:

$$187\sqrt{\frac{k_v}{F_{yw}}} = 187\sqrt{\frac{5.957}{36}} = 76.07$$

$$234\sqrt{\frac{k_v}{F_{yw}}} = 234\sqrt{\frac{5.957}{36}} = 95.19$$

Since $h/t_w = 168 > 234\sqrt{k_v/F_{yw}}$, C_v is found from AISC Equation A-G3-6:

$$C_v = \frac{44,000k_v}{(h/t_w)^2 F_{yw}} = \frac{44,000(5.957)}{(168)^2(36)} = 0.2580$$

Since $h/t_w > 187\sqrt{k_v/F_{yw}}$, AISC Equation A-G3-2, which accounts for tension-field action, will be used to determine the nominal shear strength (except for the end panels):

$$V_n = 0.6A_w F_{yw}\left(C_v + \frac{1 - C_v}{1.15\sqrt{1 + (a/h)^2}}\right)$$

$$= 0.6(23.62)(36)\left[0.2580 + \frac{1 - 0.2580}{1.15\sqrt{1 + (2.286)^2}}\right] = 263.6 \text{ kips}$$

The design shear strength is

$$\phi_v V_n = 0.90(263.6) = 237 \text{ kips}$$

From Figure 10.10, the maximum factored load shear in the middle half of the girder is 102 kips, so the shear strength is adequate where tension-field action is permitted.

For the end panels, tension-field action is not permitted, and the shear strength must be computed from AISC Equation A-G3-3:

$$V_n = 0.6A_w F_{yw} C_v = 0.6(23.62)(36)(0.2580) = 131.6 \text{ kips}$$

The design strength is

$$\phi_v V_n = 0.90(131.6) = 118 \text{ kips}$$

The maximum factored load shear in the end panel is

$$V_u = 234 \text{ kips} > 118 \text{ kips} \quad \text{(N.G.)}$$

Two options are available for increasing the shear strength: Either decrease the web slenderness (probably by increasing its thickness) or decrease the aspect ratio of each end panel by adding an intermediate stiffener. Stiffeners will be added in this example. (Decreasing the web slenderness enough to gain sufficient shear strength would bring this flexural member into the beam category, changing the analysis completely.)

The location of the first intermediate stiffener will be determined by the following strategy: First, equate the shear strength from AISC Equation A-G3-3 to the required shear strength and solve for the required value of C_v. Next, solve for k_v from Equation A-G3-6, then solve for a/h from Equation A-G3-4.

$$\phi_v V_n = \phi_v(0.6A_w F_{yw} C_v) \qquad \text{(AISC Equation A-G3-3)}$$

$$C_v = \frac{\phi_v V_n}{\phi_v(0.6A_w F_{yw})} = \frac{234}{0.90(0.6)(23.62)(36)} = 0.5096$$

$$C_v = \frac{44{,}000 k_v}{(h/t_w)^2 F_{yw}} \qquad \text{(AISC Equation A-G3-6)}$$

$$k_v = \frac{C_v(h/t_w)^2 F_{yw}}{44{,}000} = \frac{0.5096(168)^2(36)}{44{,}000} = 11.77$$

$$k_v = 5 + \frac{5}{(a/h)^2} \qquad \text{(AISC Equation A-G3-4)}$$

$$\frac{a}{h} = \sqrt{\frac{5}{k_v - 5}} = \sqrt{\frac{5}{11.77 - 5}} = 0.8594$$

The required stiffener spacing is

$$a = 0.8594h = 0.8594(63) = 54.1 \text{ in.}$$

Although a is defined as the clear spacing, we will treat it conservatively as a center-to-center spacing and place the first intermediate stiffener at 54 inches from the end of the girder. This will give a design strength approximately equal to the maximum factored load shear of 234 kips. No additional stiffeners will be needed, since the factored load shear outside of the end panels is less than the design strength of 237 kips.

The determination of the stiffener spacing can be facilitated by the use of the tables in the Numerical Values section of the Specification; this technique will be illustrated in Example 10.2.

Answer   **b.** The shear strength is inadequate. Add one intermediate stiffener 54 inches from each end of the girder.

c. Flexure–shear interaction:

Flexure–shear interaction must be checked when there is a tension field (that is, outside of the end panels) and when both of the following conditions are satisfied:

1. The factored load shear is in the range

$$0.6\phi V_n \le V_u \le \phi V_n$$

Outside of the end panels, the range is

$$0.6(237) \le V_u \le 237$$
$$142 \le V_u \le 237$$

2. The factored load bending moment is in the range

$$0.75\phi M_n \le M_u \le \phi M_n$$

For this girder,

$$0.75(3350) \le M_u \le 3350$$
$$2510 \le M_u \le 3350$$

An examination of Figure 10.10 shows that there is no combination of shear and bending moment that satisfies these conditions.

Answer   **c.** Interaction of flexure and shear does not need to be checked.

d. Bearing stiffeners:

Since bearing stiffeners are provided at each concentrated load, there is no need to check the provisions of AISC Chapter K for web yielding, web crippling, or sidesway web buckling. For both the interior and support bearing stiffeners,

$$\frac{b}{t} = \frac{7.5}{0.75} = 10.0$$

$$\frac{95}{\sqrt{F_y}} = \frac{95}{\sqrt{36}} = 15.8 > 10.0 \quad \text{(OK)}$$

Interior bearing stiffeners:

Compute the bearing strength. From Figure 10.12,

$$A_{pb} = 2at = 2(7.5 - 1)(0.75) = 9.750 \text{ in.}^2$$

FIGURE 10.12 ≡

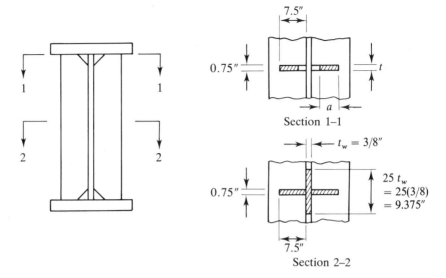

Section 1–1

Section 2–2

From AISC Equation J8-1,

$$R_n = 1.8 F_y A_{pb} = 1.8(36)(9.750) = 631.8 \text{ kips}$$
$$\phi R_n = 0.75(631.8) = 474 \text{ kips} > 60 \text{ kips} \quad \text{(OK)}$$

Check the strength of the stiffener as a compression member. Referring to Figure 10.12, we can use a length of web equal to 9.375 inches, resulting in a cross-sectional area for the "column" of

$$A = 2(0.75)(7.5) + (3/8)(9.375) = 14.77 \text{ in.}^2$$

The moment of inertia of this area about an axis in the web is

$$I = \sum (\bar{I} + A d^2)$$
$$= \frac{9.375(3/8)^3}{12} + 2\left[\frac{0.75(7.5)^3}{12} + 7.5(0.75)\left(\frac{7.5}{2} + \frac{3/8}{2} \right)^2 \right] = 227.2 \text{ in.}^4$$

and the radius of gyration is

$$r = \sqrt{\frac{I}{A}} = \sqrt{\frac{227.2}{14.77}} = 3.922 \text{ in.}$$

The slenderness ratio is

$$\frac{KL}{r} = \frac{Kh}{r} = \frac{0.75(63)}{3.922} = 12.05$$

From AISC Table 3-36 in the Numerical Values section of the Specification,

$$\phi_c F_{cr} = 30.37 \text{ ksi}$$

Section 1–1

Section 2–2

and

$$\phi_c P_n = \phi_c F_{cr} A = 30.37(14.77) = 449 \text{ kips} > 60 \text{ kips} \quad (\text{OK})$$

Bearing stiffeners at the supports:

From Figure 10.13, the design bearing strength is

$$\phi R_n = \phi(1.8 F_y A_{pb}) = 0.75(1.8)(36)[4(6.5)(0.75)]$$
$$= 948 \text{ kips} > 234 \text{ kips} \quad (\text{OK})$$

Check the stiffener–web assembly as a compression member. Referring to Figure 10.13, the moment of inertia about an axis in the plane of the web is

$$I = \sum (\bar{I} + Ad^2)$$

$$= \frac{4.5(3/8)^3}{12} + 4\left[\frac{0.75(7.5)^3}{12} + 7.5(0.75)\left(\frac{7.5}{2} + \frac{3/8}{2}\right)^2\right] = 454.3 \text{ in.}^4$$

and the area and radius of gyration are

$$A = 4.5(3/8) + 4(0.75)(7.5) = 24.19 \text{ in.}^2$$

and

$$r = \sqrt{\frac{I}{A}} = \sqrt{\frac{454.3}{24.19}} = 4.334 \text{ in.}$$

The slenderness ratio is

$$\frac{Kh}{r} = \frac{0.75(63)}{4.334} = 10.90$$

From AISC Table 3-36, $\phi_c F_{cr} = 30.41$ kips. The design strength is

$$\phi_c P_n = \phi_c F_{cr} A = 30.41(24.19) = 736 \text{ kips} > 234 \text{ kips} \quad \text{(OK)}$$

Answer   **d.** The bearing stiffeners are adequate.

10.8 ≡ DESIGN

The primary task in plate girder design is to determine the size of the web and the flanges. If a variable moment of inertia is desired, decisions must be made regarding the method of varying the flange size — that is, whether to use cover plates or different thicknesses of flange plate at different points along the length of the girder. A decision on whether to use intermediate stiffeners must be made early in the process, since this will affect the web thickness. If bearing stiffeners are needed, they must be designed. Finally, the various components must be connected by properly designed welds. The following step-by-step procedure is recommended:

1. **Select the overall depth.** As a rule of thumb, a well-proportioned girder will have a depth of one tenth to one twelfth the span length. As with any beam design, constraints on the maximum depth could establish the depth by default. Building code limitations on the depth-to-span ratio or the deflection would also influence the selection.

2. **Select a trial web size.** The web depth can be estimated by subtracting twice the flange thickness from the overall depth selected. Of course, at this stage of the design, the flange thickness must also be estimated, but the consequences of a poor estimate are minor. The web thickness t_w can then be found by using the following limitations as a guide:
   For $a/h \leq 1.5$,

   $$\frac{h}{t_w} \leq \frac{2000}{\sqrt{F_{yf}}} \qquad \text{(AISC Equation A-G1-1)}$$

   For $a/h > 1.5$,

   $$\frac{h}{t_w} \leq \frac{14{,}000}{\sqrt{F_{yf}(F_{yf} + 16.5)}} \qquad \text{(AISC Equation A-G1-2)}$$

3. **Estimate the flange size.** The required flange area can be estimated from a simple formula derived as follows:

   Let $I_x = I_{web} + I_{flanges}$

   $$\approx \frac{1}{12} t_w h^3 + 2A_f y^2 \approx \frac{1}{12} t_w h^3 + 2A_f(h/2)^2 \qquad (10.4)$$

where

A_f = cross-sectional area of one flange

y = distance from the elastic neutral axis to the centroid of the flange

The contribution of the moment of inertia of each flange about its own centroidal axis has been neglected in Equation 10.4. The section modulus can be estimated as

$$S_x = \frac{I_x}{c} \approx \frac{t_w h^3/12}{h/2} + \frac{2A_f(h/2)^2}{h/2} = \frac{t_w h^2}{6} + A_f h$$

If we assume that compression flange buckling will control the design, the required section modulus can be found from AISC Equation A-G2-2):

$$M_n = S_{xc} R_{PG} R_e F_{cr}$$

$$S_{xc} = \frac{M_n}{R_{PG} R_e F_{cr}} = \frac{M_u/\phi_b}{R_{PG} R_e F_{cr}}$$

where M_u is the maximum factored load bending moment. Equating the required section modulus to the approximate value, we have

$$\frac{M_u/\phi_b}{R_{PG} R_e F_{cr}} = \frac{t_w h^2}{6} + A_f h$$

and

$$A_f = \frac{M_u}{\phi_b h R_{PG} R_e F_{cr}} - \frac{t_w h}{6}$$

Assuming $R_{PG} = 1.0$, $R_e = 1.0$, and $F_{cr} = F_y$, the required area of one flange is

$$A_f = \frac{M_u}{0.90 h F_y} - \frac{A_w}{6} \tag{10.5}$$

where A_w is the web area. Once the required flange area is known, select the width and thickness. If the thickness used in the estimate of the web depth is retained, no adjustment in the web depth will be needed. At this point, an estimated girder weight can be computed, and M_u and A_f should be recomputed.

4. **Check the bending strength of the trial section.**

5. **Check shear.** If an end panel is being considered, or if intermediate stiffeners are not used, the shear strength can be found from AISC Equation A-G3-3, which gives the strength in the absence of a tension field. Table 9-36 or 9-50 in the Numerical Values section of the Specification can be

used for this purpose. If the tables are not used, the intermediate stiffener spacing can be determined as follows:

a. Equate the required shear strength to the shear strength given by AISC Equation A-G3-3 and solve for the required value of C_v.

b. Solve for k_v from AISC Equation A-G3-6.

c. Solve for the required value of a/h from AISC Equation A-G3-4.

If tension-field action is used, either a trial-and-error approach or AISC Table 10-36 or 10-50 can be used to obtain the required a/h. (The required cross-sectional area of the stiffeners, expressed as a percentage of the web area, is also given in these tables for certain values of h/t_w and a/h.) Establish a trial stiffener size that will satisfy the area requirement, and check the moment of inertia requirement of AISC Appendix F2.3.

6. Check for interaction of shear and bending.

7. Check the web resistance to any applied concentrated loads (web yielding, web crippling, and web sidesway buckling). If bearing stiffeners are required, the following design procedure is recommended:

a. Try a width that brings the edge of the stiffener near the edge of the flange and a thickness that satisfies the width-thickness requirement

$$\frac{b}{t} \le \frac{95}{\sqrt{F_y}}.$$

b. Compute the cross-sectional area needed for bearing strength. Compare this with the trial area and revise if necessary.

c. Check the stiffener–web assembly as a compression member.

8. Design the flange-to-web welds, stiffener-to-web welds, and any other connections (flange segments, web splices, etc.).

EXAMPLE 10.2

Design a simply supported plate girder to span 60 feet and support the service loads shown in Figure 10.14a. The maximum permissible depth is 65 inches. Use A36 steel and E70XX electrodes, and assume that the girder has continuous lateral support. The ends have bearing-type supports and are not framed.

SOLUTION

The factored loads, excluding the girder weight, are shown in Figure 10.14b. Determine the overall depth:

$$\frac{\text{span length}}{10} = \frac{60(12)}{10} = 72 \text{ in.}$$

$$\frac{\text{span length}}{12} = \frac{60(12)}{12} = 60 \text{ in.}$$

FIGURE 10.14 ≡

(a) Service Loads

(b) Factored Loads

(girder weight not included)

Use the maximum permissible depth of 65 inches.

Try a flange thickness $t_f = 1.5$ inches and a web depth of

$$h = 65 - 2(1.5) = 62 \text{ in.}$$

To determine the web thickness, first examine the limiting values of h/t_w. For this flexural member to qualify as a plate girder,

$$\frac{h}{t_w} \geq \frac{970}{\sqrt{F_{yf}}} = \frac{970}{\sqrt{36}} = 161.7$$

$$t_w \leq \frac{h}{161.7} = \frac{62}{161.7} = 0.383 \text{ in.}$$

From AISC Equations A-G1-1 and A-G1-2:

For $a/h \leq 1.5$,

$$\frac{h}{t_w} \leq \frac{2000}{\sqrt{F_{yf}}} = \frac{2000}{\sqrt{36}} = 333.3$$

$$t_w \geq \frac{62}{333.3} = 0.186 \text{ in.}$$

For $a/h > 1.5$,

$$\frac{h}{t_w} \le \frac{14{,}000}{\sqrt{F_{yf}(F_{yf} + 16.5)}} = \frac{14{,}000}{\sqrt{36(36 + 16.5)}} = 322.0$$

$$t_w \ge \frac{62}{322.0} = 0.192 \text{ in.}$$

Try a $\frac{1}{4} \times 62$ web plate.

Determine the required flange size. Referring to Figure 10.14b, the maximum factored load bending moment is

$$M_u = \frac{186.4(60)}{4} + \frac{4.040(60)^2}{8} = 4614 \text{ ft-kips}$$

From Equation 10.5, the required flange area is

$$A_f = \frac{M_u}{0.90hF_y} - \frac{A_w}{6}$$

$$= \frac{4614(12)}{0.90(62)(36)} - \frac{62(1/4)}{6} = 24.98 \text{ in.}^2$$

The girder weight can now be estimated:

Web area: $62(1/4) = 15.5$ in.^{2}
Flange area: $2(24.98) = 49.96$ in.^{2}
Total: $\overline{65.46 \text{ in.}^2}$

$$\text{weight} = \frac{65.46}{144}(490) = 222.7 \text{ lb/ft}   \quad \text{say } 250 \text{ lb/ft}$$

The adjusted bending moment is

$$M_u = 4614 + \frac{(1.2 \times 0.250)(60)^2}{8} = 4749 \text{ ft-kips}$$

and the required flange area is

$$A_f = \frac{4749(12)}{0.90(62)(36)} - \frac{62(1/4)}{6} = 25.79 \text{ in.}^2$$

If the flange thickness estimate of 1.5 inches is retained, the required width will be

$$b_f = \frac{A_f}{t_f} = \frac{25.79}{1.5} = 17.2 \text{ in.}$$

Try a $1\frac{1}{2} \times 18$ flange plate.

Figure 10.15 shows the trial section, and Figure 10.16 shows the shear and bending moment diagrams for the factored loads, which include an approximate girder weight of 250 lb/ft.

FIGURE 10.15 ≡

FIGURE 10.16 ≡

Check the flexural strength of the trial section. Referring to Figure 10-15, the moment of inertia about the axis of bending is

$$I_x = \frac{(1/4)(62)^3}{12} + 2(1.5)(18)(31.75)^2 = 59,400 \text{ in.}^4$$

and the elastic section modulus is

$$S_x = \frac{I_x}{c} = \frac{59,400}{32.5} = 1828 \text{ in.}^3$$

An examination of AISC Equations A-G2-1 and A-G2-2 shows that for a non-hybrid girder with a symmetric cross section, the flexural strength will never be controlled by tension flange yielding; therefore, only compression flange buckling will be investigated. Furthermore, since this girder has continuous lateral support, lateral-torsional buckling need not be considered. For the limit state of flange local buckling,

$$\lambda = \frac{b_f}{2t_f} = \frac{18}{2(1.5)} = 6$$

$$\lambda_p = \frac{65}{\sqrt{F_{yf}}} = \frac{65}{\sqrt{36}} = 10.83$$

Since $\lambda < \lambda_p$,

$$F_{cr} = F_{yf} = 36 \text{ ksi}$$

The following values are needed for computing the strength reduction factor R_{PG}:

$$A_w = 62\left(\frac{1}{4}\right) = 15.5 \text{ in.}^2$$

$$A_f = 18(1.5) = 27 \text{ in.}^2$$

$$a_r = \frac{A_w}{A_f} = \frac{15.5}{27} = 0.5741 < 10$$

$$\frac{h}{t_w} = \frac{62}{1/4} = 248$$

From AISC Equation A-G2-3,

$$R_{PG} = 1 - \frac{a_r}{1200 + 300a_r}\left(\frac{h}{t_w} - \frac{970}{\sqrt{F_{cr}}}\right)$$

$$= 1 - \frac{0.5741}{1200 + 300(0.5741)}\left(248 - \frac{970}{\sqrt{36}}\right) = 0.9639$$

From AISC Equation A-G2-2, the nominal flexural strength is

$$M_n = S_{xc} R_{PG} R_e F_{cr}$$
$$= 1828(0.9639)(1.0)(36) = 63,430 \text{ in.-kips} = 5286 \text{ ft-kips}$$

and the design strength is

$$\phi_b M_n = 0.90(5286) = 4757 \text{ ft-kips} > 4749 \text{ ft-kips} \quad (\text{OK})$$

Check the shear strength. The shear is maximum at the support, but tension-field action cannot be used in an end panel. Table 9-36 in the Numerical Values section of the Specification will be used to obtain the required size of the end panel. The table will be entered with $h/t_w = 248$ and

$$\frac{\phi_v V_n}{A_w} = \frac{223.4}{15.5} = 14.41 \text{ ksi}$$

This would require an a/h of less than 0.5 and is out of the range of the table, indicating that a thicker web is needed.

Try a $\frac{5}{16} \times 62$ web:

$$\frac{h}{t_w} = \frac{62}{5/16} = 198.4$$

$$A_w = 62\left(\frac{5}{16}\right) = 19.38 \text{ in.}^2$$

$$\frac{\phi_v V_n}{A_w} = \frac{223.4}{19.38} = 11.5 \text{ ksi}$$

From AISC Table 9-36, for $h/t_w = 198$ and $a/h = 0.6$, the interpolated value of $\phi_v V_n/A_w$ is 11.5 ksi. Therefore, use $a/h = 0.6$ and

$$a = 0.6h = 0.6(62) = 37.2 \text{ in.}$$

Although the required distance a is a clear spacing, the use of center-to-center distances is somewhat simpler and will be slightly conservative. Use a distance of 36 inches from the center of the end bearing stiffener to the center of the first intermediate stiffener.

Before proceeding with the shear strength analysis, determine the effect of the change in web thickness on what has already been done.

Girder weight:

Web area:	62(5/16)	= 19.38 in.²
Flange area:	2(1.5)(18)	= 54.00 in.²
Total:		73.38 in.²

$$\text{weight} = \frac{73.38}{144}(490) = 250 \text{ lb/ft} \quad (\text{same as previous assumption})$$

$$I_x = \frac{(5/16)(62)^3}{12} + 2(1.5)(18)(31.75)^2 = 60{,}640 \text{ in.}^4$$

$$S_x = \frac{I_x}{c} = \frac{60{,}640}{32.5} = 1866 \text{ in.}^3$$

$$a_r = \frac{A_w}{A_f} = \frac{19.38}{27} = 0.7178 < 10$$

From AISC Equation A-G2-3,

$$R_{PG} = 1 - \frac{a_r}{1200 + 300a_r}\left(\frac{h_c}{t_w} - \frac{970}{\sqrt{F_{cr}}}\right)$$

$$= 1 - \frac{0.7178}{1200 + 300(0.7178)}\left(198.4 - \frac{970}{\sqrt{36}}\right) = 0.9814$$

The nominal flexural strength is

$$M_n = S_{xc} R_{PG} R_e F_{cr}$$
$$= (1866)(0.9814)(1.0)(36)$$
$$= 65,930 \text{ in.-kips} = 5494 \text{ ft-kips}$$

The design strength is

$$\phi_b M_n = 0.90(5494) = 4944 \text{ ft-kips}$$

Although this is somewhat more capacity than needed, it will compensate for the weight of the stiffeners and other incidentals that we have not accounted for.

Answer   Use a $\frac{5}{16} \times 62$ web and $1\frac{1}{2} \times 18$ flanges (Figure 10.17).

FIGURE 10.17 ≡

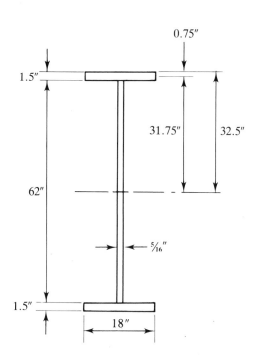

Determine the intermediate stiffener spacings needed for shear strength outside the end panels. At a distance of 36 inches from the left end, the shear is

$$V_u = 223.4 - 4.34\left(\frac{36}{12}\right) = 210.4 \text{ kips}$$

$$\frac{\phi_v V_n}{A_w} = \frac{210.4}{19.38} = 10.86 \text{ ksi}$$

Since tension-field action can be used outside the end panels, AISC Table 10-36 will be used. For $h/t_w = 200$ and $a/h = 1.6$,

$$\frac{\phi_v V_n}{A_w} = 11.2 \text{ ksi} > 10.86 \text{ ksi} \quad \text{(OK)}$$

Use $a/h = 1.6$ and

$$a = 1.6h = 1.6(62) = 99.2 \text{ in.}$$

Note that no values are given in the table when $h/t_w = 200$ and $a/h > 1.6$. This is because tension-field action is not permitted when

$$\frac{a}{h} > \left[\frac{260}{(h/t_w)}\right]^2 = \left(\frac{260}{200}\right)^2 = 1.69$$

For this reason, a maximum stiffener spacing of 99.2 inches will apply for the remainder of the girder, but to achieve a uniform spacing between the end panels, a center-to-center spacing of 81 inches, as shown in Figure 10.18, will be used.

With this reduced spacing,

$$\frac{a}{h} = \frac{81}{62} = 1.306$$

To compute the corresponding shear strength, enter Table 10-36 with $a/h = 1.3$ and $h/t_w = 200$. By interpolation,

$$\frac{\phi_v V_n}{A_w} = 12.6$$

FIGURE 10.18 ≡

36"   4 spaces @ 81"

and

$$\phi_v V_n = 12.6 A_w = 12.6(19.38) = 244.2 \text{ kips}$$

Size of intermediate stiffeners:

The cross section of the intermediate stiffeners is based on three criteria: (1) a minimum area, (2) a minimum moment of inertia, and (3) a maximum width-thickness ratio. The required area of a pair of stiffeners is a function of h/t_w and a/h, and it can be found from AISC Table 10-36 by interpolation.

For $a/h = 1.3$ and $h/t_w = 200$,

$$A_{st} = 2.3\% \text{ of the web area}$$
$$= 0.023(19.38) = 0.446 \text{ in.}^2$$

From AISC Equation A-F2-4,

$$j = \frac{2.5}{(a/h)^2} - 2 = \frac{2.5}{(1.306)^2} - 2 = -0.534 < 0.5 \quad \therefore \quad \text{use } j = 0.5$$

The required moment of inertia is

$$I_{st} = a t_w^3 j = 81(5/16)^3(0.5) = 1.24 \text{ in.}^4$$

Use a maximum value of b/t of

$$\frac{95}{\sqrt{F_y}} = \frac{95}{\sqrt{36}} = 15.8$$

Try two $\frac{1}{4} \times 4$ plates.

$$\frac{b}{t} = \frac{4}{1/4} = 16 \approx 15.8 \quad \text{(say OK)}$$

$$A_{st} \text{ provided} = 2(4)\left(\frac{1}{4}\right) = 2.0 \text{ in.}^2 > 0.446 \text{ in.}^2 \quad \text{(OK)}$$

From Figure 10.19 and the parallel axis theorem,

$$I_{st} = \sum (\bar{I} \times A d^2)$$
$$= \left[\frac{0.25(4)^3}{12} + 0.25(4)(2 + 5/32)^2\right] \times 2 \text{ stiffeners}$$
$$= 11.97 \text{ in.}^4 > 1.24 \text{ in.}^4 \quad \text{(OK)}$$

To determine the length of the stiffeners, first compute the distance between the stiffener-to-web weld and the web-to-flange weld (Figure 10.8):

$$\text{Minimum distance} = 4t_w = 4\left(\frac{5}{16}\right) = 1.25 \text{ in.}$$

$$\text{Maximum distance} = 6t_w = 6\left(\frac{5}{16}\right) = 1.875 \text{ in.}$$

FIGURE 10.19 ≡

Assuming a flange-to-web weld size of $\frac{5}{16}$ inch and 1.25 inches between welds, the approximate length of the stiffener can be taken as

$$h - \text{weld size} - 1.25 = 62 - 0.3125 - 1.25$$
$$= 60.44 \text{ in.;} \quad \text{say 60 in.}$$

Answer   Use two plates $\frac{1}{4}$ in. × 4 in. × 5 ft 0 in. for the intermediate stiffeners.

Check for interaction of flexure and shear. This must be done only where there is a tension field. The range of shear to be investigated is

$$0.6\phi V_n \le V_u \le \phi V_n$$

or

$$0.6(244.2) \le V_u \le 244.2$$
$$146.5 \le V_u \le 244.2$$

The range of bending moment is

$$0.75\phi M_n \le M_u \le \phi M_n$$

or

$$0.75(4944) \le M_u \le 4944$$
$$3708 \le M_u \le 4944$$

From Figure 10.16, $V_u = 146.5$ kips when

$$223.4 - 4.34x = 146.5 \text{ kips}$$

where

$$x = \text{distance from left end of girder} = 17.72 \text{ ft}$$

At this same location, the bending moment is

$$M_u = 223.4(17.72) - \frac{4.34(17.72)^2}{2} = 3277 \text{ ft-kips}$$

In the region where the shear is greater than 146.5 kips, the moment is less than 3708 ft-kips; therefore, interaction of flexure and shear need not be considered.

Bearing stiffeners:

Bearing stiffeners will be provided at the supports and at midspan. Since there will be a stiffener at each concentrated load, there is no need to investigate the resistance of the web to these loads. If the stiffeners were not provided, the web would need to be protected from yielding and crippling by providing enough bearing length N as required by AISC Equations K1-2 through K1-5. Sideway web buckling would not be an applicable limit state, since the girder has continuous lateral support. (This makes the unbraced length $\ell = 0$ and $(h/t_w)/(\ell/b_f) > 2.3$.)

Try a stiffener width b of 8 inches. This will give a total combined width of $2(8) + \frac{5}{16}$ (the web thickness) = 16.31 inches, slightly less than the flange width of 18 inches.

From AISC K1.9,

$$\frac{b}{t} \leq \frac{95}{\sqrt{F_y}}, \quad \text{or}$$

$$t \geq \frac{b\sqrt{F_y}}{95} = \frac{8\sqrt{36}}{95} = 0.505 \text{ in.}$$

Try two $\frac{3}{4} \times 8$ stiffeners. Assume a $\frac{5}{16}$-in. web-to-flange weld and a $\frac{1}{2}$-in. cutout in the stiffener.

Check the stiffener at the support. The bearing strength is

$$\phi R_n = 0.75(1.8 F_y A_{pb})$$
$$= 0.75(1.8)(36)(0.75)(8 - 0.5) \times 2$$
$$= 547 \text{ kips} > 223.4 \text{ kips} \quad \text{(OK)}$$

Check the stiffener as a column:

The length of web acting with the stiffener plates to form a compression member is 12 times the web thickness for an end stiffener (AISC K1.9). As shown in Figure 10.20, this length is $12(\frac{5}{16}) = 3.75$ in. Since the stiffener should be centrally located within this length, the point of support (location of the girder reaction) must be approximately $\frac{3.75}{2} = 1.875$ in. from the end of the

FIGURE 10.20 ≡

girder. Use 3 in. as shown in Figure 10.21, but base the computations on a total length of web of 3.75 in.

$$A = 2(8)\left(\frac{3}{4}\right) + \left(\frac{5}{16}\right)(3.75) = 13.17 \text{ in.}^2$$

$$I = \frac{3.75(5/16)^3}{12} + 2\left[\frac{0.75(8)^3}{12} + 8\left(\frac{3}{4}\right)\left(4 + \frac{5}{32}\right)^2\right] = 271.3 \text{ in.}^4$$

$$r = \sqrt{\frac{I}{A}} = \sqrt{\frac{271.3}{13.17}} = 4.539 \text{ in.}$$

$$\frac{KL}{r} = \frac{Kh}{r} = \frac{0.75(62)}{4.539} = 10.24$$

From AISC Table 3-36, $\phi_c F_{cr} = 30.43$ ksi. The design strength is

$$\phi_c P_n = \phi_c F_{cr} A = 30.43(13.17) = 401 \text{ kips} > 223.4 \text{ kips} \quad \text{(OK)}$$

Since the load at midspan is smaller than the reaction, use the same stiffener at midspan.

Answer   Use two plates $\frac{3}{4} \times 8$ for the bearing stiffeners.

At this point, all components of the girder have been sized. The connections of these elements will now be designed. E70 electrodes, with a design strength of $\phi F_w = 31.5$ ksi, will be used for all welds.

Flange-to-web welds:

Compute the horizontal shear flow at the flange-to-web junction:

Maximum $V_u = 223.4$ kips

$Q =$ flange area × 31.75   (see Figure 10.15)

   $= 1.5(18)(31.75) = 857.2 \text{ in.}^3$

$I_x = 60,640 \text{ in.}^4$

Maximum $\dfrac{V_u Q}{I_x} = \dfrac{223.4(857.2)}{60,640} = 3.158$ kips/in.

For the plate thicknesses being welded, the minimum weld size is $\frac{5}{16}$ in. If intermittent welds are used, their minimum length is

$$L_{min} = 4 \times \text{size} \geq 1.5 \text{ in.}$$

$$= 4\left(\frac{5}{16}\right) = 1.25 \text{ in.} \quad \therefore \quad \text{Use 1.5 in.}$$

Try $\frac{5}{16}$-in. × $1\frac{1}{2}$-in. fillet welds.

Capacity per inch $= 0.707 \times \text{size} \times \phi F_w \times 2$ welds

$= 0.707(5/16)(31.5)(2) = 13.92$ kips/in.

FIGURE 10.21 ≡

$3''$ $\longleftarrow 36'' \longrightarrow$

$60'\text{-}0''$ span

Shear capacity of base metal:

$$t(\phi F_{BM}) = t[0.90(0.60F_y)] = t(0.54F_y) = \left(\frac{5}{16}\right)(0.54)(36)$$

$$= 6.075 \text{ kips/in.} < 13.92 \text{ kips/in.}$$

Use a total weld capacity of 6.075 kips/in.

The capacity of a 1.5-in. length of a pair of welds is

$$6.075(1.5) = 9.112 \text{ kips}$$

To determine the spacing, let

$$\frac{9.112}{s} = \frac{V_u Q}{I_x}$$

where s is the center-to-center spacing of the welds in inches:

$$s = \frac{9.112}{V_u Q/I_x} = \frac{9.112}{3.158} = 2.89 \text{ in.}$$

If a center-to-center spacing of 2.75 in. is used, this will give a clear spacing of $2.75 - 1.5 = 1.25$ in. The AISC Specification gives a maximum permissible spacing of intermittent fillet welds for this application in Section B10, "Proportions of Beams and Girders." The provisions for built-up compression members (AISC E4) and built-up tension members (AISC D2) are to be used for the compression flange and tension flange connections. For compression,

$$d \le \frac{127t}{\sqrt{F_y}} \qquad \text{but no greater than 12 in.}$$

For tension,

$$d \le 24t \text{ but no greater than 12 in.}$$

where

d = clear spacing in inches

t = thickness of the thinner outside connected plate in a built-up shape

Adapting these limits to the present case,

$$\frac{127t}{\sqrt{F_y}} = \frac{127(1.5)}{\sqrt{36}} = 31.8 \text{ in.} > 12 \text{ in.}$$

$$24t = 24(1.5) = 36 \text{ in.} > 12 \text{ in.}$$

The maximum permissible clear spacing is therefore 12 in., and the required clear spacing of 1.25 in. is satisfactory.

Although the 2.75 in. center-to-center spacing can be used for the entire length of the girder, an increased spacing can be used where the shear is less than the maximum of 223.4 kips. Three different spacings will be investigated:

1. The closest required spacing of 2.75 in.
2. The maximum permissible center-to-center spacing of $12 + 1.5 = 13.5$ in.
3. An intermediate spacing of 8 in.

The 8-in. spacing can be used when

$$\frac{V_u Q}{I_x} = \frac{9.112}{s} \quad \text{or} \quad V_u = \frac{9.112 I_x}{Qs} = \frac{9.112(60,640)}{857.2(8)} = 80.58 \text{ kips}$$

Referring to Figure 10.16 and letting x be the distance from the left support,

$$V_u = 223.4 - 4.34x = 80.58 \text{ kips}$$

$$x = 32.91$$

The 13.5-in. spacing can be used when

$$V_u = \frac{9.112 I_x}{Qs} = \frac{9.112(60,640)}{857.2(13.5)} = 47.75 \text{ kips}$$

Figure 10.16 shows that the shear never gets this small, so the maximum spacing never controls.

Answer   Use $\frac{5}{16}$-in. × $1\frac{1}{2}$-in. fillet welds for the flange-to-web welds, spaced as shown in Figure 10.22.

Welds for intermediate stiffeners:

$$\text{Minimum weld size} = \frac{3}{16} \text{ in.}$$

$$\left(\text{based on the thicknesses } t_w = \frac{5}{16} \text{ in. and } t = \frac{1}{4} \text{ in.} \right)$$

$$\text{Minimum length} = 4\left(\frac{3}{16}\right) = 0.75 \text{ in.} < 1.5 \text{ in.} \qquad \text{Use 1.5 in.}$$

FIGURE 10.22 ≡

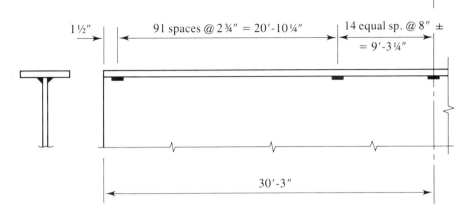

Capacity per inch for 4 welds (2 per stiffener plate):

$$0.707\left(\frac{3}{16}\right)(31.5)(4) = 16.70 \text{ kips/in.}$$

Shear capacity of base metal = 6.075 kips/in. (See previous computation for $t = \frac{5}{16}$ in.)

Use a strength of 6.075 kips/in.

From Equation 10.3, the shear to be transferred is

$$f = 0.045h\sqrt{\frac{F_y^3}{E}} = 0.045(62)\sqrt{\frac{(36)^3}{29{,}000}} = 3.539 \text{ kips/in.}$$

Use intermittent welds. The capacity of a 1.5-in. length of the 4 welds is

$$1.5(6.075) = 9.112 \text{ kips}$$

Equating the shear strength per inch to the required strength,

$$\frac{9.112}{s} = 3.539 \text{ kips/in.}$$

$$s = 2.57 \text{ in.}$$

From AISC Appendix F2.3, the maximum clear spacing is 16 times the web thickness but no greater than 10 inches.

$$16t_w = 16\left(\frac{5}{16}\right) = 5 \text{ in.}$$

Use a center-to-center spacing of 2.5 inches. This results in a clear spacing of

$$2.5 - 1.5 = 1 \text{ in.} < 5 \text{ in.}\quad\text{(OK)}$$

Answer   Use $\frac{3}{16}$-in. × $1\frac{1}{2}$-in. fillet welds for intermediate stiffeners, spaced as shown in Figure 10.23.

FIGURE 10.23 ☰

$\frac{3}{16}'' \times 1\frac{1}{2}''$ fillet welds @ $2\frac{1}{2}''$ c.c.

Section

Welds for bearing stiffeners:

$$\text{Minimum size} = \frac{5}{16} \text{ in.} \left(\text{based on the thicknesses } t_w = \frac{5}{16} \text{ in. and } t = \frac{3}{4} \text{ in.} \right)$$

$$\text{Minimum length} = 4\left(\frac{5}{16} \right) = 1.25 \text{ in.} < 1.5 \text{ in.} \qquad \text{Use 1.5 in.}$$

Use 2 welds per stiffener for a total of four. As with the intermediate stiffeners, the strength will be controlled by the base metal shear strength of 6.075 kips/in. or 9.112 kips for a 1.5-in. length.

For the end bearing stiffener, the applied load per inch is

$$\frac{\text{reaction}}{\text{length available for weld}} = \frac{223.4}{62 - 2(0.5)} = 3.662 \text{ kips/in.}$$

From $\dfrac{9.112}{s} = 3.662$, $s = 2.49$ in.

Answer   Use $\frac{5}{16}$-in. × $1\frac{1}{2}$-in. fillet welds for all bearing stiffeners, spaced as shown in Figure 10.24.

FIGURE 10.24 ☰

$\frac{3}{16}'' \times 1\frac{1}{2}''$ fillet welds @ $2\frac{1}{2}''$ c.c.

Section

The girder designed in this example is not necessarily the most economical one. Other possibilities include a girder with a thinner web and more intermediate stiffeners, and a girder with a thicker web and no intermediate stiffeners. Variables that affect economy include weight (volume of steel required) and fabricating costs. Although girders with intermediate stiffeners will usually require less steel, the savings can be offset by the additional fabrication cost. Variable flange thicknesses can also be considered. This alternative will definitely save weight, but here also, fabrication costs must be considered. A practical approach to achieving an economical design is to prepare several alternatives and compare their costs, using estimates of material and fabricating costs. *Design of Welded Structures*[53] contains many useful suggestions for the design of economical welded plate girders.

≡ PROBLEMS ≡

10.1 Determine the design flexural strength of a plate girder composed of a $\frac{3}{8}$-in. × 65-in. web and 2 in. × 14-in. flanges. Assume continuous lateral support of the compression flange. A36 steel is used.

10.2 Determine the design flexural strength of a plate girder whose web is $\frac{5}{16}$ in. × 70 in. and flanges are $\frac{3}{4}$ in. × 18 in. The compression flange has continuous lateral support, and A36 steel is used.

10.3 Determine whether the following plate girder has adequate flexural strength:

Web: $\frac{3}{8}$ in. × 65 in.

Flanges: 2 in. × 14 in.

Span length = 55 ft

Uniform service live load = 2.1 kips/ft

Lateral support of the compression flange only at the ends

A36 steel

10.4 The given plate girder is subjected to the *factored* loads shown in Figure P10.4.

FIGURE P10.4

The uniform load includes the weight of the girder. A36 steel is used. Determine the design flexural strength for the following lateral support conditions: (a) continuous lateral support and (b) lateral support at the ends and the load points.

10.5   The web of a certain girder is $\frac{5}{16}$-in. thick and $68\frac{3}{4}$-in. deep. A36 steel is used.

    a. What is the required spacing of intermediate stiffeners at the point where the factored load shear is 300 kips? Assume tension field action.

    b. What is the required cross-sectional area of double stiffeners at this point?

    c. What is the largest value of the factored load shear force V_u for which intermediate stiffeners are not required?

10.6   The girder of Problem 10.4 is provided with bearing stiffeners at the ends and at the load points. The interior bearing stiffeners consist of a pair of $\frac{5}{8}$-in. × $6\frac{1}{2}$-in. plates, clipped $\frac{1}{2}$ in. to clear the flange-to-web welds.

    a. Are these interior bearing stiffeners adequate?

    b. If the first intermediate stiffener is placed 40 inches from the end of the girder, what is the distance to the next one? Use tension field action.

    c. Determine the remaining stiffener spacings.

    d. Determine the size of the intermediate stiffeners.

10.7   The plate girder shown in Figure P10.7 is laterally supported at A, B, C, and D. A36 steel is used throughout, and the given loads are factored.

    a. Check the flexural strength.

    b. Bearing stiffeners will be placed at A, B, C, and D. Are additional stiffeners required? If so, where?

FIGURE P10.7

10.8   The details of an end bearing stiffener are shown in Figure P10.8. The stiffener plates are $\frac{9}{16}$-in. thick, and the web is $\frac{3}{16}$-in. thick. The stiffeners are clipped $\frac{1}{2}$ inch to provide clearance for the flange-to-web welds. If A36 steel is used, determine the maximum reaction that can be supported.

FIGURE P10.8

10.9   The preliminary design of a plate girder has resulted in a $\frac{5}{16}$-in. × 62-in. web. The maximum factored load bending moment is 5280 ft-kips. Based on this information, select a preliminary flange size if A36 steel is used. Assume continuous lateral support of the compression flange.

10.10 The description of a plate girder is as follows: The span length is 50 feet, the service loads consist of a uniformly distributed service live load of 5 kips/ft, continuous lateral support is provided, and A36 steel is to be used. Select the flange and web sizes. If intermediate stiffeners are required, do not design them.

10.11 A plate girder must be designed for the following conditions: a 60-foot span length, a 4 kip/ft uniform service live load, and continuous lateral support of the compression flange. Use A572 Grade 50 steel and design (a) the girder cross section, (b) intermediate stiffeners, and (c) end bearing stiffeners.

10.12 A 55-foot girder must support a uniform service live load of 3 kips/ft and a concentrated live load of 140 kips at midspan. Continuous lateral support of the compression flange will be provided. Select a cross section and compute the total girder weight for (a) A36 steel and (b) A572 steel, Grade 50. Do not design stiffeners.

10.13 A plate girder must be designed for the following conditions: a span length of 70 ft., a uniform service live load of 4 kips/ft, and two concentrated live loads of 175 kips each at the one-third points. The compression flange will be laterally braced at the ends and at the one-third points. Use A36 steel.

   a. Select the cross section.

   b. Determine the size and spacing of intermediate stiffeners.

   c. Design bearing stiffeners for the supports and points of load application.

   d. Design all welds.

10.14 Write a computer program to generate a table of shears and bending moments in a simply supported, uniformly loaded plate girder. (This table could be used in lieu of shear and moment diagrams.)

   Input: length (ft), factored load (kips/ft), web dimensions (in.), and flange dimensions (in.)

   Output: Shear and bending moment at the one-tenth points

Present your output in a table of the following form:

DISTANCE FROM LEFT SUPPORT (FT)	SHEAR (KIPS)	MOMENT (FT-KIPS)
.	.	.
.	.	.
.	.	.

10.15 Write a computer program to generate an expanded version of Table 10-36 in the Numerical Values section of the AISC Specification.

   Input: lower limit, upper limit, and increment for a/h

   Output: values of $\phi_v V_n/A_w$ for various combinations of h/t_w and a/h

Suggested procedure:

   1. Establish an outer loop in which h/t_w varies from 60 to 320 in increments of 5.

   2. Establish an inner loop in which a/h varies according to the input limits.

   3. For each combination of h/t_w and a/h, compute the following values using the equations of AISC Appendix G3:

   a. k_v (a function of a/h)

   b. C_v (a function of h/t_w and k_v)

   c. V_n (a function of F_y, C_v, and a/h)

PLASTIC ANALYSIS AND DESIGN

A.1 ≡ INTRODUCTION

The concept of plastic collapse was introduced in Section 5.2, "Bending Stress and the Plastic Moment." Failure of a structure will take place at a load that forms enough plastic hinges to create a mechanism, which will undergo uncontained displacement with any increase in the load. In a statically determinate beam, only one plastic hinge is required. As shown in Figure A.1, the hinge will form where the moment is maximum — in this case, at midspan. When the bending moment is large enough to cause the entire cross section to yield, any further increase in moment cannot be countered, and the plastic hinge has formed. This hinge is similar to an ordinary hinge except that the plastic hinge will have some moment resistance, much like a "rusty" hinge.

The plastic moment capacity, denoted by M_p, is the bending moment at which a plastic hinge forms. It is equal and opposite to the internal resisting moment corresponding to the stress distribution of Figure A.1c. The plastic moment can be computed for a given yield stress and cross-sectional shape as indicated in Figure A.2. If the stress distribution in the fully plastic condition is replaced by two equal and opposite statically equivalent concentrated forces, a couple is formed. Each of these forces has a magnitude equal to the yield stress times one half of the total cross-sectional area. The moment produced by this internal couple is

$$M_p = F_y \frac{A}{2} a = F_y Z_x$$

where A is the total cross-sectional area, a is the distance between centroids of the two half areas, and Z_x is the plastic section modulus. The factor of safety between first yielding and the fully plastic state can be expressed in terms of the section moduli as follows: From Figure A.1b, the moment causing first yield can be written as

$$M_y = F_y S_x$$

and

$$\frac{M_p}{M_y} = \frac{F_y Z_x}{F_y S_x} = \frac{Z_x}{S_x}$$

$$M_p = F_y \frac{A}{2} a = F_y Z_x$$

This ratio is a constant for a given cross-sectional shape and is called the *shape factor*. For a beam designed by allowable stress theory, it is a measure of the reserve capacity and has an average value of 1.12 for W-shapes.

In a statically indeterminate beam or frame, more than one plastic hinge will be required for the formation of a collapse mechanism. These hinges will form sequentially, although it is not always necessary to know the sequence. The analysis of statically indeterminate structures will be considered after a discussion of specification requirements.

A.2 ≡ AISC REQUIREMENTS

The AISC Specification permits plastic analysis and design to be used when the structure can remain stable, both locally and in an overall sense, up to the point of plastic collapse. Because the beam or frame could be required to undergo very large deformations as plastic hinges are being formed, lateral bracing requirements are especially severe.

To prevent local buckling, AISC B5.2 requires that the member have a compact cross-sectional shape — that is,

$$\lambda \leq \lambda_p$$

for both the web and flanges. For I-shaped members such as W- and S-shapes, the limiting width-thickness ratios from Table B5.1 are

$$\frac{b_f}{2t_f} \leq \frac{65}{\sqrt{F_y}} \quad \text{and} \quad \frac{h}{t_w} \leq \frac{640}{\sqrt{F_y}}$$

To prevent lateral buckling, AISC F1.2d limits the maximum unbraced length L_b at plastic hinge locations to L_{pd}, where, for I-shaped members,

$$L_{pd} = \frac{3600 + 2200(M_1/M_2)}{F_y} r_y \qquad \text{(AISC Equation F1-17)}$$

In this equation, M_1 is the smaller moment at the end of the unbraced length and M_2 is the larger. The ratio M_1/M_2 is positive when M_1 and M_2 bend the segment in reverse curvature and negative when they cause single curvature bending.

For compact shapes with adequate lateral bracing, M_n can be taken as M_p for use in plastic analysis. AISC F1.2d specifies, however, that in the region of the last plastic hinge to form and in regions not adjacent to plastic hinges, the usual methods must be used to compute M_n.

Other AISC Specification provisions related to plastic analysis and design are as follows:

A5.1 Plastic analysis is permitted only for $F_y \leq 65$ ksi.

C2.2 The axial force in columns caused by factored gravity and horizontal loads should not exceed $0.75\phi_c A_g F_y$.

E1.2 For columns, the slenderness parameter λ_c must not exceed $1.5K$, where K is the effective length factor.

A.3 ≡ ANALYSIS

If more than one collapse mechanism is possible, as with the continuous beam illustrated in Figure A.3, the correct one can be found and analyzed with the aid of three basic theorems of plastic analysis, given here without proof:

1. **Lower bound theorem (static theorem):** If a safe distribution of moment (one in which the moment is less than or equal to M_p everywhere) can be found, and it is statically admissible with the load (equilibrium is satisfied), then the corresponding load is less than or equal to the collapse load.

2. **Upper bound theorem (kinematic theorem):** The load that corresponds to an assumed mechanism must be greater than or equal to the collapse load. As a consequence, if all possible mechanisms are investigated, the one requiring the smallest load is the correct one.

3. **Uniqueness theorem:** If there is a safe and statically admissible distribution of moment in which enough plastic hinges form to produce a collapse mechanism, the corresponding load is the collapse load; that is, if a mechanism satisfies both the upper- and lower-bound theorems, it is the correct one.

FIGURE A.3 ≡

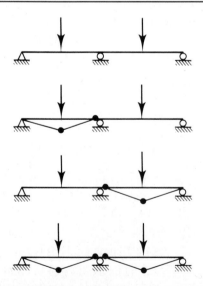

Analysis based on the lower-bound theorem is called the *equilibrium method* and will be illustrated in the following example:

≡ **EXAMPLE A.1**

Find the ultimate load for the beam in Figure A.4a by the equilibrium method of plastic analysis. Assume continuous lateral support and use A36 steel.

SOLUTION

A W30 × 99 of A36 steel is compact and, with continuous lateral support, the lateral bracing requirement is satisfied; therefore, plastic analysis is acceptable.

The loading history of the beam, from working load to collapse load, is traced in parts a through d of Figure A.4. At working loads, before yielding begins anywhere, the distribution of bending moments will be as shown in Figure A.4a, with the maximum moment occurring at the fixed ends. As the load is gradually increased, yielding begins at the supports when the bending moment reaches $M_y = F_y S_x$. Further increase in the load will cause the simultaneous formation of plastic hinges at each end at a moment of $M_p = F_y Z_x$. At this level of loading, the structure is still stable, the beam having been rendered statically determinate by the formation of the two plastic hinges. Only when a third hinge forms will a mechanism be created. This happens when the maximum positive moment attains a value of M_p. By virtue of the uniqueness theorem, the corresponding load is the collapse load, since the distribution of moment is safe and statically admissible.

At all stages of loading, the sum of the absolute values of the maximum negative and positive moments is $wL^2/8$. At collapse, this becomes

$$M_p + M_p = \frac{1}{8} w_u L^2$$

or

$$w_u = \frac{16 M_p}{L^2}$$

Since factored loads must be compared with factored strengths, $\phi_b M_p$ should have been used in the preceding equations rather than M_p. To keep notation simple, however, M_p will be used in all examples until the final step, when $\phi_b M_p$ will be substituted. In the present case, the correct result is

$$w_u = \frac{16 \phi_b M_p}{L^2}$$

For a W30 × 99,

$$M_p = F_y Z_x = \frac{36(312)}{12} = 936 \text{ ft-kips}$$

and

$$\phi_b M_p = 0.9(936) = 842.4 \text{ ft-kips}$$

FIGURE A.4 ≡

(a)   (b)

(c)   (d)

($\phi_b M_p$ can be obtained directly from the Load Factor Design Selection Table in Part 4 of the *Manual*.)

Answer

$$w_u = \frac{16(842.4)}{(30)^2} = 15.0 \text{ kips/ft.}$$

EXAMPLE A.2

If the beam of Example A.1 does not have continuous lateral support, determine where it must be braced.

SOLUTION

Since the plastic hinges at the ends form simultaneously and before the hinge forms at midspan, the maximum unbraced length should be checked with reference to the ends (that is, the last hinge to form is not subject to the bracing requirements for plastic analysis).

With respect to a hinge at the left end, assume that the brace point is at midspan. In this case, $M_1 = M_2 = M_p$, and the beam is bent in reverse curvature (the two moments are of opposite sign), so $M_1/M_2 = +1$. From AISC Equation F1-17, the maximum unbraced length is

$$L_{pd} = \frac{3600 + 2200(M_1/M_2)}{F_y} r_y = \frac{3600 + 2200(1)}{36} \quad (2.10)$$

$$= 338.3 \text{ in.} = 28.2 \text{ ft.}$$

(Note that the beam is almost satisfactory with no lateral bracing.)

With one lateral brace at midspan,

$$L_b = 15 \text{ ft.} < 28.2 \text{ ft.} \quad \text{(OK)}$$

The unbraced length under consideration encompasses the hinge at midspan, and there are no regions not adjacent to a plastic hinge, so no other design strength computations are necessary.

Answer   Use one lateral brace at midspan.

The *mechanism method* is based on the upper-bound theorem and requires that all possible collapse mechanisms be investigated. The one requiring the smallest load will control, and the corresponding load is the collapse load. The analysis of each mechanism is accomplished by application of the principle of virtual work. An assumed mechanism is subjected to virtual displacements consistent with the possible mechanism motion, and the external work is equated to the internal work. A relationship can then be found between the load and the plastic moment capacity M_p. This technique will be illustrated in the following examples.

EXAMPLE A.3

The continuous beam shown in Figure A.5 has a compact cross section with a design strength $\phi_b M_p$ of 1040 ft-kips. Use the mechanism method to find the collapse load P_u. Assume continuous lateral support.

SOLUTION   In keeping with the notation adopted in the preceding example, we will use M_p in the solution and substitute $\phi_b M_p$ in the final step.

There are two possible failure mechanisms for this beam. As indicated in Figure A.5, the two possibilities will have similar collapse mechanisms, with each segment undergoing a rigid-body motion. To investigate the mechanism in span AB, impose a virtual rotation θ at A. The corresponding rotations at the plastic hinges will be as shown in Figure A.5b, and the vertical displacement of the load will be 10θ. From the principle of virtual work,

external work = internal work
$$P_u(10\theta) = M_p(2\theta) + M_p\theta$$

(No internal work is done at A, since there is no plastic hinge.) Solving for the collapse load,

$$P_u = \frac{3M_p}{10}$$

FIGURE A.5 $\equiv$

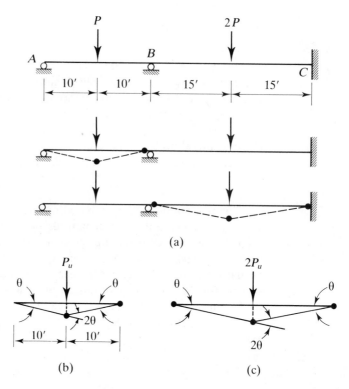

(a)

(b)   (c)

The mechanism for span BC is slightly different: All three hinges are plastic hinges. The external and internal virtual work in this case are

$$2P_u(15\theta) = M_p\theta + M_p(2\theta) + M_p\theta$$

and

$$P_u = \frac{2}{15}M_p$$

This second possibility requires the smaller load and is therefore the correct mechanism. The collapse load will be obtained using $\phi_b M_p$ in place of M_p.

Answer

$$P_u = \frac{2}{15}\phi_b M_p = \frac{2}{15}(1040) = 139 \text{ kips.}$$

$\equiv$ **EXAMPLE A.4**

Determine the collapse load P_u for the rigid frame shown in Figure A.6. Each member of the frame is a W21 × 147 with $F_y = 50$ ksi. Assume continuous lateral support.

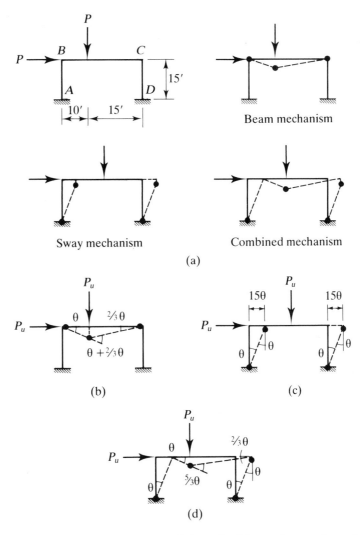

(a)

(b)   (c)

(d)

SOLUTION   A W21 × 147 is compact for $F_y = 50$ ksi, and with continuous lateral support provided, the conditions for the use of plastic analysis are satisfied.

As indicated in Figure A.6, there are three possible failure modes for this frame: a beam mechanism in member BC, a sway mechanism, and one that is a combination of the first two. The analysis of each mechanism is begun by imposing a virtual rotation of θ at one of the hinges and expressing the remaining notations and displacements as a function of this angle.

The virtual displacement of the beam mechanism is shown in Figure A.6b. From the equivalence of external and internal work,

$$P_u(10\theta) = M_p\theta + M_p\left(\frac{5}{3}\theta\right) + M_p\left(\frac{2}{3}\theta\right)$$

where M_p has been used in place of $\phi_b M_p$. Solving for P_u,

$$P_u = 0.3333 M_p$$

If axial strains in member BC are neglected, the sway mechanism will displace as shown in Figure A.6c, with the same horizontal displacement at B and C. This will make the rotations of all the plastic hinges the same:

$$P_u(15\theta) = M_p(4\theta) \qquad P_u = 0.2667 M_p$$

From Figure A.6d, the principle of virtual work for the combined mechanism gives

$$P_u(15\theta) + P_u(10\theta) = M_p\theta + M_p\left(\frac{5}{3}\theta\right) + M_p\left(\frac{2}{3}\theta + \theta\right) + M_p\theta$$

$$P_u = 0.2133 M_p \quad \text{(controls)}$$

Answer   The collapse load for the frame is $P_u = 0.2133\phi_b M_p = 0.2133(1400) = 299$ kips.

The reader will note a certain similarity between the two methods of analysis. Although the equilibrium method does not require a consideration of all possible mechanisms, it does require that the analyst recognize a mechanism when the assumed distribution of moment is consistent with one. Both methods require the assumption of failure mechanisms, but in the equilibrium method each assumption is checked for a safe and statically admissible distribution of moment, and it may not be necessary to investigate all possible mechanisms.

A.4 ≡ DESIGN

The design process is similar to analysis, except that the unknown being sought is the required plastic moment capacity M_p. The collapse load is known in advance, having been obtained by multiplying the service loads by the load factors.

≡ **EXAMPLE A.5**

The three-span continuous beam in Figure A.7 must support the gravity service loads shown. Each load consists of 25% dead load and 75% live load. Cover plates will be used in spans BC and CD to give the relative moment strengths indicated. Assume continuous lateral support and select a shape in A36 steel.

SOLUTION   The collapse loads are obtained by multiplying the service loads by the appropriate load factors. For the 45-kip service load,

$$P_u = 1.2(0.25 \times 45) + 1.6(0.75 \times 45) = 67.5 \text{ kips}$$

FIGURE A.7 ≡

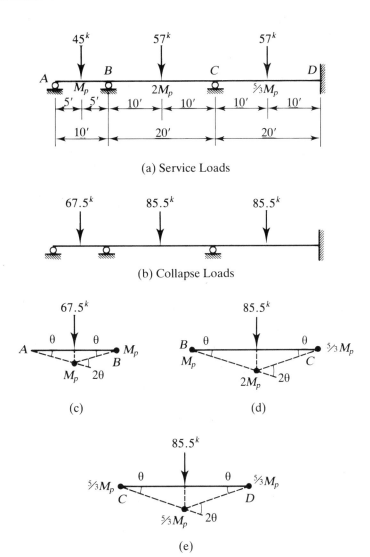

(a) Service Loads

(b) Collapse Loads

(c)

(d)

(e)

For the 55-kip service load,

$$P_u = 1.2(0.25 \times 57) + 1.6(0.75 \times 57) = 85.5 \text{ kips}$$

Three mechanisms must be investigated, one in each span. Parts c through e of Figure A.7 show each mechanism after being subjected to a virtual displacement. When a plastic hinge forms at a support where members of unequal strength meet, it will form when the bending moment equals the plastic moment capacity of the weaker member. For span AB,

external work = internal work

$$67.5(5\theta) = M_p(2\theta + \theta) \qquad M_p = 112.5 \text{ ft-kips}$$

For span BC,

$$85.5(10\theta) = M_p\theta + 2M_p(2\theta) + \frac{5}{3} M_p\theta, \qquad M_p = 128.2 \text{ ft-kips}$$

For span CD:

$$85.5(10\theta) = \frac{5}{3} M_p(\theta + 2\theta + \theta), \qquad M_p = 128.2 \text{ ft-kips}$$

The upper-bound theorem may be interpreted as follows: The value of the plastic moment corresponding to an assumed mechanism is less than or equal to the plastic moment for the collapse load. This means that the mechanism requiring the largest moment capacity is the correct one. Both of the last two mechanisms evaluated in this design problem require the same controlling value of M_p and will therefore occur simultaneously. Since this required strength is actually the required *design* strength,

Required $\phi_b M_p = 128.2$ ft-kips

From the Load Factor Design Selection Table, the lightest shape is a W16 × 31 with a design strength of $\phi_b M_p = 146$ ft-kips.

Try a W16 × 31.

Check shear (refer to Figure A.8):

Span AB:

$$\Sigma M_B = V_A(10) - 67.5(5) + 128.2 = 0$$
$$V_A = 20.93 \text{ kips}$$
$$V_B = 20.93 - 67.5 = -46.57 \text{ kips}$$

Span BC:

$$\Sigma M_B = -M_p + 85.5(10) + \left(\frac{5}{3}\right)M_p - V_C(20) = 0$$
$$V_C = \frac{85.5(10) + (2/3)M_p}{20} = \frac{855 + (2/3)(128.2)}{20} = 47.02 \text{ kips}$$
$$V_B = 85.5 - 47.02 = 38.48 \text{ kips}$$

Span CD:

$$\Sigma M_C = -\frac{5}{3} M_p + \frac{5}{3} M_p + 85.5(10) - V_D(20) = 0$$
$$V_D = 42.75 \text{ kips} = V_C$$

The maximum shear is therefore V_C from span $BC = 47.02$ kips.

FIGURE A.8 ≡

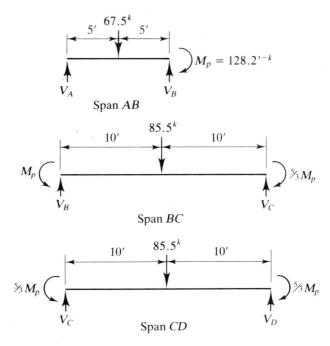

Span *AB*

Span *BC*

Span *CD*

From the Factored Uniform Load tables in Part 4 of the *Manual*, the shear design strength of a W16 × 31 is

$$\phi_v V_n = 84.9 \text{ kips} > 47.02 \text{ kips (OK)}$$

Answer   Use a W16 × 31.

A.5 ≡ CONCLUDING REMARKS

Analysis of a mechanism subjected to distributed loads presents an additional complication that has not been covered here. Any realistic problem in plastic analysis or design will certainly include this type of loading. In addition, members subjected to both axial compression and bending, as in the rigid frame of Example A.4, should be investigated for interaction of these effects.

With regard to analysis methods per se, a more formal and systematic approach to the equilibrium method exists and is discussed in considerable detail in *The Plastic Methods of Structural Analysis*.[61] A more rigorous formulation of the mechanism method is also possible. With this approach, known as the *method of inequalities*, the correct mechanism can be determined directly by linear programming techniques. For most of the structures for which plastic design would ordinarily be used, however, the mechanism method as presented in this appendix will usually be adequate.

STRUCTURAL STEEL DESIGN BASED ON ALLOWABLE STRESS

B.1 ≡ INTRODUCTION

The chief difference between allowable stress design and load and resistance factor design is in the factors of safety.* In LRFD, load factors are applied to the loads, and a resistance factor is applied to the strength; furthermore, the values of the load factors depend on the type of load and the load combination. In allowable stress design (ASD), only one factor of safety is used, and it is applied to the stress that exists at the limit state. Limit states for allowable stress design are the same as for LRFD: yielding, fracture, and buckling. The principle of allowable stress analysis and design is as follows: The stress at the limit state is divided by a factor of safety to obtain an allowable stress, and the maximum stress caused by service loads must not exceed this allowable stress. For example, for axial tension,

$$f_t = \frac{P}{A} \le F_t \tag{B.1}$$

where

f_t = computed tensile stress

P = *service* axial tensile load

F_t = allowable tensile stress

The allowable tensile stress will be either the yield stress divided by a factor of safety or the ultimate tensile stress divided by a different factor of safety. Tension members will be covered in detail in the next section.

Allowable stress design for structural steel was the norm before the introduction of the LRFD Specification in 1989. The last edition of the *ASD Specification*[62] and the corresponding *Manual of Steel Construction*[63] were published in 1989. Both of these documents are organized in the same way as their LRFD counterparts: the Specification is divided into chapters (for

---

* It is assumed that readers of this appendix are familiar with the AISC LRFD Specification and *Manual*.

example, "Chapter D, Tension Members") and the *Manual* is divided into parts (such as "Part 2, Beam and Girder Design"). The Specification is accompanied by a Commentary.

The notation of Equation B.1 is the same as that used in the Specification. A lower-case *f* is used for computed (actual) stress, and an upper-case *F* is used for allowable stress. The subscript indicates the type of stress.

Because this appendix is intended to be an introduction only, we will not use AISC Specification section numbers or equation numbers in our coverage. Although the material is taken from the Specification, the equation numbers are ours. Furthermore, when the terms *Specification* or *Manual* are used in this appendix, the allowable stress versions are implied, unless otherwise stated.

Many elements of steel design are the same for both ASD and LRFD. For example, the net area provisions for tension members are the same, including the $s^2/4g$ technique for staggered holes and the U factor for shear lag (although the ASD Specification uses the average U values and places the equation for U in the Commentary — just the reverse of the way it is done in the LRFD Specification). The definitions of compact, noncompact, and slender members are virtually the same, but the latest LRFD Specification rules are more up-to-date. As a general rule, when there is a conflict between an ASD and LRFD provision such as this, it should be resolved in favor of the LRFD Specification, which is more current.

Although there are no load factors in allowable stress design, the relative importance of various loads can still be taken into account in combinations. For example, the following load combinations for roof structures are frequently used: $D + S$, $D + W$, $D + (S/2) + W$, and $D + S + (W/3)$. In addition, the Specification permits allowable stresses to be increased by one third when wind or seismic loads are included. Many building codes also contain this provision.

The *ASD Manual* contains many tables and charts similar to the ones in the *LRFD Manual* that simplify design. Only the more important ones are discussed in this brief introduction.

B.2 ≡ TENSION MEMBERS

From Equation B.1, the computed axial tensile stress is $f_t = P/A$. The allowable stress is based on the more critical of the limit states of yielding and fracture. For yielding of the gross section, the applied stress is

$$f_t = \frac{P}{A_g} \tag{B.2}$$

where A_g is the gross cross-sectional area. The factor of safety for this limit state is 5/3, and the allowable stress is

$$F_t = \frac{F_y}{F.S.} = \frac{F_y}{5/3} = 0.60F_y \tag{B.3}$$

For fracture of the net section,

$$f_t = \frac{P}{A_e} \tag{B.4}$$

where A_e is the effective net area. The factor of safety is 2.0, resulting in an allowable stress of

$$F_t = \frac{F_u}{F.S.} = \frac{F_u}{2} = 0.50F_u \tag{B.5}$$

EXAMPLE B.1

Check the stresses in the tension member shown in Figure B.1 caused by the 50 kip service load. A36 steel and $\frac{7}{8}$-in. diameter bolts are used.

SOLUTION   Stress on the gross area: From Equations B.2 and B.3, the applied stress is

$$f_t = \frac{P}{A_g} = \frac{50}{2.48} = 20.2 \text{ ksi}$$

and the allowable stress is

$$F_t = 0.60F_y = 0.60(36) = 21.6 \text{ ksi} > 20.2 \text{ ksi} \quad \text{(OK)}$$

Stress on the net area:

$$A_n = A_g - (\text{thickness} \times \text{hole diameter})$$
$$= 2.48 - \frac{3}{8}\left(\frac{7}{8} + \frac{1}{8}\right) = 2.105 \text{ in.}^2$$

If we use the average value of U, the effective net area is

$$A_e = UA_n = 0.85A_n = 0.85(2.105) = 1.789 \text{ in.}^2$$

FIGURE B.1 ≡

L3½ × 3½ × ⅜

Section

From Equations B.4 and B.5,

$$f_t = \frac{P}{A_e} = \frac{50}{1.789} = 27.9 \text{ ksi}$$

$$F_t = 0.50F_u = 0.50(58) = 29 \text{ ksi} > 27.9 \text{ ksi} \quad \text{(OK)}$$

Answer   The member is satisfactory.

B.3 ≡ COMPRESSION MEMBERS

The stress in an axially loaded compression member is

$$f_a = \frac{P}{A_g}$$

The allowable stress, denoted by F_a, is obtained by dividing the critical buckling load by a factor of safety. The factor of safety for elastic columns (slender columns) is a constant, and the factor for inelastic columns is variable. In ASD, the compressive strength is expressed as a function of the slenderness ratio KL/r, whereas in LRFD, the strength is a function of $\lambda_c = (KL/r\pi)\sqrt{F_y/E}$. In the elastic range, the critical stress is the Euler buckling load divided by the area:

$$F_{cr} = \frac{P_{cr}}{A_g} = \frac{\pi^2 E A_g}{(KL/r)^2} \div A_g = \frac{\pi^2 E}{(KL/r)^2} \tag{B.6}$$

For the inelastic range, the proportional limit is assumed to be at $F_y/2$, and the following empirical equation is used in lieu of the tangent modulus formula:

$$F_{cr} = F_y \left[1 - \frac{(KL/r)^2}{2C_c^2} \right] \tag{B.7}$$

where C_c is the value of KL/r corresponding to a stress of $F_y/2$. Equation B.7 represents a parabola that is tangent to the Euler curve at $KL/r = C_c$ and tangent to a horizontal line at $KL/r = 0$. We can find an expression for C_c by setting the right-hand side of Equation B.6 to $F_y/2$:

$$\frac{F_y}{2} = \frac{\pi^2 E}{(KL/r)^2} = \frac{\pi^2 E}{C_c^2}$$

This results in

$$C_c = \sqrt{\frac{2\pi^2 E}{F_y}} \tag{B.8}$$

The column strength over the full range of slenderness, before the application of safety factors, is represented graphically in Figure B.2.

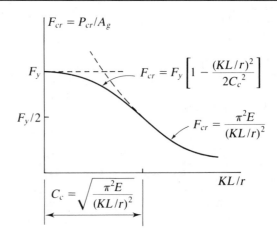

To obtain allowable compressive stresses, Equations B.6 and B.7 are divided by factors of safety. For elastic columns, the factor of safety is 23/12. For inelastic columns, the following variable factor is used:

$$F.S. = \frac{5}{3} + \frac{3(KL/r)}{8C_c} - \frac{(KL/r)^3}{8C_c^3}$$

This has a value of 5/3 at $KL/r = 0$ (the same as for yielding of a tension member) and a value of 23/12 at $KL/r = C_c$ (approximately 15% more than 5/3). Dividing the strength equations by the appropriate factors of safety, we obtain the following allowable stresses:

For $KL/r < C_c$,

$$F_a = \frac{F_y\left[1 - \dfrac{(KL/r)^2}{2C_c^2}\right]}{\dfrac{5}{3} + \dfrac{3(KL/r)}{8C_c} - \dfrac{(KL/r)^3}{8C_c^3}} \tag{B.9}$$

For $KL/r > C_c$,

$$F_a = \frac{\pi^2 E}{(KL/r)^2} \div \frac{23}{12} = \frac{12\pi^2 E}{23(KL/r)^2} \tag{B.10}$$

For members with slender cross-sectional elements, a reduction must be made in the allowable stress to account for the possibility of local buckling. This is accomplished with reduction factors from an appendix to the Specification.

≡ **EXAMPLE B.2**

Determine the allowable service load P for the compression member shown in Figure B.3.

SOLUTION

Check for slender cross-sectional elements. The width-thickness ratio limits for compression members given in the ASD Specification are the same as those in the LRFD Specification:

$$\frac{b_f}{2t_f} = 6.4 \text{ (from the properties tables in the Manual)}$$

$$\frac{95}{\sqrt{F_y}} = \frac{95}{\sqrt{36}} = 15.8 > 6.4 \quad \text{(OK)}$$

$$\frac{h}{t_w} = 25.3$$

$$\frac{253}{\sqrt{F_y}} = \frac{253}{\sqrt{36}} = 42.2 > 25.3 \quad \text{(OK)}$$

Since the stress is $f_a = P/A_g$, the corresponding load is $P = f_a A_g$, and the allowable compressive load is $F_a A_g$. From Equation B.8,

$$C_c = \sqrt{\frac{2\pi^2 E}{F_y}} = \sqrt{\frac{2\pi^2(29,000)}{36}} = 126.1$$

The maximum slenderness ratio is

$$\frac{KL}{r} = \frac{KL}{r_{min}} = \frac{KL}{r_y} = \frac{1.0(20)(12)}{2.48} = 96.77$$

FIGURE B.3 ≡

P

20'

W14 × 74
A36

This is less than C_c, so F_a is found from Equation B.9:

$$F_a = \frac{F_y\left[1 - \dfrac{(KL/r)^2}{2C_c^2}\right]}{\dfrac{5}{3} + \dfrac{3(KL/r)}{8C_c} - \dfrac{(KL/r)^3}{8C_c^3}} = \frac{36\left[1 - \dfrac{(96.77)^2}{2(126.1)^2}\right]}{\dfrac{5}{3} + \dfrac{3(96.77)}{8(126.1)} - \dfrac{(96.77)^3}{8(126.1)^3}}$$

$$= 13.38 \text{ ksi}$$

Answer $P = F_a A_g = 13.38(21.8) = 292$ kips.

Design Aids

The *ASD Manual* contains column design aids similar in form to those found in the *LRFD Manual*. Foremost among these are the tables for allowable axial loads. When these tables are entered with the effective length KL and a required service load capacity, a shape with sufficient capacity can be quickly found. Just as with the LRFD column load tables, the effective length with respect to the minimum radius of gyration r_y should be used; that is, $K_y L$ should be used. Alternatively, the tables can be entered with $K_x L/(r_x/r_y)$.

When the effective length factor K is found from the Jackson–Mooreland alignment charts, a stiffness reduction factor can be applied if the column is inelastic at failure ($KL/r < C_c$). A table for this purpose is provided in the *Manual*.

B.4 ≡ BEAMS

The maximum bending stress in a homogeneous beam that has not been stressed beyond the proportional limit is given by the *flexure formula*:

$$f_b = \frac{Mc}{I} = \frac{M}{I/c} = \frac{M}{S} \tag{B.11}$$

where

   M = maximum bending moment in the beam

   c = distance from the neutral axis to the extreme fiber

   I = moment of inertia about the axis of bending

   S = elastic section modulus

The coverage in this section will be limited to hot-rolled I- and H-shaped cross sections bent about an axis perpendicular to the web (the x-axis).

The allowable bending stress is denoted by F_b and is based on one of the limit states of yielding, local buckling, or lateral-torsional buckling. In allowable stress design, it is convenient to divide beams into two categories: laterally supported and laterally unsupported. If a beam has adequate lateral support, the allowable stress will be based on yielding if the shape is compact and will be based on local buckling if the shape is noncompact. The allowable bending stress for laterally unsupported beams will be based on lateral-torsional buckling.

Lateral Support

A beam with an unbraced length of L_b is considered to have enough lateral support to prevent lateral-torsional buckling when $L_b \le L_c$, where L_c is the smaller of

$$\frac{76b_f}{\sqrt{F_y}} \quad \text{or} \quad \frac{20,000}{(d/A_f)F_y}$$

This is,

$$L_c = \frac{76b_f}{\sqrt{F_y}} \le \frac{20,000}{(d/A_f)F_y} \tag{B.12}$$

We will use this criterion to classify beams as *laterally supported* or *laterally unsupported*.

Laterally Supported Beams

If a laterally supported beam can be stressed to the yield point without local buckling, the factor of safety is 5/3, and the allowable stress is

$$F_b = \frac{F_y}{F.S.} = \frac{F_y}{5/3} = 0.60F_y$$

This corresponds to a shape whose flange width-thickness ratio is at the upper limit for noncompactness; that is, $b_f/2t_f = 95/\sqrt{F_y}$. (This is different from the LRFD limit, but it will be used here since it is incorporated in an AISC equation for allowable stress.) If the shape is compact, the fully plastic condition can be reached without local buckling, and an additional 10% is permitted. The allowable stress for this case is

$$F_b = 1.10(0.60F_y) = 0.66F_y$$

For noncompact shapes, AISC uses a linear transition between $0.60F_y$ and $0.66F_y$, based on the value of $b_f/2t_f$. (All hot-rolled I- and H-shapes in the *Manual* have compact webs.) This results in the following equation for the allowable stress:

$$F_b = F_y \left[0.79 - 0.002 \frac{b_f}{2t_f} \sqrt{F_y} \right]$$

Figure B.4 shows the relationship between flange width-thickness ratio and allowable stress for laterally supported beams. Slender shapes are dealt with in an appendix to the Specification, but none of the hot-rolled I- and H-shapes in the *Manual* are slender.

In summary, the allowable stresses for laterally supported beams are as follows:

If the shape is compact,

$$F_b = 0.66F_y \tag{B.13}$$

If the shape is noncompact,

$$F_b = F_y \left[0.79 - 0.002 \frac{b_f}{2t_f} \sqrt{F_y} \right] \tag{B.14}$$

Laterally Unsupported Beams

The strength of laterally unsupported beams is based on the limit state of lateral-torsional buckling. In ASD, this is treated as having two components: uniform warping and nonuniform warping. Uniform warping is elastic and the limit stress is given by

$$f_u = \frac{0.65E}{L_b d/A_f} \tag{B.15}$$

where

d = overall depth of beam

A_f = area of compression flange

FIGURE B.4 ≡

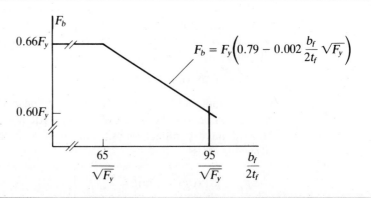

Nonuniform warping can be either inelastic or elastic. For elastic warping, the failure stress is

$$f_{nu} = \frac{\pi^2 E}{(L_b/r_y)^2} \tag{B.16}$$

For inelastic warping, the following empirical equation, similar to the one for compression members, is used:

$$f_{nu} = \frac{10}{9} F_y \left[1 - \frac{(L_b/r_y)^2}{2C^2} \right] \tag{B.17}$$

where

    C = maximum value of L_b for which the nonuniform warping is inelastic (if $L_b > C$, the warping is elastic)

$$= 3\pi \sqrt{\frac{E}{5F_y}}$$

The buckling stresses given by equations B.15–B.17 are subject to an upper limit of F_y. Figure B.5 shows the uniform warping stress as a function of L_b, and Figure B.6 shows the nonuniform warping stress.

To obtain the AISC equations for allowable bending stress based on lateral-torsional buckling, the following adjustments are made to the foregoing equations:

1. All failure stresses are divided by a factor of safety of 5/3.

2. The radius of gyration r_y is replaced by r_T, the radius of gyration about the weak axis for a portion of the cross section consisting of the compression flange and one third of the compressed part of the web. This value does not differ appreciably from r_y and is tabulated in the *ASD Manual*.

3. All equations are expressed in terms of the ratio L_b/r_T.

FIGURE B.5 ≡

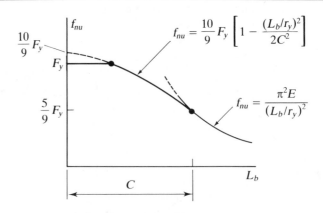

4. The factor C_b is incorporated to account for variation of bending moment over the unbraced length (the warping equations are based on uniform moment).

5. Although the lateral-torsional buckling strength derives from both the uniform and nonuniform warping components, AISC conservatively uses only one component — the larger one.

The AISC equations for allowable bending stress for laterally unsupported beams can be summarized as follows:

For $\dfrac{L_b}{r_T} < \sqrt{\dfrac{102{,}000 C_b}{F_y}}$,

$$F_b = 0.60 F_y$$

For $\sqrt{\dfrac{102{,}000 C_b}{F_y}} \le \dfrac{L_b}{r_T} \le \sqrt{\dfrac{510{,}000 C_b}{F_y}}$, use the larger of

$$F_b = \left[\frac{2}{3} - \frac{F_y (L_b/r_T)^2}{1{,}530{,}000 C_b} \right] F_y \le 0.60 F_y \quad \text{(inelastic nonuniform warping)} \quad \text{(B.18)}$$

and

$$F_b = \frac{12{,}000 C_b}{L_b d/A_f} \le 0.60 F_y \quad \text{(uniform warping)} \quad \text{(B.19)}$$

For $\dfrac{L_b}{r_T} > \sqrt{\dfrac{510{,}000 C_b}{F_y}}$, use the larger of

$$F_b = \frac{170{,}000 C_b}{(L_b/r_T)^2} \le 0.60 F_y \quad \text{(elastic nonuniform warping)} \quad \text{(B.20)}$$

and

$$F_b = \frac{12,000C_b}{L_b d/A_f} \leq 0.60F_y \quad \text{(uniform warping)} \tag{B.19}$$

The ASD Specification gives an equation for the modifier C_b that is different from the one in the LRFD Specification, but either equation can be used. Note that while the flexural strength according to LRFD is directly proportional to C_b, this is not the case for the allowable stresses given by equations B.18–B.20. This somewhat complicates allowable stress design of beams.

Shear

The shearing stress is computed as the maximum service load shear force divided by the web area:

$$f_v = \frac{V}{A_w} \approx \frac{V}{t_w d}$$

The allowable shearing stress is based on shear yielding and is taken as two-thirds of the allowable tensile stress on the gross section, or

$$F_v = \frac{2}{3} F_t = \frac{2}{3}(0.60F_y) = 0.40F_y \tag{B.21}$$

EXAMPLE B.3

A W16 × 100 is used for a simply supported, uniformly loaded beam that is laterally braced only at its ends. If A36 steel is used, determine the maximum service load bending moment that can be resisted for span lengths of (a) 10 ft, (b) 15 ft, and (c) 40 ft.

SOLUTION

Determine L_c. From Equation B.12,

$$\frac{76b_f}{\sqrt{F_y}} = \frac{76(10.42)}{\sqrt{36}} = 132.0 \text{ in.} = 11.0 \text{ ft}$$

$$\frac{20,000}{(d/A_f)F_y} = \frac{20,000}{\dfrac{16.97}{10.42(0.985)}(36)} = 336.0 \text{ in.} = 28.0 \text{ ft}$$

The smaller value controls, ∴ $L_c = 11.0$ ft.

a. 10-ft span:

$$L_b = 10 \text{ ft} < L_c \quad \therefore \quad \text{the beam is laterally supported.}$$

Since a W16 × 100 is a compact shape for A36 steel, the allowable stress from Equation B.13 is

$$F_b = 0.66F_y = 0.66(36) = 23.76 \text{ ksi}$$

The maximum bending stress for a moment M is given by Equation B.11 as $f_b = M/S$, so the maximum moment occurs when the stress f_b is equal to the allowable stress F_b:

$$M = F_b S = 23.76(175) = 4158 \text{ in.-kips} = 346 \text{ ft-kips}$$

Answer   a. Maximum moment = 346 ft-kips.

b. 15-ft span:

$$L_b = 15 \text{ ft} > L_c = 11.0 \text{ ft} \quad \therefore \quad \text{the beam is laterally unsupported.}$$

$r_T = 2.81$ in.   (This value is given in the properties tables in the *ASD Manual*.)

$$\frac{L_b}{r_T} = \frac{15(12)}{2.81} = 64.06$$

For a simply supported, uniformly loaded beam with lateral bracing at the ends, $C_b = 1.14$ (this was calculated with the LRFD Specification equation, but the ASD equation could be used).

Determine the limits for L_b/r_T:

$$\sqrt{\frac{102{,}000C_b}{F_y}} = \sqrt{\frac{102{,}000(1.14)}{36}} = 56.8$$

$$\sqrt{\frac{510{,}000C_b}{F_y}} = \sqrt{\frac{510{,}000(1.14)}{36}} = 127$$

Since $56.8 < L_b/r_T < 127$, use equations B.18 and B.19:

$$F_b = \left[\frac{2}{3} - \frac{F_y(L_b/r_T)^2}{1{,}530{,}000C_b}\right]F_y \leq 0.60F_y$$

$$= \left[\frac{2}{3} - \frac{36(64.06)^2}{1{,}530{,}000(1.14)}\right]36 = 20.95 \text{ ksi}$$

or

$$F_b = \frac{12{,}000C_b}{L_b d/A_f} \leq 0.60F_y$$

$$= \frac{12{,}000(1.14)}{(15 \times 12)(16.97)/(10.42 \times 0.985)} = 45.97 \text{ ksi}$$

The larger value is greater than $0.60F_y = 0.60(36) = 21.6$ ksi, so use

$$F_b = 0.60F_y = 21.6 \text{ ksi}$$

The maximum bending moment is

$$M = F_b S = 21.6(175) = 3780 \text{ in.-kips} = 315 \text{ ft-kips}$$

Answer   **b.** Maximum moment = 315 ft-kips.

c. 40-ft span:

$$\frac{L_b}{r_T} = \frac{40(12)}{2.81} = 170.8 > \sqrt{\frac{510,000C_b}{F_y}} = 127$$

Use equations B.20 and B.19:

$$F_b = \frac{170,000C_b}{(L_b/r_T)^2} = \frac{170,000(1.14)}{(170.8)^2} = 6.643 \text{ ksi} < 0.60F_y$$

or

$$F_b = \frac{12,000C_b}{L_b d/A_f} = \frac{12,000(1.14)}{(40 \times 12)(16.97)/(10.42 \times 0.985)} = 17.24 \text{ ksi} < 0.60F_y$$

Use $F_b = 17.24$ ksi. The maximum moment is

$$M = F_b S = 17.24(175) = 3017 \text{ in.-kips} = 251 \text{ ft-kips}$$

Answer   **c.** Maximum moment = 251 ft-kips.

Design Aids

Most of the design aids for beams in the *LRFD Manual* have their counterparts in the *ASD Manual*. These include design charts that give allowable bending moment as a function of unbraced length for shapes normally used for beams. These curves are based on $C_b = 1.0$, but they cannot be used directly for other values of C_b, since the allowable bending stress F_b is not directly proportional to C_b.

B.5 ≡ BEAM-COLUMNS

Structural members subjected to both bending and axial stress are analyzed with interaction equations incorporating ratios of actual stress to allowable stress. The ASD Specification equations are variations of

$$\frac{f_a}{F_a} + \frac{f_{bx}}{F_{bx}} + \frac{f_{by}}{F_{by}} \le 1.0$$

where x and y denote the axes of bending. Two equations are used in the Specification: one that is evaluated with a bending stress based on the absolute maximum moment in the member, and one with a bending stress based on the maximum end moment. A single amplification factor is used; there are no separate factors to account for the sway and nonsway components. This amplification factor has the form

$$\frac{C_m}{1 - \dfrac{f_a}{F'_e}}$$

where C_m is defined as follows:

Members subject to sidesway:

$$C_m = 0.85$$

Members not subject to sidesway: If there are no transverse loads on the member,

$$C_m = 0.6 - 0.4(M_1/M_2) \tag{B.22}$$

where M_1 and M_2 are the bending moments at the ends of the member, with M_1 the smaller in absolute value. The ratio M_1/M_2 is positive if the member is bent in reverse curvature and is negative for single curvature bending.

For members restrained against sidesway and subjected to transverse loads,

$$C_m = 0.85 \text{ if the ends are restrained against rotation}$$
$$= 1.0 \text{ if the ends are not restrained}$$

The factor F'_e is the Euler buckling stress divided by a factor of safety of 23/12:

$$F'_e = \frac{12\pi^2 E}{23(KL_b/r_b)^2} \tag{B.23}$$

The subscript b refers to the axis of bending. If bending about the x-axis is being considered, $F'_e = F'_{ex}$ and $KL_b/r_b = K_x L/r_x$. Similarly, for F'_{ey}, use $K_y L/r_y$.

The following interaction equations must be checked:
If $f_a/F_a \leq 0.15$, moment amplification is not required, and

$$\frac{f_a}{F_a} + \frac{f_{bx}}{F_{bx}} + \frac{f_{by}}{F_{by}} \leq 1.0 \tag{B.24}$$

If $f_a/F_a > 0.15$, *both* of the following equations must be checked:

$$\frac{f_a}{F_a} + \frac{C_{mx} f_{bx}}{\left(1 - \dfrac{f_a}{F'_{ex}}\right) F_{bx}} + \frac{C_{my} f_{by}}{\left(1 - \dfrac{f_a}{F'_{ey}}\right) F_{by}} \leq 1.0 \tag{B.25}$$

and

$$\frac{f_a}{0.60F_y} + \frac{f_{bx}}{F_{bx}} + \frac{f_{by}}{F_{by}} \leq 1.0 \qquad\qquad (B.26)$$

Equation B.25 is a stability check, and the maximum bending moments, regardless of where they occur, must be used when computing f_{bx} and f_{by}. Equation B.26, which does not use amplification factors, is a stress check, and the maximum *end* moments should be used to compute f_{bx} and f_{by}. Note that $0.60F_y$ is used in Equation B.26 in lieu of F_a since yielding, rather than buckling, is the limit state. For the same reason, the member can be considered as laterally supported when computing F_{bx} for Equation B.26, but the actual lateral bracing conditions must be accounted for when evaluating Equation B.25.

The purpose of the factor C_{mx} in Equation B.25 is to account for the x-axis moment gradient in the member. In a laterally unsupported member, the factor C_b, used in computing F_{bx}, also accounts for this moment gradient, so the Specification requires that C_b be taken as unity when computing F_{bx} for Equation B.25 *for members braced against joint translation.*

≡ EXAMPLE B.4

The beam-column shown in Figure B.7 is part of a braced frame. Bending is about the x-axis, and lateral bracing is provided at the ends. Assume $K_x = K_y = 1.0$ and review this member for compliance with the AISC Specification.

SOLUTION   Determine which interaction equation(s) to use. The axial compressive stress is

$$f_a = \frac{P}{A_g} = \frac{100}{14.4} = 6.944 \text{ ksi}$$

Compute the allowable compressive stress. The maximum slenderness ratio is

$$\frac{K_y L}{r_y} = \frac{1.0(15)(12)}{2.54} = 70.87$$

FIGURE B.7 ≡

100^k

$35^{'\text{-}k}$

$15'$

W10 × 49
A36

$60^{'\text{-}k}$

From Equation B.8,

$$C_c = \sqrt{\frac{2\pi^2 E}{F_y}} = \sqrt{\frac{2\pi^2(29,000)}{36}} = 126.1$$

Since $KL/r < C_c$, compute the allowable compressive stress with Equation B.9:

$$F_a = \frac{F_y\left[1 - \dfrac{(KL/r)^2}{2C_c^2}\right]}{\dfrac{5}{3} + \dfrac{3(KL/r)}{8C_c} - \dfrac{(KL/r)^3}{8C_c^3}} = \frac{36\left[1 - \dfrac{(70.87)^2}{2(126.1)^2}\right]}{\dfrac{5}{3} + \dfrac{3(70.87)}{8(126.1)} - \dfrac{(70.87)^3}{8(126.1)^3}} = 16.34 \text{ ksi}$$

$$\frac{f_a}{F_a} = \frac{6.944}{16.34} = 0.4250 > 0.15 \quad \therefore \quad \text{Check Equations B.25 and B.26.}$$

$$f_{bx} = \frac{M_x}{S_x} = \frac{60(12)}{54.6} = 13.19 \text{ ksi}$$

$$f_{by} = 0$$

Compute the allowable bending stress. From Equation B.12,

$$\frac{76b_f}{\sqrt{F_y}} = \frac{76(10.00)}{\sqrt{36}} = 126.7 \text{ in.} = 10.6 \text{ ft}$$

$$\frac{20,000}{(d/A_f)F_y} = \frac{20,000}{\dfrac{9.98}{0.560(10.00)}(36)} = 311.7 \text{ in.} = 26.0 \text{ ft}$$

The smaller value controls, so $L_c = 10.6$ ft. This is less than the unbraced length of $L_b = 15$ ft, so the member will be treated as laterally unsupported. Since this member is braced against sidesway, use $C_b = 1.0$.

$$\sqrt{\frac{102,000C_b}{F_y}} = \sqrt{\frac{102,000(1.0)}{36}} = 53.2$$

$$\sqrt{\frac{510,000C_b}{F_y}} = \sqrt{\frac{510,000(1.0)}{36}} = 119$$

$$\frac{L_b}{r_T} = \frac{15(12)}{2.74} = 65.69 \quad (r_T \text{ is tabulated in the } Manual.)$$

Since $53.2 < L_b/r_T < 119$, use the larger of the values computed with equations B.18 and B.19, subject to an upper limit of

$$0.60F_y = 0.60(36) = 21.6 \text{ ksi}$$

From Equation B.18,

$$\left[\frac{2}{3} - \frac{F_y(L_b/r_T)^2}{1,530,000C_b}\right]F_y = \left[\frac{2}{3} - \frac{36(65.69)^2}{1,530,000(1.0)}\right]36 = 20.34 \text{ ksi}$$

From Equation B.19,

$$\frac{12{,}000 C_b}{L_b d / A_f} = \frac{12{,}000(1.0)}{(15 \times 12)(9.98)/(0.560 \times 10.00)} = 37.4 \text{ ksi}$$

The larger value is greater than $0.60 F_y$, so

$$F_{bx} = 0.60 F_y = 21.6 \text{ ksi}$$

Check Equation B.26 first. In this equation, which is a check on the stress condition at the support, the allowable bending stress may be computed as if the member had full lateral support of its compression flange. A W10 × 49 is compact for A36 steel, so the allowable stress can be taken as $0.66 F_y$. Bending is about the x-axis, so the terms involving y-axis bending will be omitted.

$$\frac{f_a}{0.60 F_y} + \frac{f_{bx}}{F_{bx}} = \frac{6.944}{0.60(36)} + \frac{13.19}{0.66(36)} = 0.877 < 1.0 \quad \text{(OK)}$$

Check Equation B.25. From Equation B.22,

$$C_m = 0.6 - 0.4 \frac{M_1}{M_2} = 0.6 - 0.4\left(-\frac{35}{60}\right) = 0.8333$$

The slenderness ratio for use in computing F'_{ex} is

$$\frac{KL_b}{r_b} = \frac{K_x L}{r_x} = \frac{1.0(15)(12)}{4.35} = 41.38$$

and

$$F'_{ex} = \frac{12\pi^2 E}{23(K_x L/r_x)^2} = \frac{12\pi^2(29{,}000)}{23(41.38)^2} = 87.21 \text{ ksi}$$

$$\frac{f_a}{F_a} + \frac{C_{mx} f_{bx}}{\left(1 - \dfrac{f_a}{F'_{ex}}\right) F_{bx}} = 0.4250 + \frac{0.8333(13.19)}{\left(1 - \dfrac{6.944}{87.21}\right) 21.6} = 0.978 < 1.0 \quad \text{(OK)}$$

Answer   The W10 × 49 is satisfactory.

Design Aids

Other than the tables and charts for column and beam design, the principal *Manual* design aid for beam-columns is a table of constants for use in making a preliminary member selection.[42] These constants enable the designer to transform the bending moment into an equivalent axial load that can be combined with the actual load to obtain a total effective axial load. The column allowable load tables can then be entered with this effective axial load and a trial shape can be selected for evaluation.

B.6 ≡ CONCLUDING REMARKS

Although allowable stress design is rapidly being supplanted by load and resistance factor design, it is still authorized by AISC and will be in use for some time to come. Those interested in pursuing ASD beyond the brief introduction in this appendix are referred to *Design of Steel Structures*,[64] which incorporates the latest AISC Specification provisions.

REFERENCES

1 *The BOCA National Building Code* (11th ed.). Chicago, IL: Building Officials and Code Administrators International, 1990.

2 *Uniform Building Code.* Whittier, CA: International Conference of Building Officials, 1991.

3 *Standard Building Code.* Birmingham, AL: Southern Building Code Congress International, 1988.

4 *Minimum Design Loads for Buildings and Other Structures*, ASCE 7-88 (formerly ANSI A58.1). New York: American Society of Civil Engineers, 1990.

5 *Load and Resistance Factor Design Specification for Structural Steel Buildings.* Chicago, IL: American Institute of Steel Construction, 1993.

6 *Standard Specifications for Highway Bridges* (15th ed.). Washington, DC: The American Association of State Highway and Transportation Officials, 1992.

7 *Specifications for Steel Railway Bridges.* Chicago, IL: American Railway Engineering Association, 1992.

8 *Load and Resistance Factor Design Specification for Cold-Formed Steel Members.* Washington, DC: American Iron and Steel Institute, 1991.

9 Murphy, G. *Properties of Engineering Materials* (3rd ed.). Scranton, PA: International Textbook Co., 1957.

10 Tall, L. (ed.). *Structural Steel Design.* New York: Ronald Press Co., 1964.

11 *1992 Annual Book of ASTM Standards.* Philadelphia, PA: American Society for Testing and Materials.

12 "ASTM A6, Standard Specification for General Requirements for Rolled Steel Plates, Shapes, Steel Piling, and Bars for Structural Use." *1992 Annual Book for ASTM Standards*, Vol. 1.04. Philadelphia, PA: American Society for Testing and Materials.

13 *Manual of Steel Construction: Load and Resistance Factor Design* (2nd ed.). Chicago, IL: American Institute of Steel Construction, 1994.

14 Ravindra, M. K., and Galambos, T. V. "Load and Resistance Factor Design for Steel." *Journal of the Structural Division*, ASCE 104 (September 1978): 1337–1353.

15   Yura, J. A., Galambos, T. V., and Ravindra, M. K. "The Bending Resistance of Steel Beams." *Journal of the Structural Division*, ASCE 104 (September 1978): 1355–1370.

16   Bjorhovde, R., Galambos, T. V., and Ravindra, M. K. "LRFD Criteria for Steel Beam-Columns." *Journal of the Structural Division*, ASCE 104 (September 1978): 1371–1387.

17   Cooper, P. B., Galambos, T. V., and Ravindra, M. K. "LRFD Criteria for Plate Girders." *Journal of the Structural Division*, ASCE 104 (September 1978): 1389–1407.

18   Hansell, W. C., Galambos, T. V., Ravindra, M. K., and Viest, I. M. "Composite Beam Criteria in LRFD." *Journal of the Structural Division*, ASCE 104 (September 1978): 1409–1426.

19   Fisher, J. W., Galambos, T. V., Kulak, G. L., and Ravindra, M. K. "Load and Resistance Factor Design Criteria for Connectors." *Journal of the Structural Division*, ASCE 104 (September 1978): 1427–1441.

20   Ravindra, M. K., Cornell, C. A., and Galambos, T. V. "Wind and Snow Load Factors for Use in LRFD." *Journal of the Structural Division*, ASCE 104 (September 1978): 1443–1457.

21   Galambos, T. V., and Ravindra, M. K. "Properties of Steel for Use in LRFD." *Journal of the Structural Division*, ASCE 104 (September 1978): 1459–1468.

22   *Building Code Requirements for Reinforced Concrete* (ACI 318-89). Detroit, MI: American Concrete Institute, 1989.

23   *Load and Resistance Factor Design Specification for Structural Joints Using ASTM A325 or A490 Bolts.* Chicago, IL: Research Council on Structural Connections of the Engineering Foundation, 1988.

24   McGuire, W. *Steel Structures.* Englewood Cliffs, NJ: Prentice-Hall, 1968.

25   Munse, W. H., and Chesson, E., Jr. "Riveted and Bolted Joints: Net Section Design," *Journal of the Structural Division*, ASCE, Vol. 89, No. ST1 (Feb. 1963) 107–126.

26   Cochrane, V. H. "Rules for Riveted Hole Deduction in Tension Members," *Engineering News Record*, Nov. 16, 1922.

27   *Cable Roof Structures*, Bethlehem Steel, Nov. 1968.

28   Lothars, J. E. *Design in Structural Steel* (3rd ed.). Englewood Cliffs, NJ: Prentice-Hall, 1972.

29   Timoshenko, Stephen P. *History of Strength of Materials.* New York: McGraw-Hill, 1953.

30   Shanley, F. R. "Inelastic Column Theory," *Journal of Aeronautical Sciences*, Vol. 14, No. 5 (May 1947), 261.

31   Bleich, Friedrich, *Buckling Strength of Metal Structures*. New York: McGraw-Hill, 1952.

32   Galambos, Theodore V. (ed.). *Guide to Stability Design Criteria for Metal Structures* (4th ed.). Structural Stability Research Council. New York: Wiley-Interscience, 1988.

33   Johnston, Bruce G. (ed.). *Guide to Stability Design Criteria for Metal Structures* (3rd ed.). Structural Stability Research Council. New York: Wiley-Interscience, 1976.

34   Yura, Joseph A. "The Effective Length of Columns in Unbraced Frames," *Engineering Journal*, AISC, Vol. 8, No. 2 (1971), 37–42.

35   Disque, Robert O. "Inelastic K-Factor for Column Design," *Engineering Journal*, AISC, Vol. 10, No. 2 (1973), 33–35.

36   Timoshenko, Stephen P., and Gere, James M. *Theory of Elastic Stability* (2nd ed.). New York: McGraw-Hill, 1961.

37   Higdon, A., Ohlsen, E. H., and Stiles, W. B. *Mechanics of Materials*. New York: John Wiley & Sons, 1960.

38   Darwin, David. *Design of Steel and Composite Beams with Web Openings. AISC Design Guide No. 2*. Chicago, IL: American Institute of Steel Construction, 1990.

39   *Standard Specifications, Load Tables, and Weight Tables for Steel Joists and Joist Girders*. Myrtle Beach, SC: Steel Joist Institute, 1992.

40   *Torsional Analysis of Steel Members*. Chicago, IL: American Institute of Steel Construction, 1983.

41   Yura, Joseph, A. *Elements for Teaching Load and Resistance Factor Design*. Chicago, IL: American Institute of Steel Construction, 1988.

42   Burgett, Lewis B. "Selection of a 'Trial' Column Section." *Engineering Journal*, AISC, Vol. 10, No. 2 (1973), 54–59.

43   Ad Hoc Committee on Serviceability. "Structural Serviceability: A Critical Appraisal and Research Needs." *Journal of Structural Engineering*, American Society of Civil Engineers, 112(12) (1986), 2646–2664.

44   *High-Strength Bolting for Structural Joints*. Bethlehem Steel, March, 1969.

45   Kulak, G. L., Fisher, J. W., and Struik, J. H. A., *Guide to Design Criteria for Bolted and Riveted Joints* (2nd ed.). New York: John Wiley & Sons, 1987.

46   Bickford, John H. *An Introduction to the Design and Behavior of Bolted Joints*. New York: Marcel Dekker, 1981.

47   *Structural Welding Code (ANSI/AWS D1.1-92)*. Miami: American Welding Society, 1992.

48   Crawford, S. H., and Kulak, G. L. "Eccentrically Loaded Bolted Connections." *Journal of the Structural Division*, ASCE, Vol. 97, No. ST3 (March 1971), 765–783.

49   Butler, L. J., Pal, S., and Kulak, G. L. "Eccentrically Loaded Welded Connections." *Journal of the Structural Division*, ASCE, Vol. 98, No. ST5 (May 1972), 989–1005.

50   Kulak and Timler, "Tests on Eccentrically Loaded Fillet Welds," Department of Civil Engineering, University of Alberta, Edmonton, December 1984.

51   Krishnamurthy, N. "A Fresh Look at Bolted End-Plate Behavior and Design." *Engineering Journal*, AISC, Vol. 15, No. 2 (1978), 39–49.

52   Hendrick, A., and Murray, T. M. "Column Web Compression Strength at End-Plate Connections," *Engineering Journal*, AISC, Vol. 21, No. 3 (1984), 161–169.

53   Blodgett, O. W. *Design of Welded Structures*. Cleveland, OH: James F. Lincoln Arc Welding Foundation, 1966.

54   *Manual of Steel Construction, Volume II — Connections*. Chicago, IL: American Institute of Steel Construction, 1993.

55   *Detailing for Steel Construction*. Chicago, IL: American Institute of Steel Construction, 1983.

56   Ollgaard, J. G., Slutter, R. G., and Fisher, J. W. "Shear Strength of Stud Connectors in Lightweight and Normal-Weight Concrete." *Engineering Journal*, AISC, Vol. 8, No. 2 (1971), 55–64.

57   Task Group 20, Structural Stability Research Council. "A Specification for the Design of Steel-Concrete Composite Columns." *Engineering Journal*, AISC, Vol. 16, No. 4 (1979), 101–115.

58   Zahn, C. J. "Plate Girder Design Using LRFD." *Engineering Journal*, AISC, Vol. 24, No. 1 (1987), 11–20.

59   Basler, K. "Strength of Plate Girders in Shear." *Journal of the Structural Division*, ASCE, Vol. 87, No. ST7 (October 1961), 151–197.

60   Salmon, C. G., and Johnson, J. E. *Steel Structures, Design and Behavior* (3rd ed.). New York: Harper & Row, 1990.

61   Neal, B. G. *The Plastic Methods of Structural Analysis* (3rd ed.). London: Chapman & Hall, 1977.

62   *Specification for Structural Steel Buildings: Allowable Stress Design and Plastic Design*. Chicago, IL: American Institute of Steel Construction, 1989.

63   *Manual of Steel Construction: Allowable Stress Design*. Ninth Edition, Chicago, IL: American Institute of Steel Construction, 1989.

64   Gaylord, Edwin H., Gaylord, Charles N., and Stallmeyer, James E. *Design of Steel Structures* (3rd ed.). New York: McGraw-Hill, 1992.

ANSWERS TO SELECTED PROBLEMS

NOTE Answers are not given to problems that have more than one correct answer or in which the answer reveals the solution technique or procedure (this includes most design problems), or to computer programming assignments.

CHAPTER 3: TENSION MEMBERS

3.1	97.2 kips
3.2	73.8 kips
3.3	98 kips < 113 kips; member adequate.
3.4	$D = 137$ kips, $L = 68.4$ kips
3.5	**(a)** 1680 kips; **(b)** 1970 kips
3.6	237 kips
3.8	60.8 kips
3.9	2.16 in.^{2}
3.10	5 in.
3.11	7 in.
3.12	381 kips (with U from equation)
3.13	8.38 in.
3.14	**(a)** 286 kips (for $U = 0.90$) or 282 kips (for $U = 0.85$); **(b)** 334 kips (for $U = 0.90$) or 316 kips (for $U = 0.85$)
3.15	162 kips
3.16	286 kips (for $U = 0.90$) or 280 kips (for $U = 0.85$)
3.17	$2\sqrt{gd}$
3.18	186 kips
3.19	5.08 in.^{2}
3.20	129 kips
3.21	Required $A_g = 3.89$ in.^{2}, required $A_n = 3.41$ in.^{2}
3.22	Required $A_g = 4.93$ in.^{2}, required $A_n = 5.36$ in.^{2}
3.23	Required $A_g = 9.20$ in.^{2}, required $A_e = 6.85$ in.^{2}
3.24	Required $A_g = 5.62$ in.^{2}, required $A_e = 4.18$ in.^{2}

3.25	Required $A_g = 2.22$ in.^{2}, required $A_n = 1.95$ in.^{2}
3.26	Required $A_g = 8.02$ in.^{2}, required $A_n = 7.03$ in.^{2}
3.27	Required $A_g = 4.17$ in.^{2}, required $A_n = 3.65$ in.^{2}
3.28	Required diameter $= 2.65$ in.
3.29	Required diameter $= 2.28$ in.
3.30	Required diameter $= 0.128$ in.
3.33	7.94 kips at peak, 7.22 kips elsewhere

☰ CHAPTER 4: COMPRESSION MEMBERS ☰

4.1	340 kips
4.2	**(a)** 338 kips; **(b)** 204 kips; **(c)** 93.0 kips
4.3	172 kips
4.4	370 kips
4.5	200 kips
4.6	84.0 kips
4.7	198 kips
4.8	1040 kips
4.9	94.4 kips
4.19	531 kips
4.20	804 kips
4.21	1060 kips
4.22	171 kips
4.23	736 kips
4.24	736 kips
4.27	282 kips
4.28	**(a)** 1.78; **(b)** 533 kips
4.29	226 kips
4.30	113 kips
4.32	18,000 kips
4.33	860 kips
4.34	183 kips
4.35	11,900 kips
4.36	44.2 kips
4.37	132 kips

☰ CHAPTER 5: BEAMS ☰

5.1	**(a)** $Z = 1000$ in.^{3}, $M_p = 3000$ ft-kips
5.2	**(a)** noncompact; **(b)** compact; **(c)** compact; **(d)** noncompact

5.5	1570 ft-kips
5.6	$w_L = 2.23$ kips/ft, $w_D = 0.891$ kips/ft
5.7	23.6 ft
5.8	Yes. $\phi_b M_n = 109$ ft-kips > 94.5 ft-kips
5.9	No. $\phi_b M_n = 397$ ft-kips < 406 ft-kips
5.11	1.22
5.12	244 ft-kips
5.13	204 ft-kips
5.14	29.1 kips
5.15	$w_D = 84.8$ lb/ft, $w_L = 254$ lb/ft
5.16	Satisfactory. $\phi_b M_n = 606$ ft-kips > 398 ft-kips, $\phi_v V_n = 266$ kips > 48.8 kips
5.17	$\phi_b M_n = 28.8$ ft-kips, $\phi_v V_n = 26.8$ kips
5.18	No. $\phi_v V_n = 84.9$ kips < 108 kips
5.37	**(a)** 32.0 in.^{3}; **(b)** 36.1 in.^{3}
5.44	Satisfactory. $0.588 < 1.0$
5.46	**(a)** Satisfactory. $0.792 < 1.0$; **(b)** Unsatisfactory. $1.42 > 1.0$
5.47	Satisfactory. $0.702 < 1.0$
5.48	Inadequate. $1.23 > 1.0$

☰ CHAPTER 6: BEAM-COLUMNS ☰

6.1	Satisfactory. $0.524 < 1.0$
6.2	Yes. $0.968 < 1.0$
6.3	Yes. $0.529 < 1.0$
6.4	Satisfactory. $0.537 < 1.0$
6.5	Yes. Result of interaction equation $= 1.00$.
6.6	Yes. $0.543 < 1.0$
6.7	0.368; use 1.0
6.8	0.963; use 1.0
6.9	**(a)** 0.608; use 1.0; **(b)** 0.623; use 1.0
6.10	**(a)** 0.853; use 1.0; **(b)** 1.002
6.11	**(a)** 1.08; **(b)** 1.06
6.12	Yes. $0.892 < 1.0$
6.13	Satisfactory. $0.962 < 1.0$
6.14	Satisfactory. $0.893 < 1.0$
6.15	Yes. $0.898 < 1.0$
6.16	No. $1.30 > 1.0$
6.17	Satisfactory. $0.497 < 1.0$
6.18	Load combination A4-2: $0.884 < 1.0$ (OK). Load combination A4-4: $0.723 < 1.0$ (OK)

$\equiv$ **CHAPTER 7: SIMPLE CONNECTIONS** $\equiv$

7.1	32.5 kips
7.2	42.4 kips
7.3	86.6 kips
7.4	2
7.5	10
7.6	**(a)** 12; **(b)** 15.5 in.; **(c)** 5.62 in.; use $5\frac{5}{8}$ in.
7.8	**(a)** 130 kips; **(b)** 218 kips, 123 kips; **(c)** 218 kips
7.9	**(a)** 20; **(b)** 8
7.10	65.0 kips, 30.7 kips
7.11	49.3 kips
7.12	112 kips
7.13	2
7.15	Adequate. Bolt tension controls: $B = 29.8$ kips > 25 kips.
7.16	Adequate. Angle thickness controls: Required $t_f = 0.329$ in. < 0.5 in.
7.17	Satisfactory. Shear: $\phi R_n = 15.9$ kips > 8.94 kips. Bearing: $\phi R_n = 47.4$ kips > 8.94 kips. Tension: $\phi R_n = 26.0$ kips > 17.9 kips.
7.19	Satisfactory. Shear: $\phi R_n = 27.1$ kips > 5.8 kips. Bearing: $\phi R_n = 55.3$ kips > 5.8 kips. Tension: $\phi R_n = 40.6$ kips > 17.4 kips.
7.21	72.9 kips
7.22	134 kips
7.23	89.1 kips
7.24	64.8 kips
7.25	8 in.
7.26	20.1 in.
7.27	9 in.
7.28	$\frac{1}{4}$ in.
7.29	4.29 in.
7.30	18 in.

$\equiv$ **CHAPTER 8: ECCENTRIC CONNECTIONS** $\equiv$

8.1	43.7 kips
8.2	1 in.
8.3	9.08 kips
8.4	1.68P in.-kips (for P in kips)
8.5	Inadequate. Shear: $\phi R_n = 28.3$ kips < 36.6 kips.
8.6	$1\frac{1}{4}$ in.

8.7   35.7 kips

8.8   25.3 kips

8.9   11.5 kips

8.10 20.5 kips

8.11 $\frac{3}{4}$ in.

8.12 30 kips

8.13 Satisfactory. Shear: $\phi R_n = 15.9$ kips/bolt > 11.3 kips. Bearing: $\phi R_n = 55.6$ kips/bolt > 11.3 kips. Tension: $\phi M_n = 1770$ in.-kips > 1020 in.-kips. Slip-critical load per bolt $= 7.51$ kips > 7.5 kips.

8.14 Inadequate. Shear: $\phi R_n = 28.3$ kips/bolt > 26.8 kips (OK). Bearing: $\phi R_n = 53.8$ kips/bolt > 26.8 kips (OK). Tension: $\phi M_n = 1410$ in.-kips < 1530 in.-kips (N.G.).

8.15 Satisfactory. Shear: $\phi R_n = 15.9$ kips/bolt > 15.0 kips. Bearing: $\phi R_n = 47.4$ kips/bolt > 15.0 kips. Tension: $\phi M_n = 1350$ in.-kips > 900 in.-kips.

8.16 Satisfactory. *Beam-to-tee connection:* Shear: $\phi R_n = 15.9$ kips/bolt > 14.2 kips. Bearing: $\phi R_n = 27.8$ kips/bolt > 14.2 kips. *Tee-to-column connection:* Shear: $\phi R_n = 15.9$ kips/bolt > 9.47 kips. Tension: $\phi M_n = 794$ in.-kips > 170 in.-kips.

8.17 Satisfactory. Shear: $\phi R_n = 28.3$ kips/bolt > 27.0 kips. Bearing: $\phi R_n = 58.5$ kips/bolt > 27.0 kips. Tension: $\phi M_n = 1080$ in.-kips > 324 in.-kips.

8.19 0.502 in. Use $\frac{9}{16}$ in.

8.20 Unsatisfactory. $\phi R_n = 11.1$ kips/in. < 11.7 kips/in.

8.21 Unsatisfactory. $\phi R_n = 4.18$ kips/in. < 4.59 kips/in.

8.22 5.55 kips/in.

8.23 3.54 kips/in.

8.26 Unsatisfactory. $\phi R_n = 6.96$ kips/in. < 10.2 kips/in.

8.27 23.6 kips

8.28 3.25 kips/in.

8.29 0.202 in. Use $\frac{1}{4}$ in.

8.31 0.168 in. Use $\frac{3}{16}$ in.

8.32 59.5 kips

8.33 1.18 kips/in.

8.34 1.09 kips/in.

8.35 **(a)** Unsatisfactory. $\phi R_n = 6.96$ kips/in. < 9.52 kips/in. **(b)** Unsatisfactory. $\phi R_n = 6.96$ kips/in. < 11.5 kips/in.

8.36 65.3 ft-kips

8.37 Stiffeners are required opposite both flanges.

8.38 Stiffeners are not required.

8.40 Adequate. Slip-critical load $= 81.8$ kips > 20 kips. Shear: $\phi R_n = 21.6$ kips/bolt > 3.75 kips. Bearing: $\phi R_n = 39.7$ kips/bolt > 3.75 kips. Tension: $\phi R_n = 40.6$ kips/bolt > 10.1 kips.

8.41 Adequate. Slip-critical load $= 81.8$ kips > 45 kips. Shear: $\phi R_n = 21.6$ kips/bolt > 8.44 kips. Bearing: $\phi R_n = 51.2$ kips/bolt > 8.44 kips. Tension: $\phi R_n = 40.6$ kips/bolt > 9.12 kips.

CHAPTER 9: COMPOSITE CONSTRUCTION

9.1 4.30 in. from top of slab

9.2 5440 in.[4]

9.3 17.6 ksi (tension)

9.4 (a) 1.99 ksi; (b) 21.8 ksi

9.5 354 ft-kips

9.6 908 ft-kips

9.7 412 ft-kips

9.8 1390 ft-kips

9.9 509 ft-kips

9.10 Adequate. Before concrete cures: $\phi_b M_n = 272$ ft-kips > 37.5 ft-kips. After concrete cures: $M_u = 90.6$ ft-kips $< \phi_b M_n$ of steel shape, $\phi_v V_n = 124$ kips > 14.5 kips.

9.11 Adequate. Before concrete cures: $\phi_b M_n = 766$ ft-kips > 76.7 ft-kips. After concrete cures: $\phi_b M_n = 1450$ ft-kips > 1080 ft-kips, $\phi_v V_n = 288$ kips > 145 kips.

9.12 Adequate. Before concrete cures: $\phi_b M_n = 245$ ft-kips > 56.8 ft-kips. After concrete cures: $\phi_b M_n = 412$ ft-kips > 327 ft-kips, $\phi_v V_n = 126$ kips > 43.6 kips.

9.13 Adequate. Before concrete cures: $\phi_b M_n = 146$ ft-kips > 47.1 ft-kips. After concrete cures: $\phi_b M_n = 264$ ft-kips > 185 ft-kips, $\phi_v V_n = 84.9$ kips > 27.4 kips.

9.14 (a) Adequate. Before concrete cures: $\phi_b M_n = 146$ ft-kips > 64.9 ft-kips. After concrete cures: $\phi_b M_n = 261$ ft-kips > 227 ft-kips, $\phi_v V_n = 84.9$ kips > 30.3 kips. (b) 70

9.15 (a) Adequate. Before concrete cures: $\phi_b M_n = 751$ ft-kips > 131 ft-kips. After concrete cures: $\phi_b M_n = 1180$ ft-kips > 611 ft-kips, $\phi_v V_n = 256$ kips > 61.1 kips.

9.16 58

9.21 (a) 1.25 in. (slab and beam, before concrete cures), 1.90 in. (after concrete cures, before creep occurs); (b) 1.91 in.

9.22 0.473 in.

9.23 Dead load deflection $= 0.0889$ in. Live load deflection $= 0.393$ in. Total deflection $= 0.482$ in.

9.24 Dead load deflection (without creep) $= 0.525$ in. Dead load deflection (with creep) $= 0.532$ in. Live load deflection $= 0.283$ in. Total deflection (without creep) $= 0.808$ in. Total deflection (with creep) $= 0.815$ in.

9.27 (a) 460 ft-kips; (b) 426 ft-kips

9.28 **(a)** Satisfactory. Before concrete cures: $\phi_b M_n = 362$ ft-kips > 83.6 ft-kips. After concrete cures: $\phi_b M_n = 603$ ft-kips (with full composite action) > 532 ft-kips, $\phi_v V_n = 181$ kips > 53.2 kips.

9.29 **(a)** Satisfactory. Before concrete cures: $\phi_b M_n = 348$ ft-kips > 162 ft-kips. After concrete cures: $\phi_b M_n = 635$ ft-kips (with full composite action) > 485 ft-kips, $\phi_v V_n = 166$ kips > 55.5 kips.

CHAPTER 10: PLATE GIRDERS

10.1 5550 ft-kips

10.2 2890 ft-kips

10.3 Adequate. $\phi_b M_n = 1660$ ft-kips > 1390 ft-kips.

10.4 **(a)** 7730 ft-kips; **(b)** 7730 ft-kips

10.5 **(a)** 67 in.; **(b)** 0.62 in.^{2} (total); **(c)** 52.7 kips

10.6 **(a)** Adequate. Bearing: $\phi R_n = 364$ kips > 200 kips. Column strength: $\phi_c P_n = 352$ kips > 200 kips. **(b)** 75.9 in.

10.7 **(a)** Satisfactory. $\phi_b M_n = 4000$ ft-kips > 3090 ft-kips. **(b)** Yes. Stiffeners are required at 58 in. and 125 in. from each end.

10.8 446 kips

INDEX